Inners and Stability
of Dynamic Systems

E. I. Jury

Professor of Electrical Engineering
University of California, Berkeley

A WILEY-INTERSCIENCE PUBLICATION

John Wiley & Sons

New York · London · Sydney · Toronto

Library of Congress Cataloging in Publication Data

Jury, Eliahu Ibraham, 1923–
  Inners and stability of dynamic systems.

  "A Wiley-Interscience publication."
  Includes bibliographical references.
  1. System analysis.   2. Stability.   3. Matrices.
I. Title.

QA402.J87   620′.72   74–13036
ISBN 0-471-45335-8

Printed in the United States of America

10 9 8 7 6 5 4 3 2 1

To Joyce

Preface

The stability problem is very important in many disciplines such as Engineering, Physics, Chemistry, Computer Sciences, Biology, and Mathematics. It has been studied and researched since the middle of the last century. During this time, several books on the subject have been published in both the mathematical and engineering literatures. The present treatise can be considered as one of a continuing series of publications on the problem of stability; however, it differs from the earlier published books in the following three points: First, the problem of stability of linear time-invariant continuous and discrete systems and its related topics is considered from the notion of *inners* of a square matrix. This notion, which is discussed thoroughly, makes it possible to unify most of the stability criteria in one framework. Such a unification is advantageous in clarifying and organizing the known criteria and in relieving the burden of remembering the detailed form of each of them.

Second, the content of this text goes beyond the stability problem. The problems of positivity and nonnegativity are thoroughly discussed, as well as the evaluation of complex integrals which arise in communication and control theories. Though the latter problems are discussed in other texts, this is the first to include all the above topics in one unified approach, which is again based on the inners.

Third, to my knowledge, this is the first text devoted to stability and other associated problems which also includes the computational algorithms. The use of the computer is becoming an important aspect in the theory of stability in view of the widespread use of digital computers for both analysis and design of complex systems.

Having listed the above unique features of this treatise, the contents are delineated in the following: In Chapter 1, the inners notion is introduced and defined, and the historical background related to the inners is documented. The chapter is of an introductory nature and relates to matrix theory; a background in elementary matrix theory is a prerequisite.

In Chapter 2, the various stability criteria for both continuous and discrete time systems are presented in a unified approach based on the notion of positive innerwise matrices. Also treated are problems connected with relative stability, aperiodicity, and critical constraints. Finally, conditions on root distribution of polynomials are discussed in terms of inners.

Chapter 3 presents the various criteria developed or discussed in Chapter 2 in terms of positive definite symmetric matrices. This approach also offers a unified formulation of various criteria and represents the counterpart of the unified approach based on the inners. The connection between positive definite symmetric matrices and positive innerwise matrices for the root-clustering problems is established. In addition, the stability criteria based on the A matrix and a generalized Lyapunov theorem are given. The material of this and earlier chapters presents an exhaustive account of stability and related criteria which has been studied during the last century.

Chapter 4 gives the inner representation of complex integrals which arise in the evaluation of the total square integral (or sum) of a signal arising in problems of communication, control systems, or digital filtering. The connection between the stability criteria developed in Chapters 2 and 3 and the evaluation of these complex integrals is delineated. Again, the inners approach offers a simplified method for the integral evaluation. Much research has been done during the last three decades on this problem. The discussion presented covers most of this work and establishes the connection between the various approaches.

In Chapter 5, an exhaustive study is made of positive and strictly positive real functions and matrices. In recent years this material has become of importance in diverse areas such as absolute stability of nonlinear systems, optimality, and sensitivity of dynamic systems. Methods for testing positivity and nonnegativity, which are connected with the inners notion in Chapter 1, are fully discussed. In Section 5.4, necessary and sufficient conditions for stability of nonuniformly distributed parameter systems are given.

Stability conditions of two- and multidimensional filters are fully discussed in Chapter 6. This material is fairly new and, to my knowledge, has not been discussed in any text. This extension of the stability concepts to higher-dimensional systems will become of major interest and importance in the future. In the development of the material of this chapter, the positivity test developed in Chapter 5 is much emphasized. Again, the inners play a role in testing the stability conditions.

Chapter 7 dwells in depth on the computational algorithm for evaluating the inners determinant. The double triangularization algorithm developed offers a unified computation method for dealing with all the problems discussed in the preceding six chapters. Computer examples are presented which cover many areas of research. This detailed account supplements the theoretical material of the text and, hopefully, will enhance its future use and importance.

The three appendices present various proofs of the stability criteria, which were left out in the text. Also included is a table of Chebyshev functions calculated from the inners algorithm.

This text is written mainly for engineers, physicists, and computer specialists. It is not my intention to compete with the existing mathematical books written on some of the same topics. Hence, mathematical rigor and detailed proofs are often sacrificed for the sake of engineering applications. However, the ideas presented might be of use to applied mathematicians in putting the theory into a rigorous and elegant formulation. It is hoped that such an interest will be stimulated by this book.

The material of this text can be taught as a two-unit graduate course on stability theory. It is also of use as a reference research monograph and as a supplemental text for many courses in control, communication, digital filters, and computer science.

In the preparation of this manuscript, I have been aided and encouraged by many colleagues and students. In particular, I am grateful to Professor D. D. Šiljak for his constructive review and valuable suggestions in the writing of this text. His research efforts on positivity and nonnegativity conditions constitute the bulk of Chapter 5. During the last few years, I have been fortunate to collaborate on research problems with Professor B. D. O. Anderson. Our joint research efforts are integrated in various sections of this text, in particular, the material of Chapter 6, which was critically reviewed by Professor Anderson. For this and other research contributions, I am very grateful and thankful to him.

Thanks are also due to my research students who carefully scrutinized the manuscript. Among them, I wish to offer my thanks to Dr. Ahn and to P. S. Kamat, S. Gutman, and N. A. Pendergrass. Thanks are also due to Professor N. Bose.

The painstaking efforts and patience of Ms. Doris Simpson, who typed the manuscript, are gratefully appreciated.

Most of the research material presented in the text was developed through research grants awarded by the National Science Foundation. This generous support is gratefully acknowledged.

I also wish to thank Professor Tom Everhart, Chairman of the Department of Electrical Engineering and Computer Sciences at the University of California at Berkeley. His support and encouragement have contributed significantly to the early completion of this text. I am very grateful to him and to my colleagues in the department for their support.

E. I. JURY

Berkeley, California

Contents

Inners:
Definitions and Theorems

The name *inners* is given to the square submatrices that arise in a square $n \times n$ matrix. Previously these submatrices were referred to in the literature [1,2] as dotted frames, innermost frames of array, central minors, and symmetrically situated principal minors [3]. Those frames of array that are called inners play an important role in the theory of polynomials. Historically, Trudi [4] was perhaps the first to study the determinants of these inners, which he called *disencumbered remainders*. Later, Bôcher [5] referred to these determinants as *subresultants*. Also, in the literature, the name *bigradients* [6] was adopted because of the pattern of two sets of sloping elements.

The importance of these determinants, as will be mentioned in detail in this book, lies in obtaining the "greatest common divisor of two polynomials," in obtaining condition on the stability of dynamic systems, in the root distribution of polynomials, and, as we shall see, in many other applications of system theory.

The purpose of this chapter is to list the newly introduced definitions of inners, to present some of their properties, and to mention the historical background and use of these terms in the past.

1.1 INNERS, DISENCUMBERED REMAINDERS, BIGRADIENTS, AND SUBRESULTANTS

In the following material, the definition of the inners is introduced. It will be shown that when these inners take a special form, their determinants have been known for a long time and play an important role in the theory of polynomials.

DEFINITION 1.1 [7]. *Let $\Delta = \Delta_N$ be an $N \times N$ matrix. Form matrix Δ_{N-2}, which is $N - 2 \times N - 2$ from Δ, by deleting rows 1 and N and columns 1 and N; Δ_{N-2} is called an inner. Now repeat this process on Δ_{N-2} to form Δ_{N-4}. Continue until it ends forming $\Delta_1, \Delta_3, \Delta_5, \ldots, \Delta_{N-2}$ for N-odd and $\Delta_2, \Delta_4, \ldots, \Delta_{N-2}$ for N-even. The appropriate set is called the inners of the matrix Δ.*

1

Remark. If N is even (larger than or equal to four), the number of inners is $[(N - 2)/2]$. The inners $\Delta_2, \Delta_4, \ldots$ are designated as the first, second, $\ldots$ inners, respectively. If N is odd, the number of inners is $[(N - 1)/2]$. The first, second, $\ldots$ inners are $\Delta_1, \Delta_3, \ldots$, respectively. Note that in this case the first inner, Δ_1, is a one-element matrix—that is, a scalar.

Example 1.1. Let $N = 6$. The inners of a 6×6 matrix are formed as follows:

$$
\Delta_6 = \begin{bmatrix}
a_{11} & a_{12} & a_{13} & a_{14} & a_{15} & a_{16} \\
a_{21} & a_{22} & a_{23} & a_{24} & a_{25} & a_{26} \\
a_{31} & a_{32} & a_{33} & a_{34} & a_{35} & a_{36} \\
a_{41} & a_{42} & a_{43} & a_{44} & a_{45} & a_{46} \\
a_{51} & a_{52} & a_{53} & a_{54} & a_{55} & a_{56} \\
a_{61} & a_{62} & a_{63} & a_{64} & a_{65} & a_{66}
\end{bmatrix}
\tag{1.1}
$$

The two inners are Δ_2 and Δ_4.

Example 1.2. Let $N = 5$. The inners of a 5×5 matrix are formed as follows:

$$
\Delta_5 = \begin{bmatrix}
a_{11} & a_{12} & a_{13} & a_{14} & a_{15} \\
a_{21} & a_{22} & a_{23} & a_{24} & a_{25} \\
a_{31} & a_{32} & a_{33} & a_{34} & a_{35} \\
a_{41} & a_{42} & a_{43} & a_{44} & a_{45} \\
a_{51} & a_{52} & a_{53} & a_{54} & a_{55}
\end{bmatrix}
\tag{1.2}
$$

The two inners are $\Delta_1 = a_{33}$ and Δ_3.

HISTORICAL BACKGROUND

Let $A(z)$ and $B(z)$ be the following polynomials:

$$
A(z) = z^n + a_{n-1}z^{n-1} + \cdots + a_0
\tag{1.3}
$$

$$
B(z) = b_m z^m + b_{m-1}z^{m-1} + \cdots + b_0
\tag{1.4}
$$

where z is a complex variable and a_i, b_i are real or complex. We assume that $m \leq n$. A basic classical result is that the determinant of the $(m + n)$-order *Sylvester matrix* [5],

$$\Delta_{n+m} = \begin{bmatrix} 1 & a_{n-1} & a_{n-2} & \cdots & a_0 & 0 & \cdots & 0 \\ 0 & 1 & a_{n-1} & \cdots & a_1 & a_0 & \cdots & 0 \\ \vdots & & & & & & & \\ 0 & 0 & \cdots & 1 & a_{n-1} & \cdots & \cdots & a_0 \\ \vdots & \vdots & \cdots & b_m & b_{m-1} & b_{m-2} & \cdots & b_0 \\ 0 & b_m & b_{m-1} & \cdots & \cdots & b_1 & \cdots & 0 \\ b_m & b_{m-1} & b_{m-2} & \cdots & \cdots & b_0 & \cdots & 0 \end{bmatrix} \begin{array}{l} \\ \left.\rule{0pt}{1.2cm}\right\} m \text{ rows} \\ \\ \left.\rule{0pt}{1.2cm}\right\} n \text{ rows} \\ \end{array} \qquad (1.5)$$

is nonzero if and only if $A(z)$ and $B(z)$ are relatively prime† [that is, no common zeros exist between $A(z)$ and $B(z)$]. This determinant is called the *resultant*‡ $R[A, B]$ of A and B.

The way in which the Sylvester matrix is formed can best be illustrated if we assume $n = 4$, $m = 3$ as follows:

$$\Delta_7 = \begin{bmatrix} 1 & a_3 & a_2 & a_1 & a_0 & 0 & 0 \\ 0 & 1 & a_3 & a_2 & a_1 & a_0 & 0 \\ 0 & 0 & 1 & a_3 & a_2 & a_1 & a_0 \\ 0 & 0 & 0 & b_3 & h_2 & b_1 & b_0 \\ 0 & 0 & b_3 & b_2 & b_1 & b_0 & 0 \\ 0 & b_3 & b_2 & b_1 & b_0 & 0 & 0 \\ b_3 & b_2 & b_1 & b_0 & 0 & 0 & 0 \end{bmatrix} \qquad (1.6)$$

Remarks

1. The determinants associated with the inners defined earlier are called bigradients or subresultants.

2. The Sylvester matrix§ has a certain pattern of left triangle of zeros, whereas the definition of the innerwise matrix has no such pattern. It is more general and includes the Sylvester pattern as a special case.

3. If $m < n$ and the coefficients a's and b's are real, the remainders of the Sturmian division process on the two polynomials may be replaced by determinants that are called by Trudi disencumbered remainders. The matrices of these remainders have the same form as the innerwise matrices of the Sylvester matrix except that the last column of any of the inners may be a function of a polynomial. This point is illustrated in Example 1.3.

† There exist alternate ways of testing for relative primeness that involve checking the singularity of lower-dimension matrices [3].
‡ Also, in special cases of certain polynomials, referred to as dialytic eliminant [8].
§ This matrix is also referred to as the "Chevron" matrix [17].

Example 1.3 [1]. Let

$$A(z) = a_0 + a_1 z + a_2 z^2 + a_3 z^3 + a_4 z^4 \qquad (1.7)$$

$$B(z) = b_0 + b_1 z + b_2 z^2 + b_3 z^3 \qquad (1.8)$$

where $a_4 \neq 0$, $b_3 \neq 0$.

Trudi's disencumbered remainders are obtained by using the Sturmian division process† on the two equations $A(z)$ and $B(z)$. They are given by the following two determinants:

$$\begin{vmatrix} a_4 & a_3 & a_2 z^2 + a_1 z + a_0 \\ 0 & b_3 & b_2 z^2 + b_1 z + b_0 \\ b_3 & b_2 & b_1 z^2 + b_0 z \end{vmatrix} \qquad (1.9)$$

$$\begin{vmatrix} a_4 & a_3 & a_2 & a_1 & a_0 z \\ 0 & a_4 & a_3 & a_2 & a_1 z + a_0 \\ 0 & 0 & b_3 & b_2 & b_1 z + b_0 \\ 0 & b_3 & b_2 & b_1 & b_0 z \\ b_3 & b_2 & b_1 & b_0 & 0 \end{vmatrix} \qquad (1.10)$$

and the determinant of the following matrix:

$$\begin{bmatrix} a_4 & a_3 & a_2 & a_1 & a_0 & 0 & 0 \\ 0 & a_4 & a_3 & a_2 & a_1 & a_0 & 0 \\ 0 & 0 & a_4 & a_3 & a_2 & a_1 & a_0 \\ 0 & 0 & 0 & b_3 & b_2 & b_1 & b_0 \\ 0 & 0 & b_3 & b_2 & b_1 & b_0 & 0 \\ 0 & b_3 & b_2 & b_1 & b_0 & 0 & 0 \\ b_3 & b_2 & b_1 & b_0 & 0 & 0 & 0 \end{bmatrix} \qquad (1.11)$$

The coefficients of z^2, z, and z^0 are given, respectively, by the determinants of the second and third inners of the preceding matrix as well as by the determinant of the matrix.

It is possible to form bigradients that are themselves polynomials.‡ Such

† This division process can be expressed as follows:

$$q_i(z) = q_{i+1}(z) x_i(z) - q_{i+2}(z) \qquad i = 0, 1, 2, \dots$$

where $q_0(z) = A(z)$, $q_1(z) = B(z)$, and $x_i(z)$ is a constant or linear in z obtained in a form to cancel the highest-order terms in z in the above division process. The remainders are $k_i q_i(z)$ where the k_i's are arbitrary constants.

‡ See Reference 22 for an interesting application.

polynomial bigradients play an important role in obtaining the greatest common divisor of specified degree of two polynomials. This topic is briefly explained as follows [6]:

Replace the last column of Δ_{i+j} inner by

$$[z^{i-1}A(z), z^{i-2}A(z), \ldots, z^{j-2}B(z), z^{j-1}B(z)]'$$

where $'$ is the transpose.

The greatest common divisor (gcd) of $A(z)$ and $B(z)$ is $|\Delta_{m-j+n-j}(z)|$ and has degree j where the first nonzero term in the sequence $|\Delta_{m-k+n-k}|$, $k = 0, 1, 2, \ldots$ occurs when $k = j$.

1.2   POSITIVE INNERWISE AND NEGATIVE INNERWISE MATRICES

If the determinants of all the inners, as well as of the matrix, are positive, we designate this matrix as positive innerwise or (pi). If all the determinants are negative, we designate this matrix as negative innerwise or (ni).

The definition of a positive innerwise matrix plays an important role in stability and root-clustering problems, as will be discussed in detail in later chapters. For such problems the innerwise matrix takes a special form of left triangle of zeros as in the Sylvester matrix form. There are other applications of positive innerwise matrix where the matrix form is of a more general nature. It is because of such more general applications that we do not refer to the inner matrices determinants as subresultants, bigradients, or disencumbered remainders. One such application is presented in the following theorem, and others will be presented here as well as in later chapters.

THEOREM 1.1† **[7].**   *For any matrix (symmetric or nonsymmetric) to be positive definite, it is necessary that it be positive innerwise.*

The proof can be obtained by contradiction. Suppose that the matrix is positive definite (pd) but not (pi); then one of the inners determinants has negative sign or zero, which implies that the inner is not (pd). Hence the matrix cannot be positive definite. This necessary condition can be utilized in connection with Sylvester's theorem for checking the positive definiteness of a symmetric matrix. The nonsymmetric matrix is positive definite if and only if the symmetric matrix constructed by the sum of its transpose with itself is positive definite. Thus we can apply Sylvester's theorem to check the positive definiteness of a newly constructed symmetric matrix (i.e., all the leading principal minors of this matrix are positive).

If the matrix is not symmetric, we can also apply Sylvester's theorem to the

† This theorem can be readily deduced from the definition of a positive definite matrix as given by Gantmacher [13].

matrix formed from the original matrix plus its transpose. The newly formed matrix is symmetric. The application of Sylvester's theorem follows from the fact that if the matrix is (pd), the formed matrix is also (pd) and vice versa.

Example 1.4. Let the matrix A associated with a quadratic polynomial form be as follows:

$$A = \begin{bmatrix} 1 & 2 & 5 & 6 \\ 1 & 3 & 5 & 12 \\ 2 & 4 & 6 & 8 \\ & & \Delta_2 & \\ 4 & 8 & 9 & 10 \end{bmatrix} \tag{1.12}$$

and consider $A + A'$ to give

$$A + A' = \begin{bmatrix} 2 & 3 & 7 & 10 \\ 3 & 6 & 9 & 20 \\ 7 & 9 & 12 & 17 \\ & & \Delta_3 & \\ 10 & 20 & 17 & 20 \end{bmatrix} \tag{1.13}$$

In applying Sylvester's theorem to (1.13), we ascertain by calculating the third leading principal minor $|\Delta_3|$, which is negative, that A is not (pd). However, this conclusion can be readily ascertained from calculating $|\Delta_2|$, the first inner determinant of (1.12). This offers a simplification in the calculations by checking first whether A is (pi) or not. Furthermore, if the given symmetric matrix is $A + A'$, then by checking its first inner determinant (which is negative), we ascertain that it is not (pd).

The definition of a negative innerwise matrix is utilized in obtaining conditions on the root distribution of polynomials, as will be explained in Chapter 2.

1.3 NULL-INNERWISE, NONNULL-INNERWISE, AND SEMI-INNERWISE SUBMATRICES

DEFINITION 1.2. *If the determinants of the inners of Δ_N as well as of Δ_N are zero, we designate it as null innerwise. If none of the determinants is zero, we designate it as nonnull innerwise. If some of the inners determinants are zero, we designate it as a semi-innerwise submatrix.*

The definitions of null-innerwise matrices or submatrices occur in the critical cases of root distribution of polynomials. The importance of their definitions will become apparent in following chapters. However, we present an application to nonnull-innerwise matrix in this section.

THEOREM 1.2.   *The necessary and sufficient condition for the rational function*

$$Z(s) \triangleq \frac{g(s)}{f(s)} = \frac{b_{n-1}s^{n-1} + b_{n-2}s^{n-2} + \cdots + b_1 s + b_0}{a_n s^n + a_{n-1}s^{n-1} + \cdots + a_1 s + a_0}$$

$$\text{with } a_n \neq 0; \quad (1.14)$$

to have a certain continued fraction expansion is that the matrix Δ_{2n-1} given below be nonnull innerwise.

$$\Delta_{2n-1} \triangleq \begin{bmatrix} a_n & a_{n-1} & \cdots & a_2 & a_1 & a_0 & \cdots & 0 & 0 \\ 0 & a_n & & a_3 & a_2 & a_1 & \cdots & 0 & 0 \\ 0 & 0 & & a_n & a_{n-1} & a_{n-2} & \cdots & & a_0 \\ 0 & & & 0 & b_{n-1} & b_{n-2} & \cdots & & b_0 \\ 0 & 0 & & b_{n-1} & b_{n-2} & b_{n-3} & \cdots & & 0 \\ 0 & b_{n-1} & & b_{n-2} & & & \cdots & & 0 \\ b_{n-1} & b_{n-2} & \cdots & & b_0 & 0 & \cdots & 0 \end{bmatrix} \quad (1.15)$$

The proof readily follows from Wall [10] and the rearrangement of his matrix in inner form. It should be noted that Wall's matrix is presented in minor array form, whereby the odd minors need be checked. The inner-minor array transformation will be discussed in the next section.

1.4   INNER-MINOR ARRAY MATRIX TRANSFORMATIONS [11]

In many applications the subresultants are obtained as the leading principal minor arrays of a matrix, such as in Hurwitz's matrix. Also, many criteria on root clustering can be expressed in minor form. Hence an inner-minor array matrix transformation that preserves the determinant's values of the matrices of same dimensions is needed. The following transformation gives equivalent inner determinants and either odd or even leading principal minors.

$$M = P \mathscr{I} P' \quad (1.16)\dagger$$

where M = minor matrix form $N \times N$

$\mathscr{I}$ = inner matrix form $N \times N$

P = permutation matrix of dimension N whose elements are either zero or unity‡

P' = permutation matrix transpose

† It should be noted that the minors are defined as determinants in the literature.
‡ Note that $|P| = 1$.

For N-even equal to four,

$$P = \begin{bmatrix} 0 & 0 & 1 & 0 \\ 0 & 1 & 0 & 0 \\ 0 & 0 & 0 & 1 \\ 1 & 0 & 0 & 0 \end{bmatrix} = P_{3241} \qquad (1.17)$$

The subscript 3 relates to column *three* and row *one*, whose element is unity. Similarly, the subscript two related to column *two* and row *two*, whose element is unity, and so forth. The rest of the elements are zero. It may be noted that matrix premultiplication by P changes the row location, while postmultiplication by P' changes the column location in this matrix [11].

For N equal to three,

$$P = \begin{bmatrix} 0 & 1 & 0 \\ 1 & 0 & 0 \\ 0 & 0 & 1 \end{bmatrix} = P_{213} \qquad (1.18)$$

Therefore the permutation transformation for both odd and even matrix dimensions is

$$P_1, P_{21}, P_{213}, P_{3241}, P_{32415}, P_{435261}, P_{4352617}, \ldots$$

Similarly, one can obtain the inners from the minor arrays as follows:

$$\mathscr{I} = P'MP \qquad (1.19)$$

Equation (1.19) is evident by noting (1.16) and $P' = P^{-1}$, where P^{-1} is the inverse of P.

One can easily verify that another type of permutation transformation exists, which can be expressed for both even and odd N, as follows:

$$P_1, P_{12}, P_{231}, P_{2314}, P_{34251}, P_{342516}, P_{531246}, \ldots$$

One can also obtain a general relationship for generating the permutation matrix P for N-even and N-odd. These relationships are given below (and can easily be verified).

$$p_i = n \pm (-1)^i [i/2]$$
$$N\text{-odd}, \quad i = 1, 2, \ldots, N = 2n - 1 \quad (1.20)$$

where $N \times N$ is the dimension of the matrix, and

$$p_i = n \pm (-1)^i [i/2]$$
$$N\text{-even}, \quad i = 1, 2, \ldots, N = 2n \quad (1.21)$$

The notation $[i/2]$ indicates the minimum integer value in the fraction.

For the case of root clustering in the left half of the complex plane, as in stability of linear continuous systems (Routh–Hurwitz criterion), the stability matrix takes a special form, as known in the literature. To obtain the corresponding innerwise matrix, we premultiply the Hurwitz matrix by the following premutation matrix: For n-even [the permutation matrix dimension is $(n - 1) \times (n - 1)$ where n is the degree of the polynomial to be tested for stability],

$$
P = \begin{bmatrix}
0 & 1 & 0 & 0 & \cdots & 0 & 0 & 0 & 0 \\
0 & 0 & 0 & 1 & \cdots & 0 & 0 & 0 & 0 \\
\cdot & \cdot & \cdot & & & \cdot & \cdot & \cdot & \cdot \\
0 & 0 & 0 & 0 & 0 & 0 & 0 & 0 & 1 \\
0 & 0 & 0 & \cdot & \cdots & \cdot & 1 & 0 & 0 \\
\cdot & \cdot & \cdot & \cdot & \cdot & \cdot & \cdot & \cdot & \cdot
\end{bmatrix}
\tag{1.22}
$$

the pattern is†

$$
\begin{aligned}
p_{12} &= 1, \quad p_{24} = 1, \quad p_{35} = 1 \\
p_{n/2-1,n-2} &= 1, \quad p_{n/2,n-1} = 1 \\
p_{n/2+1,n-3} &= 1, \quad p_{n/2+2,n-5} = 1, \ldots
\end{aligned}
$$

When $n = 4$ and $n = 6$, the pattern is

$$
n = 4 \qquad\qquad\qquad n = 6
$$

$$
P = \begin{bmatrix} 0 & 1 & 0 \\ 0 & 0 & 1 \\ 1 & 0 & 0 \end{bmatrix}
\qquad
P = \begin{bmatrix}
0 & 1 & 0 & 0 & 0 \\
0 & 0 & 0 & 1 & 0 \\
0 & 0 & 0 & 0 & 1 \\
0 & 0 & 1 & 0 & 0 \\
1 & 0 & 0 & 0 & 0
\end{bmatrix}
$$

For n-odd [the permutation matrix dimension is $(n - 1) \times (n - 1)$],

$$
P = \begin{bmatrix}
1 & 0 & 0 & \cdots & 0 & 0 & 0 \\
0 & 0 & 1 & \cdots & 0 & 0 & 0 \\
\cdot & \cdot & \cdot & & \cdot & \cdot & \cdot \\
0 & \cdot & \cdot & \cdots & \cdot & 0 & 1 \\
0 & \cdot & \cdot & \cdots & 1 & 0 & 0 \\
\cdot & \cdot & \cdot & \cdots & \cdot & \cdot & \cdot
\end{bmatrix}
$$

† In this pattern the first subscript in p_{jk} refers to row location and the second to column location.

the pattern is

$$p_{11} = 1, \quad p_{23} = 1$$
$$p_{(n-1)/2,(n+1/2)} = 1, \quad p_{(n-1)/2+1,n-1} = 1$$
$$p_{(n-1)/2+2,n-3} = 1, \ldots$$

For example, $n = 5$; the permutation matrix is

$$P = \begin{bmatrix} 1 & 0 & 0 & 0 \\ 0 & 0 & 1 & 0 \\ 0 & 0 & 0 & 1 \\ 0 & 1 & 0 & 0 \end{bmatrix}$$

The innerwise matrix has the unified form of left triangles as shown earlier.

Finally, in order to show that, for the general problem of root clustering, only the innerwise matrix offers a unifying form which has a pattern that can be utilized advantageously for computational purposes (as will be discussed in Chapter 7), the following example is discussed. This example can be generalized for any degree polynomial and for any region of root clustering in the complex plane.

Example 1.5. Let

$$F(z) = a_4 z^4 + a_3 z^3 + a_2 z^2 + a_1 z + a_0 \qquad a_4 > 0 \quad (1.23)$$

The stability condition in the open left half-plane is [9]

(a) $a_k > 0$, $\qquad\qquad\qquad\qquad\qquad\qquad k = 0, 1, 2, 3, 4$ (1.24)

(b) In the Hurwitz matrix,

$$\Delta_3 = \begin{bmatrix} a_3 & a_1 & 0 \\[-2pt] {\scriptstyle\Delta_1} & & \\ a_4 & a_2 & a_0 \\ 0 & a_3 & a_1 \end{bmatrix} \qquad (1.25)$$

$|\Delta_1| > 0, |\Delta_3| > 0.$

The inner form of the preceding matrix can be obtained as discussed above by using the premultiplying matrix.

$$\Delta_3^1 = \begin{bmatrix} 0 & 1 & 0 \\ 0 & 0 & 1 \\ 1 & 0 & 0 \end{bmatrix} \begin{bmatrix} a_3 & a_1 & 0 \\ a_4 & a_2 & a_0 \\ 0 & a_3 & a_1 \end{bmatrix} = \begin{bmatrix} a_4 & a_2 & a_0 \\ 0 & a_3 & a_1 \\ a_3 & a_1 & 0 \end{bmatrix} \qquad (1.26)$$

The stability condition is that Δ_3^1 be positive innerwise (pi) plus condition (a) above.

For the same polynomial, the stability condition within the unit circle [7] is

$$F(1) > 0 \qquad F(-1) > 0 \tag{1.27}$$

and $\Delta_3^{\pm}$ is positive innerwise.

For instance, to obtain Δ_3^+, we have

$$\Delta_3^+ = X_3 + Y_3 = \begin{bmatrix} a_4 & a_3 & a_2 \\ 0 & a_4 & a_3 \\ 0 & 0 & a_4 \end{bmatrix}$$

$$+ \begin{bmatrix} 0 & 0 & a_0 \\ 0 & a_0 & a_1 \\ a_0 & a_1 & a_2 \end{bmatrix}$$

$$= \begin{bmatrix} a_4 & a_3 & a_2 + a_0 \\ 0 & a_4 + a_0 & a_3 + a_1 \\ a_0 & a_1 & a_4 + a_2 \end{bmatrix} \tag{1.28}$$

Δ_3^- can be obtained similarly.

The innerwise matrix Δ_3^+ can be transformed directly into a minor array form whereby the inner Δ_1^+ becomes the leading principal minor array and the determinant values of the matrices are the same. The transformed array is given by

$$\Delta_3^+ = \begin{bmatrix} a_4 + a_0 & 0 & a_3 + a_1 \\ a_3 & a_4 & a_2 + a_0 \\ a_1 & a_0 & a_4 + a_2 \end{bmatrix} \tag{1.29}$$

The preceding matrix is obtained by exchanging the second column by the first and the second row by the first row in (1.28).

Remark. In comparing the inner form for the left half-plane and for the unit circle, the first element of the second row in both cases is zero. This is the unifying feature. The minor array form has no such unifying pattern. Furthermore, there exists no inner-minor array transformation that makes the first element of the third row in Δ_3^+ zero, as in the left half-plane. This is also valid for any n and for more general regions in the complex plane.

1.5 EXAMPLES

In this section we will consider two examples in which the matrices are generated in terms of their inners. Starting in the first row (or row and column) of the matrix, the entries of the last inner are generated by a certain rule and, similarly, the entries of all the other inners. Hence the inners play an important role in the generation of the elements of certain matrices.

Example 1.6. This example is based on an easy way of illustrating the construction of a symmetric matrix [14] that occurs in the stability of linear discrete systems. This matrix is given as follows:

$$b_{ij} = \sum_{p=0}^{min(i,j)} (a_{i-p}a_{j-p} - a_{n+p-i}a_{n+p-j}) \qquad (1.30)$$

for $i, j = 0, 1, 2, \ldots, n - 1$, with $a_0 = 1$, and where the a_i's are the coefficients of the following characteristic polynomial of nth degree:

$$P(z) = a_0 z^n + a_1 z^{n-1} + a_2 z^{n-2} + \cdots + a_n$$

$$a_0 \equiv 1 \quad (1.31)$$

The procedure for generating the matrix B is as follows:

1. Form the first coefficient array from $P(z)$ as shown in (1.31a). The second row is the reverse of the first row. Obtain the entries of the third row as indicated by the arrows in (1.31a). These constitute the elements of the first row of B. Since B is symmetric in both, the diagonals, the entries of the first column, and the last row of the columns are easily obtained.

$$
\begin{array}{c}
1 \qquad\qquad a_1 \quad a_2 \quad \cdots \quad a_{n-1} \qquad a_n \\
a_n \qquad a_{n-1} \quad \cdots \quad a_1 \qquad 1 \\
\hline
1 - a_n^2 \quad a_1 - a_n a_{n-1} \quad \cdots \quad a_{n-1} - a_1 a_n
\end{array}
\qquad (1.31a)
$$

2. To get the row corresponding to the last inner, add to the terms at the tails of the arrows in (1.32) the term (written below) obtained from the coefficient array.

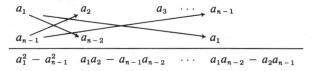

$$
\begin{array}{c}
a_1 \qquad\qquad a_2 \qquad\qquad a_3 \quad \cdots \quad a_{n-1} \\
a_{n-1} \qquad a_{n-2} \qquad\qquad a_1 \\
\hline
a_1^2 - a_{n-1}^2 \quad a_1 a_2 - a_{n-1} a_{n-2} \quad \cdots \quad a_1 a_{n-2} - a_2 a_{n-1}
\end{array}
$$

This coefficient array is obtained from the coefficient array of step 1 by omitting the outer entries. The rows and columns of the last inner are then filled by utilizing the symmetry as in step 1.

3. Succeeding inners are obtained in the same manner as in step 2. When n is odd, the first inner is obtained from a coefficient array of two elements in each row.

$$
B = \begin{bmatrix}
1 - a_n^2 & a_1 - a_n a_{n-1} & \cdots & a_{n-3} - a_n a_3 & a_{n-2} - a_n a_2 & a_{n-1} - a_1 a_n \\
a_1 - a_n a_{n-1} & 1 - a_n^2 + (a_1^2 - a_{n-1}^2) & a_1 - a_n a_{n-1} + (a_1 a_2 - a_{n-1} a_{n-2}) & \cdots & a_{n-3} - a_n a_3 + (a_1 a_{n-2} - a_2 a_{n-1}) & a_{n-2} - a_n a_2 \\
& & \text{First inner} & & & \vdots \\
a_{n-2} - a_n a_2 & & & & & a_1 - a_n a_{n-1} \\
a_{n-1} - a_1 a_n & a_1 - a_n a_{n-1} & \cdots & & & 1 - a_n^2
\end{bmatrix}
\tag{1.32}
$$

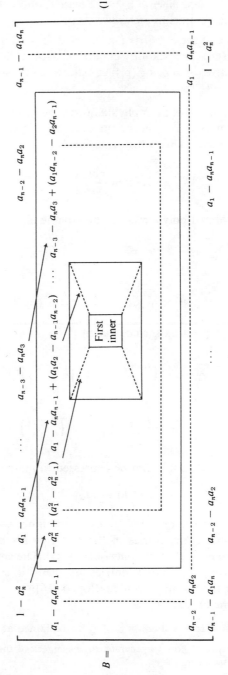

13

When n is even, the first inner is a 2×2 matrix, obtained from the coefficient arrays of the three elements in each row.

The significance of the B matrix (as will be discussed in Chapter 3), is to be positive definite (pd) for a stability condition. Hence the determinants of the inners being positive (or B being (pi)) constitute, as shown earlier, a *necessary* condition for B to be (pd).

Example 1.7† [15]. This example illustrates the mechanism involved in the bilinear transformation from inside the unit circle in the z-plane to the open left half of the s-plane. This transformation is

$$z = \frac{s + 1}{s - 1} \tag{1.33}$$

where the transformed polynomial in the z-plane is

$$a_n z^n + a_{n-1} z^{n-1} + \cdots + a_1 z + a_0 \tag{1.34}$$

and in the s-plane is

$$b_n s^n + b_{n-1} s^{n-1} + \cdots + b_1 s + b_0 \tag{1.35}$$

The relationship expressing the coefficient is

$$b_{n-K} = \sum_{j=0}^{n} a_{n-j} \left[\binom{n-j}{K} - \binom{j}{1}\binom{n-j}{K-1} + \binom{j}{2}\binom{n-j}{K-2} \right.$$

$$\left. - \cdots + (-1)^{K-1}\binom{j}{K-1}\binom{n-j}{1} + (-1)^K\binom{j}{K} \right] \tag{1.36}$$

The foregoing relationship can be expressed in matrix form as

$$b] = Qa] \tag{1.37}$$

The matrix Q can be generated by using the inner's similarity to the first example. In this case, the entries of the first row are all one, and the first columns are the binomial coefficients of $(s + 1)^n$. The entries of the last inner are generated from the known entries by using a certain combinatorial relationship that has recently been developed [16].

To illustrate the procedure, let $n = 5$.

1. Form the first row and column of Q as explained above.

† This example illustrates how to generate all the inners of the matrix in succession, starting from the last inner.

$$Q = \begin{bmatrix} 1 & 1 & 1 & 1 & 1 & 1 \\ 5 & 5-(1+1) & 1 & -1 & -3 & -5 \\ & =3 & & & & \\ 10 & 2 & -2 & -2 & 2 & 10 \\ 10 & -2 & -2 & 2 & 2 & -10 \\ 5 & -3 & 1 & 1 & -3 & 5 \\ 1 & -1 & 1 & -1 & 1 & -1 \end{bmatrix} \tag{1.38}$$

2. The entries of the last row are the same as the first except for the sign change, which should be $(+, -, +, -, +, \ldots)$. Similarly for the last column.

3. The entries of the last inner row are obtained in a combinatorial procedure as indicated in the arrow; that is, $5 - (1 + 1) = 3$, $3 - (1 + 1) = 1$, $1 - (1 + 1) = -1, \ldots$.

4. The entries of the first inner are similarly obtained following step 3.

Remark. A certain symmetry in the coefficients can be utilized. In this case, one needs to calculate only the entries of the 2×2 matrix as shown in dotted lines. The other entries can readily be written by inspection (to be discussed in Chapter 7).

In concluding this chapter we may mention that although many applications of the inners in system theory will be thoroughly discussed in the remaining chapters,† it is worthwhile to mention two applications in different fields—namely, quantum physics [18] and sparse matrix theory [19].

In the field of quantum physics, the innerwise matrix arises in calculating the energy-level patterns for rigid asymmetric rotors used in obtaining the rotational energies of the molecules [18,20]. The innerwise matrix in this case is used to obtain a transformation to certain symmetrized basis functions and is referred to in the literature as the Wang [21] symmetrizing transformation.

In the field of sparse matrix theory, the innerwise matrix is often arrived at in order that no zero-block matrices can become nonzero because of roundoff error in the process of the Gauss elimination method of computation. In these cases, one would like to determine the permutation matrices P and Q such that [19]

$$PAQ = \tilde{A} \tag{1.39}$$

where $\tilde{A}$ is an innerwise matrix with left triangle of zeros and A is a given sparse matrix whose inverse is to be evaluated.

† See also References 23–26.

REFERENCES

[1] A. T. FULLER, "Conditions for aperiodicity in linear systems," *Brit. J. Appl. Phys.*, **6**, 195–198 (June 1955).

[2] A. T. FULLER, "Stability criteria for linear systems and realizability criteria for RC networks," *Proc. Camb. Phil. Soc.*, **53**, 878–896 (1957).

[3] S. BARNETT, 'Discussion on "'inners,' Approach to some problems of system theory," by E. I. Jury,' *IEEE Trans. on Automatic Control*, **AC-17** (1), 175–176 (Feb. 1972).

[4] N. TRUDI, *Teoria di Determinanti e Loro Applicazioni*. Napoli: Libreria Scientifica e Industriale, d. B. Pellerano, 1862.

[5] M. BÔCHER, *Introduction to Higher Algebra*. New York: Dover, 1964 (reprint of 1907 edition).

[6] A. S. HOUSEHOLDER, "Bigradients and the problem of Routh–Hurwitz," *SIAM Rev.*, **10**, 56–66 (1968).

[7] E. I. JURY, "'Inners,' approach to some problems of system theory," *IEEE Trans. on Automatic Control*, **AC-16** (3), 233–240 (June 1971).

[8] B. M. BROWN, "On the distribution of the zeros of a polynomial," *The Quarterly Journal of Mathematics, Oxford Series*, **16** (63), 241–256 (Sept. 1968).

[9] S. BARNETT, "Matrices, polynomials, and linear time-invariant systems," *IEEE Trans. on Automatic Control*, **AC-18** (1), 1–9 (Feb. 1973).

[10] H. S. WALL, *Analytic Theory of Continued Fractions*. New York: Van Nostrand, 1948, pp. 161–166.

[11] R. K. SHARMA and E. I. JURY, "Inner–minor transformations," *IEEE Trans. on Automatic Control*, **AC-18** (1), 75–76 (Feb. 1973).

[12] J. H. WILKINSON, *The Algebraic Eigenvalue Problem*. Oxford: Clarendon Press, 1965, pp. 44–45.

[13] F. R. GANTMACHER, *The Theory of Matrices*. New York: Chelsea, 1959, Vol. 1, p. 306.

[14] I. G. SARMA and M. A. PAI, "A note on the Liapunov matrix equation for linear discrete systems," *IEEE Trans. on Automatic Control (Correspondence)*, **AC-13**, 119–121 (Feb. 1968).

[15] E. I. JURY, "Remarks on the mechanics of bilinear transformation," *IEEE Trans. on Audio and Electroacoustics*, **AU-21** (4), 380–382 (Aug. 1973).

[16] H. M. POWER, "Further comments on the mechanics of the bilinear transformation," *IEEE Trans. on Education*, pp. 114–115 (Aug. 1970).

[17] A. ROWE, "The generalized resultant matrix," *J. Inst. Math. Appl.*, **9**, 390–396 (1972).

[18] G. W. KING, R. M. HAINES, and P. C. CROSS, "The asymmetric rotors," I. "Calculation and symmetry classification of energy levels," *J. Chem. Phys.*, **11**, 27–42 (Jan. 1943).

[19] R. P. TEWARSON, "Computations with sparse matrices," *SIAM Rev.*, **12** (4), 527–543 (Oct. 1970).

[20] M. TINKHAM, *Group Theory and Quantum Mechanics*. New York: McGraw-Hill, 1964.

[21] S. C. WANG, "On the assymetrical top in quantum mechanics," *Phys. Rev.*, *Second Series*, **34** (2), 243–252 (July 15, 1929).

[22] M. URUSKI and M. S. PEIKARSKI, "Synthesis of a network containing a cascade of commensurate transmission lines and lumped elements," *Proc. IEE* (*London*), **119** (2), 153–160 (Feb. 1972).

[23] M. P. EPSTEIN, "The use of resultants to locate extreme values of polynomials," *SIAM J. Appl. Math.*, **16** (1), 62–70 (1968).

[24] P. C. MÜLLER, "Solution of the matrix equations $AX + XB = -Q$ and $S^T X + XS = -Q$," *SIAM J. Appl. Math.*, **18** (3), 682–687 (May 1970).

[25] R. E. HARTWIG, "The resultant and the matrix equation $AX = XB$," *SIAM J. Appl. Math.*, **22** (4), 538–544 (June 1972).

[26] R. E. HARTWIG, "Resultants and the solution of $AX - XB = -C^*$," *SIAM J. Appl. Math.*, **23** (1), 104–117 (July 1972).

Stability and Positive Innerwise Matrices

The problem of stability of linear time-invariant dynamic systems (both continuous and discrete) has been the subject of investigation by many mathematicians, physicists, and engineers during the last century. Numerous authoritative books have been written on this and related topics. The aim of this chapter is to represent the various stability and other related criteria in a unified form in terms of positive innerwise matrices. These various criteria are well developed in the literature; consequently, in order to keep within the limits of this manuscript, most of the known proofs are not repeated and only new ones will be discussed. Authoritative references are given to the well-known methods of proof as well as to other forms of stability criteria.

As an extension of the various stability criteria, conditions on the root clustering of polynomials in certain regions in the complex plane with applications are also presented. Finally, conditions on the root distribution of polynomials in terms of innerwise matrices are discussed in the last section. The application of the root-distribution problem to positivity and non-negativity conditions will be presented in Chapter 5.

2.1 STABILITY OF LINEAR TIME-INVARIANT CONTINUOUS SYSTEMS (ROOT CLUSTERING IN THE OPEN LEFT HALF OF THE COMPLEX PLANE)

It is well known that the condition of stability of linear time-invariant continuous systems can be represented, in terms of the roots of the characteristic equation, to be in the open left half of the s-plane. The characteristic equation is a real polynomial with degree n. For relative stability (as will be discussed in the next section), the condition is represented in terms of a complex polynomial having its roots in the open left half-plane. Hence in this section we will start from complex polynomials and obtain the conditions under which the roots occur in the open left half-plane. We will obtain Hurwitz's and Liénard–Chipart's criteria as a special case.

$$
\Delta_{2n-1}=
\left[
\begin{array}{ccccccccccccc}
1 & -a''_{n-1} & 1 & 0 & 0 & 0 & 0 & 0 & \cdots & \cdots & 0 & 0 \\
0 & 1 & -a''_{n-1} & 1 & 0 & 0 & 0 & 0 & \cdots & \cdots & 0 & 0 \\
0 & 0 & 1 & -a''_{n-1} & 1 & 0 & 0 & 0 & \cdots & \cdots & 0 & 0 \\
0 & 0 & 0 & 1 & -a''_{n-1} & 1 & 0 & 0 & \cdots & \cdots & 0 & 0 \\
0 & 0 & 0 & 0 & \boxed{-a'_{n-1}} & -a''_{n-2} & a'_{n-2} & -a''_{n-3} & \cdots & \cdots & * & 0 \\
0 & 0 & 0 & 0 & 1 & \boxed{-a''_{n-2}} & -a'_{n-2} & a''_{n-3} & \cdots & \cdots & ** & 0 \\
0 & 0 & 0 & 0 & 0 & a'_{n-1} & \boxed{a'_{n-2}} & -a''_{n-3} & a'_{n-3} & -a''_{n-4} & \cdots & 0 \\
0 & 0 & 0 & 0 & 0 & -a''_{n-2} & \boxed{-a''_{n-3}} & -a'_{n-3} & a''_{n-4} & \cdots & \cdots & 0 \\
0 & 0 & 0 & 0 & 0 & -a'_{n-3} & a''_{n-4} & \boxed{a'_{n-4}} & -a''_{n-5} & -a'_{n-5} & \cdots & 0 \\
\vdots & & & & & & & & & & \vdots & \vdots \\
0 & 0 & 0 & 0 & 0 & -a'_{n-6} & -a''_{n-6} & -a'_{n-6} & -a''_{n-7} & \cdots & 0 & 0 \\
a'_{n-1} & -a''_{n-2} & & & & & & & & & 0 & 0
\end{array}
\right]
\tag{2.2}
$$

(Innermost boxes: $\Delta_1 \subset \Delta_3 \subset \Delta_5 \subset \Delta_7$.)

$*$ $(-1)^{(n+1)/2}a_0''$ when n is odd; $(-1)^{n/2}a_0'$ when n is even

$**$ $(-1)^{(n-1)/2}a_0'$ when n is odd; $(-1)^{n/2}a_0''$ when n is even

Let $F(s)$ be represented as follows:

$$F(s) = s^n + (a'_{n-1} + ja''_{n-1})s^{n-1} + (a'_{n-2} + ja''_{n-2})s^{n-2} + \cdots$$
$$+ a'_0 + ja''_0 \qquad (2.1)$$

The necessary and sufficient conditions for the roots of $F(s) = 0$ to lie in the open left half-plane are represented (following Marden [1])† in terms of a $2n - 1 \times 2n - 1$ matrix Δ_{2n-1}. This matrix is to be positive innerwise (pi) and is given as in equation 2.2 on the previous page.

Remarks

1. An alternate form of the necessary and sufficient conditions for the roots of $F(s)$ to be in the open left half of the s-plane (especially when the coefficient of s^n is not unity) can be represented in terms of a Δ_{2n} matrix to be positive innerwise (pi). This is determined from the Hurwitz determinantal form (and called the generalized Routh–Hurwitz criteria).‡ This case is given as follows. Let $F(js)$ be given as

$$F(js) = (b_n + jc_n)s^n + (b_{n-1} + jc_{n-1})s^{n-1} + \cdots + (b_1 + jc_1)s + b_0 + jc_0,$$
$$c_n \neq 0 \quad (2.3)$$

Then Δ_{2n} can be represented as

$$\Delta_{2n} = \begin{bmatrix} c_n & c_{n-1} & c_{n-2} & \cdots & 0 & \cdots & 0 \\ 0 & & & & & & \\ 0 & c_n & c_{n-1} & c_{n-2} & c_{n-3} & & \\ \vdots & 0 & c_n & c_{n-1} & c_{n-2} & \cdots & c_1 & c_0 \\ 0 & & & & & & \\ \vdots & 0 & b_n & b_{n-1} & b_{n-2} & \cdots & b_1 & b_0 \\ 0 & b_n & b_{n-1} & b_{n-2} & b_{n-3} & & \\ 0 & & & & & & \\ 0 & & & & & & \\ b_n & b_{n-1} & b_{n-2} & \cdots & 0 & \cdots & 0 \end{bmatrix} \qquad (2.4)$$

2. If the constant term in $F(s)$ is real, the necessary and sufficient conditions for the roots to be in the open left half-plane can be obtained by using algorithms suggested by Duffin [2]. In this algorithm a simple necessary condition can be conveniently utilized (as will be shown in the next section).

† See Exercise, p. 178, and (40.1), p. 179 (when expressed in inner form), of Reference 1.
‡ See Gantmacher [19].

Example 2.1. Let $n = 5$ in (2.1). The innerwise matrix of (2.2) is given by

$$
\Delta_9 = \begin{bmatrix}
1 & -a''_4 & -a'_3 & a''_2 & a'_1 & -a''_0 & 0 & 0 & 0 \\
0 & 1 & -a''_4 & -a'_3 & a''_2 & a'_1 & -a''_0 & 0 & 0 \\
0 & 0 & 1 & -a''_4 & -a'_3 & a''_2 & a'_1 & -a''_0 & 0 \\
0 & 0 & 0 & 1 & -a''_4 & -a'_{33} & a''_2 & a'_1 & -a''_0 \\
0 & 0 & 0 & 0 & a'_4 & -a''_2 & -a'_2 & a''_1 & a'_0 \\
0 & 0 & 0 & a'_4 & -a''_3 & -a'_2 & a''_1 & a'_1 & 0 \\
0 & 0 & a'_4 & -a''_3 & -a'_2 & a''_1 & a'_0 & 0 & 0 \\
0 & a'_4 & -a''_3 & -a'_2 & a''_1 & a'_0 & 0 & 0 & 0 \\
a'_4 & -a''_3 & -a'_2 & a''_1 & a'_0 & 0 & 0 & 0 & 0
\end{bmatrix}
\quad (2.5)
$$

Now, if we let $F(s)$ in (2.1) be a real polynomial with $a'_i = a_i$ (and $a''_i = 0$), the matrix in (2.2), being positive innerwise, becomes equivalent to about two half-sized matrices being positive innerwise. These two matrices for odd and even inners can be combined into one minor form matrix. This matrix is known to be the *Hurwitz matrix* [3], and the stability condition reduces to Hurwitz's matrix being positive (i.e., all leading principal minors are positive). This point can easily be explained by the following example.

Example 2.2. Let $F(s)$ be

$$F(s) = s^5 + a_4 s^4 + a_3 s^3 + a_2 s^2 + a_1 s + a_0$$

From (2.5) we have

$$
\Delta_9 = \begin{bmatrix}
1 & 0 & -a_3 & 0 & a_1 & 0 & 0 & 0 & 0 \\
0 & 1 & 0 & -a_3 & 0 & a_1 & 0 & 0 & 0 \\
0 & 0 & 1 & 0 & -a_3 & 0 & a_1 & 0 & 0 \\
0 & 0 & 0 & 1 & 0 & -a_3 & 0 & a_1 & 0 \\
0 & 0 & 0 & 0 & a_4 & 0 & -a_2 & 0 & a_0 \\
0 & 0 & 0 & a_4 & 0 & -a_2 & 0 & a_0 & 0 \\
0 & 0 & a_4 & 0 & -a_2 & 0 & a_0 & 0 & 0 \\
0 & a_4 & 0 & -a_2 & 0 & a_0 & 0 & 0 & 0 \\
a_4 & 0 & -a_2 & 0 & a_0 & 0 & 0 & 0 & 0
\end{bmatrix}
\quad (2.6)
$$

The preceding matrix can be written as two innerwise matrices.

$$\Delta_4 = \begin{bmatrix} a_4 & a_2 & a_0 & 0 \\ 0 & a_4 & a_2 & a_0 \\ 0 & 1 & a_3 & a_1 \\ 1 & a_3 & a_1 & 0 \end{bmatrix} \qquad \Delta_5 = \begin{bmatrix} 1 & a_3 & a_1 & 0 & 0 \\ 0 & 1 & a_3 & a_1 & 0 \\ 0 & 0 & a_4 & a_2 & a_0 \\ 0 & a_4 & a_2 & a_0 & 0 \\ a_4 & a_2 & a_0 & 0 & 0 \end{bmatrix} \qquad (2.7)$$

The innerwise matrices can then be combined into one matrix.

$$\Delta_{5_H} = \begin{bmatrix} a_4 & a_2 & a_0 & 0 & 0 \\ 1 & a_3 & a_1 & 0 & 0 \\ 0 & a_4 & a_2 & a_0 & 0 \\ 0 & 1 & a_3 & a_1 & 0 \\ 0 & 0 & a_4 & a_2 & a_0 \end{bmatrix} \qquad (2.8)$$

This matrix is recognized as the Hurwitz matrix for $n = 5$. If the coefficient of s^n is $a_n > 0$, then, instead of one, the entry a_n is substituted in the preceding matrices.

It should be noted that Δ_4 and Δ_5, being positive innerwise, are equivalent to Δ_{5_H} being positive, and vice versa.

We can also obtain the Hurwitz matrix from (2.4) when $c_i = a_i$ and b_i''s $= 0$.

It is of interest to note from (2.8) that $\Delta_{5_H} = a_0 \Delta_{4_H}$. This relationship also holds for the general case; that is, $\Delta_{n_H} = a_0 \Delta_{n-1_H}$. It will be utilized here and in later chapters.

Furthermore, a necessary condition for stability is that all the a_i''s are positive. Based on this condition and Hurwitz determinants, Liénard and Chipart [4] were able to simplify the Hurwitz criterion significantly. They showed that, for stability, the sign of either the odd or the even Hurwitz determinants must be positive. This simplified criterion is presented in an inner form as follows [5].

LIÉNARD–CHIPART CRITERION

Let $F(s)$ be the real polynomial

$$F(s) = a_n s^n + a_{n-1} s^{n-1} + \cdots + a_1 s + a_0$$
$$\text{with } a_n > 0 \quad (2.9)$$

The necessary and sufficient conditions for the roots of $F(s)$ to lie in the open left half-plane can be given as follows [5]:

1. The a_i's (or half of them) be positive.

2.   The following Δ_{n-1} matrices for n-odd or n-even be positive innerwise, respectively.

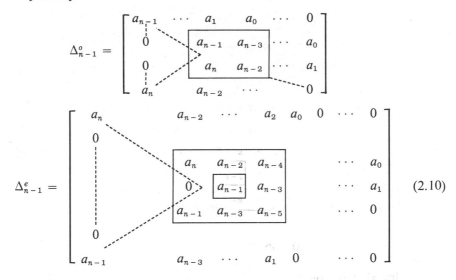

$$\Delta_{n-1}^o =
\begin{bmatrix}
a_{n-1} & \cdots & a_1 & a_0 & \cdots & 0 \\
0 & & a_{n-1} & a_{n-3} & \cdots & a_0 \\
0 & & a_n & a_{n-2} & \cdots & a_1 \\
a_n & & a_{n-2} & \cdots & & 0
\end{bmatrix}$$

$$\Delta_{n-1}^e =
\begin{bmatrix}
a_n & & a_{n-2} & \cdots & a_2 & a_0 & 0 & \cdots & 0 \\
0 & & & & & & & & \\
& & a_n & a_{n-2} & a_{n-4} & & & \cdots & a_0 \\
& & 0 & a_{n-1} & a_{n-3} & & & \cdots & a_1 \\
& & a_{n-1} & a_{n-3} & a_{n-5} & & & \cdots & 0 \\
0 & & & & & & & & \\
a_{n-1} & & a_{n-3} & \cdots & a_1 & 0 & & \cdots & 0
\end{bmatrix} \qquad (2.10)$$

A newly derived proof of the Liénard–Chipart criterion, based on the Hermite symmetric matrix, will be presented in the next chapter. We may note that the Liénard–Chipart criterion can be represented in several forms [19]. The one given above is computationally the simplest.

2.2   RELATIVE STABILITY AND APERIODICITY OF CONTINUOUS SYSTEMS [6]

In this section significant problems in the study of dynamic systems will be discussed. Relative stability is important in obtaining an acceptable transient response of the linear continuous systems. Mathematically, this problem is represented by finding the necessary and sufficient conditions for the roots of the system's characteristic equation to lie in a certain sector in the left half of the s-plane. This sector is defined by a certain damping ratio ζ. The second problem of aperiodicity arises in obtaining a response that has no oscillations or that has oscillations of a finite number only. Mathematically, this situation is represented by obtaining the necessary and sufficient conditions for all the roots of the characteristic equation to be distinct and on the negative real axis. This would also include stability. Both problems are dealt with below.

Consider the characteristic equation of the linear system in the form

$$F(s) = \sum_{k=0}^{n} a_k s^k = 0 \qquad\qquad a_n > 0 \quad (2.11)$$

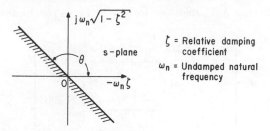

Figure 2.1. Relative stability boundary in s-plane.

If the complex variable s is substituted by a new variable

$$s = We^{j(\theta - \pi/2)} \qquad (2.12)$$

the left half of the s-plane that is to the left of the straight line in Figure 2.1 is mapped into the left half of the W-plane as shown in Figure 2.2. After substituting (2.12) into $F(s)$ of (2.11), one obtains

$$F_1(W) = \sum_{k=0}^{n} a_k e^{jk(\theta - \pi/2)} W^k = 0 \qquad (2.13)$$

Now consider

$$F_1(jW) = \sum_{k=0}^{n} a_k e^{jk\theta} W^k = 0 \qquad (2.14)$$

The coefficients $a_k e^{jk\theta}$ can be developed in the form

$$a_k e^{jk\theta} = b_k + jc_k \qquad (2.15)$$

In checking the conditions for the roots of (2.13) to lie in the left half of the W-plane, we can apply the Hurwitz form (or the generalized Routh–Hurwitz criterion discussed earlier) to $F_1(jW)$. This is represented in (2.4) as being positive innerwise (pi), where

$$b_k = (-1)^k a_k T_k(\zeta)$$
$$T_k(\zeta) = \text{Chebyshev function of the first kind}$$

$$c_k = (-1)^{k+1} a_k \sqrt{1 - \zeta^2}$$
$$U_k(\zeta) = \text{Chebyshev function of second kind}$$

with $\qquad T_0(\zeta) = 1, \quad T_1(\zeta) = \zeta, \quad U_0(\zeta) = 0, \quad U_1(\zeta) = 1.$

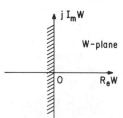

Figure 2.2. Relative stability boundary in W-plane.

The damping ratio ζ determines the relative stability of the sector angle in the s-plane.

If we use the form (2.1) for $F_1(W)$, the relative stability condition can be more conveniently represented in terms of the Δ_{2n-1} matrix of (2.2) as being positive innerwise (pi).

Example 2.3.   Let $F(s)$ be

$$F(s) = s^3 + 4.464s^2 + 7.464s + 4 \tag{2.16}$$

We want to check whether or not all roots of the preceding equation have the relative damping coefficient

$$\zeta = \sin^{-1}\left(\theta - \frac{\pi}{2}\right) = \frac{1}{\sqrt{2}} = 0.707 \tag{2.17}$$

Form $F_1(jW)$ from (2.14) as follows:

$$F_1(jW) = (b_3 + jc_3)W^3 + (b_2 + jc_2)W^2 + (b_1 + jc_1)W + (b_0 + jc_0) \tag{2.18}$$

Now

$$b_k = (-1)^k a_k T_k(0.707) \qquad c_k = (-1)^{k+1} a_k \sqrt{1 - \tfrac{1}{2}} U_k(0.707) \tag{2.19}$$

From tables of Chebyshev functions we have

$$\begin{aligned}
T_0(0.707) &= 1, \quad T_1(0.707) = 0.707, \quad T_2 = 0, \quad T_3 = 0.707 \\
U_0(0.707) &= 0, \quad U_1(0.707) = 1, \quad U_2 = 1.414, \quad U_3 = 1.00
\end{aligned} \tag{2.20}$$

Hence

$$\begin{aligned}
F_1(jW) &= (0.707 + j0.707)W^3 + (0 + 4.464j)W^2 \\
&\quad + (-5.28 + 5.28j)W + 4
\end{aligned} \tag{2.21}$$

The conditions for relative stability [noting (2.4)] reduce to the following matrix Δ_6 to be positive innerwise.

$$\Delta_6 = \begin{bmatrix}
0.707 & -4.464 & 5.28 & 0 & 0 & 0 \\
0 & 0.707 & -4.464 & 5.28 & 0 & 0 \\
0 & 0 & 0.707 & -4.64 & 5.28 & 0 \\
0 & 0 & 0.707 & 0 & -5.28 & 4 \\
0 & 0.707 & 0 & -5.28 & 4 & 0 \\
0.707 & 0 & -5.28 & 4 & 0 & 0
\end{bmatrix} \tag{2.22}$$

Calculating the determinants of the inners, we obtain

$$|\Delta_2| = 0.707 \times 4.464 > 0 \tag{2.23}$$

$$|\Delta_4| = 0.5 \times 4.464 \times 12.36 > 0 \tag{2.24}$$

$$|\Delta_6| = 0.5 \times 4.464 \times 0.707 \times 12.36 \times 6.10 > 0 \tag{2.25}$$

Hence the polynomial in (2.10) is relatively stable.

We can also check relative stability by first obtaining $F_1(jW)$, normalizing the coefficient of W^3 to unity, and then applying (2.2) in order to determine whether Δ_5 is positive innerwise or not. In this example it is (as can be easily verified) positive innerwise.

Remarks

1. It should be noted that a necessary condition for relative stability (same as stability) is that all the a_i''s in (2.11) should be positive [7].

2. In the equation of $F_1(jW)$ the constant term is always real and equals a_0. The reason is that $U_0(\zeta) \equiv 0$ and $T_0(\zeta) = 1$. As a consequence of this property, one can apply the recently developed necessary condition for left half-plane roots by Duffin [2]. This is stated and follows.

Let

$$\begin{aligned} F_2(W) &= A_0 + A_1 W + A_2 W^2 + \cdots + A_n W^n \\ &\quad + j(B_1 W + B_2 W^2 + \cdots + B_n W^n) \end{aligned} \tag{2.26}$$

where A_j and B_j are real. A necessary condition for the roots of the foregoing polynomial to be in the open left half of the W-plane is

$$A_j A_{j+1} + B_j B_{j+1} > 0 \qquad j = 0, 1, 2, \ldots, n-1 \tag{2.27}$$

Note that for $j = 0$, the preceding equation reduces to

$$A_0 A_1 > 0 \tag{2.28}$$

In order to apply the above necessary condition to the example, we obtain $F_2(W)$ as follows:

$$\begin{aligned} F_2(W) = F_1(jW \times (-j)) &= (0.707j - 0.707)W^3 + (0 + 4.464j)W^2 \\ &\quad + (5.28 + 5.28j)W + 4 \end{aligned} \tag{2.29}$$

The necessary condition for relative stability is

$$A_0 A_1 = 4 \times 5.28 > 0 \tag{2.30}$$

$$A_2 A_3 + B_2 B_3 = 0 + 5.28 \times 4.464 > 0 \tag{2.31}$$

$$A_2 A_3 + B_2 B_3 = 0 + 4.464 \times 0.707 > 0 \tag{2.32}$$

As expected, the necessary condition is satisfied.

Relative stability for complex polynomials of practical use has been also obtained.[†]

The condition for aperiodicity (aperiodic and stable system) can be obtained from Sturm sequences, as was developed by Fuller [8]. It is represented in the following:

Given a real polynomial in the form of (2.9), the necessary and sufficient conditions for all the roots of (2.9) to be distinct and negative real are

1. All the a_i's > 0. $\qquad\qquad\qquad\qquad\qquad\qquad\qquad$ (2.33)
2. The following $2n - 1 \times 2n - 1$ matrix Δ_{2n-1} is positive innerwise.

$$
\Delta_{2n-1} =
\begin{bmatrix}
a_n & \cdots & a_3 & a_2 & a_1 & a_0 & 0 & \cdots & 0 \\
\vdots & \cdots & \vdots & \vdots & \vdots & \vdots & \vdots & \cdots & \vdots \\
0 & \cdots & a_n & a_{n-1} & a_{n-2} & a_{n-3} & a_{n-4} & \cdots & 0 \\
0 & \cdots & 0 & a_n & a_{n-1} & a_{n-2} & a_{n-3} & \cdots & a_0 \\
0 & \cdots & 0 & 0 & na_n & (n-1)a_{n-1} & (n-2)a_{n-2} & \cdots & a_1 \\
0 & \cdots & 0 & na_n & (n-1)a_{n-1} & (n-2)a_{n-2} & (n-3)a_{n-3} & \cdots & 0 \\
0 & \cdots & na_n & (n-1)a_{n-1} & (n-2)a_{n-2} & (n-3)a_{n-3} & (n-4)a_{n-4} & \cdots & 0 \\
\vdots & \cdots & \vdots & \vdots & \vdots & \vdots & \vdots & \cdots & \vdots \\
na_n & \cdots & 3a_3 & 2a_2 & a_1 & 0 & 0 & \cdots & 0
\end{bmatrix}
$$

$$(2.34)$$

The proof of the above condition can be readily ascertained from the next two theorems.

THEOREM 2.1 (FULLER [8]). *From (2.3), form the following characteristic equation of some fictitious system.*

$$G(s) = F(s^2) + sF'(s^2) = 0 \qquad\qquad (2.35)$$

where $\qquad\qquad\qquad F'(s) = \dfrac{dF(s)}{ds}$ $\qquad\qquad\qquad\qquad$ (2.36)

If the system represented in (2.9) is aperiodic, then all the roots of the fictitious system lie in the open left half-plane. Hence, by applying the Liénard–Chipart criterion to $G(s)$, (2.33) and (2.34) are readily ascertained. Furthermore, one can also apply the Hurwitz criteria to $G(s)$ and thus obtain the minor array form represented by Meerov.[‡]

[†] B. D. O. Anderson, N. K. Bose, and E. I. Jury, "A simple test for zeros of a complex polynomial in a sector." *IEEE Trans. on Automatic Control* (Aug. 1974), pp. 437–438.
[‡] M. V. Meerov, "Criteria of aperiodic systems," *Isvest. Akad. Nauk, SSSR, OTN.*, (12), 1169 (1945).

THEOREM 2.2 (ROMANOV [9]). *The polynomial $F(s)$ in (2.9) with real and positive coefficients has all its n distinct roots on the negative real axis if and only if*

$$\hat{F}(s) = F(js) + j\frac{dF(js)}{d(js)} \tag{2.37}$$

has all its roots in the open left half-plane.

Since the polynomial $\hat{F}(s)$ is complex, we can apply relationship (2.5) to require that all its roots are in the open left half-plane. The condition for this is exactly (2.34) being (pi).

A proof and discussion of Theorem 2.2 will be the subject of Section 3.3. At this point it may be mentioned that (2.37) is not unexpected, since the aperiodicity condition can be obtained from the relative stability, discussed in this section, as a limiting case when $\zeta \to 1$ or $\theta = \pi$. A verification of the aperiodicity condition from this limiting case of relative stability is an interesting topic to be tackled. Another limiting case is to show that Liénard–Chipart criteria can be obtained from the limit $\zeta \to 0$ in the relative stability.

2.3 STABILITY AND APERIODICITY OF LINEAR TIME-INVARIANT DISCRETE SYSTEMS

a. Stability Condition

The condition of stability of linear time-invariant discrete (or sampled-data) systems can be presented in terms of the roots of the system characteristic equation to be inside the unit circle in the z-plane. In this case, the characteristic equation is given in terms of a real polynomial with degree n.

Similar to the stability criteria for the continuous case, we will present the root clustering inside the unit circle, first, for the complex polynomial and then for the real polynomial. This, in turn, will lead to Schur–Cohn, determinantal, and simplified determinantal stability criteria.

Let $F(z)$ be represented as

$$F(z) = a_n z^n + a_{n-1} z^{n-1} + \cdots + a_1 z + a_0 \tag{2.38}$$

SCHUR–COHN CRITERION [10]†

The necessary and sufficient condition for the roots of $F(z) = 0$ in (2.38) with complex coefficients a_k to lie inside the unit circle is given by the following matrix Δ_{2n} to be positive innerwise (pi),

† See also I. Schur, "Über Potenzreihen die in Innern des Einheitkreises beschränkt sind," *S. Für Math.*, **147**, 205–232 (1917).

$$
\Delta_{2n} =
\begin{bmatrix}
a_n & a_{n-1} & & \cdots & & a_1 & 0 & 0 & \cdots & 0 & a_0 \\
0 & \vdots & a_n & a_{n-1} & a_{n-2} & 0 & 0 & & & a_0 & \vdots \\
\vdots & \vdots & 0 & a_n & a_{n-1} & 0 & a_0 & & & a_1 & \vdots \\
\vdots & \vdots & 0 & 0 & a_n & a_0 & a_1 & & & a_2 & \vdots \\
\vdots & \vdots & 0 & 0 & \bar{a}_0 & \bar{a}_n & \bar{a}_{n-1} & \bar{a}_{n-2} & & & \vdots \\
\vdots & \vdots & 0 & \bar{a}_0 & \bar{a}_1 & 0 & \bar{a}_n & \bar{a}_{n-1} & & & \vdots \\
0 & \vdots & \bar{a}_0 & \bar{a}_1 & \bar{a}_2 & 0 & 0 & \bar{a}_n & & & \vdots \\
\bar{a}_0 & \bar{a}_1 & & \cdots & & \bar{a}_{n-1} & 0 & 0 & \cdots & & \bar{a}_n
\end{bmatrix}
\tag{2.39}
$$

where $\bar{a}_k$ is a complex conjugate of a_k.

DETERMINANTAL CRITERION [11]

When the coefficients in (2.38) are real, then the root-clustering condition within the unit circle can be further simplified, similar to the Hurwitz criterion, in the following way.

The necessary and sufficient condition for the roots of $F(z) = 0$ (with $a_n > 0$) to lie inside the unit circle (or, equivalently, for the linear time-invariant discrete system to be stable) is given by

$$
F(1) > 0 \qquad (-1)^n F(-1) > 0 \tag{2.40}
$$

and the matrices $\Delta_{n-1}^{\pm} = X_{n-1} \pm Y_{n-1}$ are positive innerwise (pi), where

$$
X_{n-1} =
\begin{bmatrix}
a_n & a_{n-1} & a_{n-2} & \cdots & a_2 \\
0 & a_n & a_{n-1} & \cdots & a_3 \\
0 & 0 & a_n & \cdots & a_4 \\
\vdots & \vdots & \vdots & & \vdots \\
0 & 0 & 0 & \cdots & a_n
\end{bmatrix}
$$

$$
Y_{n-1} =
\begin{bmatrix}
0 & 0 & & \cdots & a_0 \\
\vdots & \vdots & & & \vdots \\
0 & 0 & a_0 & \cdots & a_{n-4} \\
0 & a_0 & a_1 & \cdots & a_{n-3} \\
a_0 & a_1 & a_2 & \cdots & a_{n-2}
\end{bmatrix}
\tag{2.41}
$$

The matrices $\Delta_{n-1}^{\pm}$ for n-odd are

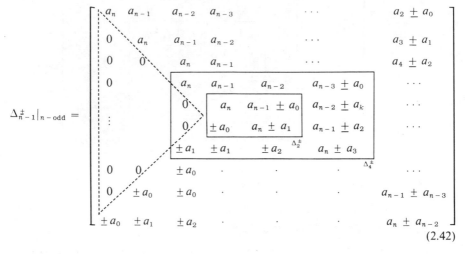

$$(2.42)$$

Remark. The inner determinants sign (2.42) also yield information on the root distribution within the unit circle. Hence the preceding criterion can be further simplified for root clustering within the unit circle, similar to the Liénard–Chipart criterion for the continuous case, as follows.

SIMPLIFIED DETERMINANTAL CRITERION [12]†

The necessary and sufficient condition for the roots of $F(z) = 0$ (with $a_n > 0$) to lie inside the unit circle is given by:

1. The matrix Δ_{n-1}^{-} is positive innerwise (pi).

2. For n-odd, $n \triangleq 2m - 1$, either

$$B_{2i} > 0 \qquad B_{2m-1} > 0 \qquad \text{or} \qquad B_{2i+1} > 0 \qquad B_0 > 0$$
$$i = 0, 1, \ldots, m - 1 \quad (2.43)$$

where

$$B_i \triangleq \sum_{r=0}^{2m-1} \left[\sum_j (-1)^{r+i-j+1} a_r \binom{r}{j} \binom{2m-1-r}{i-j} \right] \qquad (2.44)$$

Note that

$$B_{2m-1} = F(1) \qquad \text{and} \qquad B_0 = -F(-1) \qquad (2.45)$$

or n-even, $n \triangleq 2m$, either

$$B_{2i} > 0 \qquad i = 0, 1, \ldots, m \quad (2.46)$$

† See also B. D. O. Anderson and E. I. Jury, "On the reduced Hermité and reduced Schur–Cohn matrix relationship," *Int. J. Control*, Vol. 19, No. 5, pp. 877–890 (1974).

or
$$B_{2i+1} > 0, \quad B_0 > 0, \quad B_{2m} > 0$$
$$i = 0, 1, \ldots, m - 1 \quad (2.47)$$

where
$$B_i = \sum_{r=0}^{2m} \left[\sum_j (-1)^{r+i-j} a_r \binom{r}{j} \binom{2m-r}{i-j} \right] \quad (2.48)$$

[The summation over j is governed by $\max(0, 2m - r - i) \le j \le \min(i, r)$.]
Note that

$$B_0 = F(-1) \qquad B_{2m} = F(1)$$

and the coefficients B_i are obtained from the bilinear transformation of the polynomial $F(z)$ to $F_1(W)$, whose roots are in the open left half of the W-plane. A computational method for obtaining the B_i from the coefficients a_i will be presented in Section 7.3.

Example 2.4. Let $F(z)$ be given as follows:

$$F(z) = a_5 z^5 + a_4 z^4 + a_3 z^3 + a_2 z^2 + a_1 z + a_0$$
$$a_5 > 0 \quad (2.49)$$

By applying the bilinear transformation $z = (W + 1)/(W - 1)$, the polynomial $F(z)$ becomes

$$F_1(W) = B_5 W^5 + B_4 W^4 + B_3 W^3 + B_2 W^2 + B_1 W + B_0$$
$$(2.50)$$

The stability condition for $F(z)$ is given by:

1. B_0, B_2, B_4, B_5 are positive or B_0, B_1, B_3, B_5 are positive.

2. The following matrix Δ_4^- is positive innerwise (pi):

$$\Delta_4^- = \begin{bmatrix} a_5 & a_4 & a_5 & a_2 - a_0 \\ 0 & a_5 & a_4 - a_0 & a_3 - a_1 \\ 0 & -a_0 & a_5 - a_1 & a_4 - a_2 \\ -a_0 & -a_1 & -a_2 & a_5 - a_3 \end{bmatrix} \quad (2.51)$$

Note that

$$B_0 = -F(-1) \qquad B_5 = F(1) \quad (2.52)$$
$$B_4 = 5a_5 + 3a_4 + a_3 - a_2 - 3a_1 - 5a_0 \quad (2.53)$$
$$B_2 = 10a_5 - 2a_4 - 2a_3 + 2a_2 + 2a_1 - 10a_0 \quad (2.54)$$

Similarly, one can readily obtain B_1 and B_3 from (2.44).

APERIODICITY CONDITION [13]

An aperiodicity condition in discrete systems is obtained when all the roots of the characteristic equation are distinct and lie on the real axis in the interval $(0, 1)$ in the z-plane. Such an aperiodicity condition yields only a finite number of maxima and minima in the system response (less than n, the order of the system). This condition is important when excessive oscillations are not desirable, as, for example, in instrumentation systems.

In order to obtain the necessary and sufficient condition for aperiodicity, we can apply either one of the following transformations to $F(z) = 0$.

1. *Bilinear Transformation* [13]. This transformation involves mapping the real segment $(0, 1)$ of the z-plane onto the negative real axis of the W-plane. It is given by the transformation

$$z = \frac{W}{W - 1} \quad \text{or} \quad W = \frac{z}{z - 1} \qquad (2.55)$$

If (2.55) is substituted into (2.38) for $F(z) = 0$, we obtain

$$F(W) = B_n W^n + B_{n-1} W^{n-1} + \cdots + B_1 W + B_0 = 0 \qquad (2.56)$$

The relationship between the a_i's and the B_i's is given by

$$B_{n-k} = \sum_{r=k}^{n} \binom{r}{k} a_{n-r} (-1)^k \qquad (2.57)$$

In order to obtain the aperiodicity condition for the discrete case, we can apply the aperiodicity condition developed earlier for the continuous case to (2.56).

2. *A Nonlinear Transformation* [14]. This transformation maps the real segment $(0, 1)$ in the z-plane onto the periphery of the unit circle $|W| = 1$ in the W-plane. This transformation is given as follows:

$$z = \frac{1}{4} \frac{(W + 1)^2}{W} = \frac{1}{4} \left(\sqrt{W} + \frac{1}{\sqrt{W}} \right)^2 \qquad (2.58)$$

The aperiodicity condition reduces to the requirement that all the roots of the following polynomial in W should be distinct and should lie on the unit circle in the W-plane:

$$F_1(W) = W^n \sum_{i=0}^{n} a_i \left(\frac{1}{4} \right)^i \left(\sqrt{W} + \frac{1}{\sqrt{W}} \right)^{2i} = F_{2n}(W) \qquad (2.59)$$

and also $F(1) \neq 0$.

The preceding polynomial $F_{2n}(W)$ is a special form and of degree $2n$, and it can be expressed as

$$F_{2n}(W) = F_n(W) + W^n F_n^*(W) \qquad (2.60)$$

where $F_n^*(W)$ is the inverse polynomial of $F_n(W)$ and equals $W^n F_n(1/W)$.

The coefficients of $F_1(W)$—that is, the b_i—are related to a_i by the following formulas.

$$b_i = \sum_{l=0}^{i} a_{n-l} \left(\frac{1}{4}\right)^{n-l} \binom{2n-2l}{i-l}$$

$$i = 0, 1, \ldots, n-1 \quad (2.61)$$

$$b_n = \frac{1}{2} \sum_{l=0}^{n} a_{n-l} \left(\frac{1}{4}\right)^{n-l} \binom{2n-2l}{n-l} \qquad (2.62)$$

THEOREM 2.3 [14]. *The discrete system characteristic polynomial $F(z)$ given in (2.9) is aperiodic if and only if the roots (which are distinct) of $F_n^*(W) = 0$ lie inside the unit circle and $F(1) \neq 0$.*

Proof [14]:   Assume that all the roots of $F_n(W) = 0$—that is, the distinct n roots—lie outside the unit circle. The distribution of the roots of $F_{2n}(W)$ can be obtained by forming the polynomial

$$H_{2n}(W) = F_n(W) + k_{2n} W^n F_n^*(W) \qquad (2.62a)$$

Using Rouche's theorem [1], if $|k_{2n}| < 1$, all the roots (which are distinct) of H_{2n} lie outside the unit circle. If $|k_{2n}| > 1$, they lie inside the unit circle. The distribution of the roots are continuous functions of the polynomial coefficients; then for $|k_{2n}| = 1$—that is, when $H_{2n} = F_{2n}(W)$—all the roots of $F_{2n}(W)$ are distinct and should lie on the unit circle. This constitutes the aperiodicity condition. Note that the condition for the roots of $F_n(W) = 0$ to lie outside the unit circle is identical to that of $F_n^*(W) = 0$ to lie inside the unit circle, similar to the stability condition.

Example 2.5.   Let $F(z)$ be given as

$$F_4(z) = 0.252 - 56.392z + 322.784z^2 - 522.24z^3 + 256z^4 \quad (2.63)$$

$$F_4(1) \neq 0 \qquad (2.64)$$

Using (2.61) and (2.62), we obtain

$$F_1(W) = 1 - 0.16W - 0.786W^2 + 0.448W^3 + 0.2W^4 \quad (2.65)$$

or

$$F_1^*(W) = 0.2 + 0.448W - 0.786W^2 - 0.16W^3 + W^4 \quad (2.66)$$

The condition for aperiodicity requires that all the roots that are distinct of $F_1^*(W)$ lie inside the unit circle. By applying the stability condition to this example, it can be ascertained that it is aperiodic.

Remark. The coefficients b_i and b_n indicated in (2.61) and (2.62) can easily be generated from the a_i's by using a certain combinatorial rule, which is presented in Section 7.3.

2.4 CONDITIONS ON ROOT CLUSTERING OF POLYNOMIALS

Here we will present additional useful criteria for root clustering that are not discussed in earlier sections of this chapter. Included are root clustering within the unity-shifted unit circle in the $\zeta = z - 1$ plane and root clustering on the real and imaginary axes in the complex plane. Furthermore, features of root clustering for the rational polynomials that occur in the physical realization of RC and LC networks are also given. In the latter cases, the poles and zeros interlace along the negative real axis and the imaginary axis, respectively.

a. Root Clustering within the Unity-Shifted Unit Circle [15]

In the study of discrete systems it is often desirable to present the formulation of stability in the $\zeta = z - 1$ plane, where the roots of the characteristic equation $F(z) = 0$ given in (2.38) should lie within the shifted circle, as shown in Figure 2.3. The necessary and sufficient condition for such root

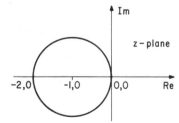

Figure 2.3. Unity-shifted unit circle.

clustering is formulated by Jury and Tschauner [15]. It is expressed in inners form as follows.

1. For n-even,

$$F(0) > 0 \qquad F(-2) > 0 \tag{2.67}$$

and the following $n - 1 \times n - 1$ matrix should be positive innerwise, in addition to the coefficients $A_{m,v}$ (or half of them) given below being positive.

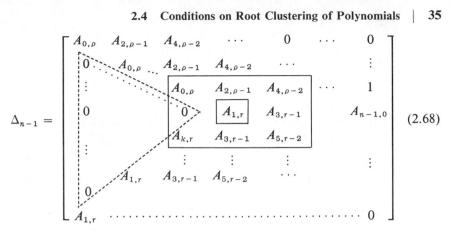

$$\Delta_{n-1} = \begin{bmatrix} A_{0,\rho} & A_{2,\rho-1} & A_{4,\rho-2} & \cdots & 0 & \cdots & 0 \\ 0 & A_{0,\rho} & A_{2,\rho-1} & A_{4,\rho-2} & \cdots & & \vdots \\ \vdots & & A_{0,\rho} & A_{2,\rho-1} & A_{4,\rho-2} & \cdots & 1 \\ 0 & & 0 & A_{1,r} & A_{3,r-1} & & A_{n-1,0} \\ & & A_{k,r} & A_{3,r-1} & A_{5,r-2} & & \\ \vdots & & \vdots & \vdots & \vdots & & \vdots \\ & A_{1,r} & A_{3,r-1} & A_{5,r-2} & \cdots & & \\ 0 & & & & & & 0 \\ A_{1,r} & \cdots & \cdots & \cdots & \cdots & \cdots & 0 \end{bmatrix} \quad (2.68)$$

where
$$\rho = r = \frac{n-2}{2}$$

and
$$A_{m,v} = \sum_{\mu=0}^{m} (-1)^\mu \binom{\mu+v}{\mu} a_{m-\mu}$$

$$m = 0, 1, 2, \ldots, n, \quad v = 0, 1, 2, \ldots, \rho \text{ or } r \quad (2.69)$$

2. For n-odd,
$$F(0) > 0 \qquad F(-2) < 0 \qquad\qquad (2.70)$$

and the following $n - 1 \times n - 1$ matrix should be positive innerwise, in addition to the coefficients $A_{m,v}$ (or half of them) given below being positive.

$$\Delta_{n-1} = \begin{bmatrix} A_{1,r} & A_{3,r-1} & \cdots & \cdots & 0 \\ 0 & & \vdots & & \\ \vdots & & A_{1,r} & A_{3,r-1} & \cdots & 1 \\ & 0 & & & & \\ 0 & & & & & \\ \vdots & & 0 & & & \\ & & & & & \\ 0 & & A_{0,\rho} & A_{2,\rho-1} & A_{n-1,0} \\ A_{0,\rho} & A_{2,\rho-1} & \cdots & \cdots & 0 \end{bmatrix} \quad (2.71)$$

where
$$r = \frac{n-3}{2} \qquad \rho = r + 1 \qquad\qquad (2.72)$$

and
$$A_{m,v} = \sum_{\mu=0}^{m} (-1)^{\mu} \binom{\mu+v}{\mu} a_{m-\mu} \qquad A_{0,v} = a_0 \quad (2.73)$$

where $\quad m = 0, 1, 2, \ldots, n \quad$ and $\quad v = 0, 1, 2, \ldots, p$ or r

b. Necessary and Sufficient Condition for All the Roots of a Real Polynomial to Be Distinct and to Lie on the Real Axis [17]

The necessary and sufficient condition that (2.9) have all its roots distinct and lying on the real axis is that Δ_{2n-1} in (2.34) be positive innerwise (pi).

System application of this condition can be found in distributed parameters systems. In particular, the existence and uniqueness of the solution of the partial differential equation are guaranteed as in lossless transmission lines [16].

c. Necessary and Sufficient Condition for Root Clustering on the Imaginary Axis [11]

For a polynomial

$$\hat{F}(z) = a_n z^{2n} + a_{n-1} z^{2n-2} + \cdots + a_1 z^2 + a_0 \qquad (2.74)$$

to have all its roots distinct and on the imaginary axis, it is necessary and sufficient that:

1. All the coefficients of $\hat{F}(z)$ be positive.
2. The matrix Δ_{2n-1} given in (2.34) be positive innerwise (pi).

Application of the above conditions will be discussed in detail in Chapter 5 in connection with positive real functions.

d. Physical Realization of an RC Passive Network [17]

The necessary and sufficient conditions for a rational function

$$Z(s) = \frac{g(s)}{f(s)} = \frac{b_{n-1}s^{n-1} + b_{n-2}s^{n-2} + \cdots + b_1 s + b_0\dagger}{a_n s^n + a_{n-1}s^{n-1} + \cdots + a_1 s + a_0} \qquad (2.75)$$

to be realizable as the driving-point impedance of an RC network are formulated as follows:

1. All the a_i's and b_i's be positive.
2. The matrix Δ of dimension $2n - 1 \times 2n - 1$ be positive innerwise (pi).

† In cases where the degrees of the numerator and denominator are the same, $Z(s) = r + Z_1(s)$, where $Z_1(s)$ is of the form (2.75) and $r > 0$.

$$\Delta \triangleq = \begin{bmatrix} a_n & a_{n-1} & \cdots & a_2 & a_1 & a_0 & \cdots & 0 & 0 \\ 0 & a_n & & a_3 & a_2 & a_1 & \cdots & 0 & 0 \\ \vdots & \vdots & & \vdots & \vdots & \vdots & & & \vdots \\ 0 & 0 & \cdots & a_n & a_{n-1} & a_{n-2} & \cdots & a_1 & a_0 \\ 0 & 0 & \cdots & 0 & b_{n-1} & b_{n-2} & \cdots & b_1 & b_0 \\ 0 & 0 & \cdots & b_{n-1} & b_{n-2} & b_{n-3} & \cdots & b_0 & 0 \\ \vdots & & & \vdots & \vdots & \vdots & & & \vdots \\ 0 & b_{n-1} & \cdots & b_2 & b_1 & b_0 & \cdots & 0 & 0 \\ b_{n-1} & b_{n-2} & \cdots & b_1 & b_0 & 0 & \cdots & 0 & 0 \end{bmatrix} \qquad (2.76)$$

e.   Physical Realization of an LC Passive Network [17]

The necessary and sufficient conditions for

$$Z_1(s) = \frac{b_{n-1}s^{2n-1} + b_{n-2}s^{2n-3} + \cdots + b_0 s}{a_n s^{2n} + a_{n-1}s^{2n-2} + \cdots + a_0} \qquad (2.77)$$

to be realizable as the driving-point impedance of an LC passive network are the same as condition 1 and 2 of part (d).

2.5   CRITICAL CONSTRAINTS ON STABILITY, RELATIVE STABILITY, AND APERIODICITY FOR CONTINUOUS AND DISCRETE SYSTEMS

It is known that if one or more parameters (e.g., gain) for an initially stable feedback system are changed, stability ceases when either the Hurwitz determinant $|\Delta_{n-1}|$ (the last determinant of the innerwise matrix) or the constant coefficient of the characteristic equation changes sign and becomes negative. The usefulness of this fact is important in system design when one is interested in obtaining the maximum parameter (gain) for stability limit. In this case, one needs to solve only two equations, one of $(n-1)$th order at most and the other of first order, thus avoiding the solution of all other inner determinants equations. These two equations are referred to as *critical constraints*. Such critical constraints also appear in the stability within the unit circle, shifted unit circle, relative stability in the s-plane, relative stability inside the unit circle, and in aperiodicity conditions for both continuous and discrete systems. The objective of this section is to introduce the critical constraints for all cases.

a. Stability Constraints of Continuous Systems [18]†

It is known that if the continuous system characteristic equation is described by

$$F(s) = b_n s^n + b_{n-1} s^{n-1} + b_1 s + b_0 \qquad b_n > 0 \quad (2.78)$$

then the last Hurwitz (or innerwise) determinant is given by

$$|\Delta_n| = b_0 |\Delta_{n-1}| \qquad (2.79)$$

If a parameter of an initially stable system is changed such that a real root (simple or of odd multiple order) passes from the open left half to the right half of the s-plane, then b_0 changes sign immediately. Hence one critical condition can readily be ascertained as

$$b_0 \geq 0 \qquad (2.80)$$

Furthermore, according to Orlando's formula [19], the Hurwitz determinant $|\Delta_{n-1}|$ can be expressed as

$$|\Delta_{n-1}| = (-1)^{n(n-1)/2} b_n^{n-1} \prod_{\substack{i < k}}^{1 \cdots n} (s_i + s_k) \qquad (2.81)$$

where s_i, s_k are the roots of $F(s) = 0$.

If

$$s_k = \bar{s}_i \text{ (conjugate)}, \qquad (2.82)$$

then $\qquad s_i + s_k = 2 \operatorname{Re} [s_i] \qquad (2.83)$

Here $|\Delta_{n-1}|$ changes sign when a pair of complex roots (or odd number of pairs with the same real part) passes from the left half-plane to the right half-plane. Thus the other critical condition where stability ceases is given by

$$|\Delta_{n-1}| \geq 0 \qquad (2.84)$$

Furthermore, for the case of complex roots of even multiplicity, $|\Delta_{n-1}| = 0$ is also critical (although it does not change sign).

In the case of conditional stability, the critical equations b_0 and $|\Delta_{n-1}| = 0$ give more than one critical value of the parameter. Thus special investigation is required to ascertain these critical values. For example, $|\Delta_{n-1}|$ may become positive again after being negative, which may be due either to the fact that the complex roots moved back to the open left half-plane or to the fact that another pair of roots moved to the right half-plane.

---

† See also R. A. Frazer and W. J. Duncan, "On the criteria for the stability of small motion," *Proc. Roy. Soc. A*, **124,** 642 (1929).

b.   Relative Stability Constraints (Continuous Case)†

Similar to the stability constraints discussed in part (a), the critical constraints for relative stability can be obtained from (2.11) and (2.2) as follows:

$$a_0 \geq 0 \qquad (2.85)$$

and
$$|\Delta_{2n-1}| \geq 0 \qquad (2.86)$$

The preceding constraints yield the parameter value when complex roots move out of the damping line (shown in Figure 2.1) or when real roots move to the right half-plane. This fact is evident, since relative stability is transformed to that of open left half-plane roots by using the transformation in (2.12).

c.   Aperiodicity Constraints (Continuous Case) [18]

It was indicated earlier in Section 2.2 that the criterion for aperiodicity is the same as the criterion of stability of the system described by $F(s^2) + sF'(s^2) = 0$. Therefore, the same reasoning as in case (a) can be applied here. The corresponding critical constraints are obtained from (2.9) and (2.34).

$$a_0 \geq 0 \qquad (2.87)$$

and
$$|\Delta_{2n-1}| \geq 0 \qquad (2.88)$$

In this case, when a parameter changes for an initially aperiodic system, the determinant $|\Delta_{2n-1}|$ is the first to become zero when two roots become complex. When the roots remain real but move outside the specified section on its real axis, then the critical condition is

$$a_0 \geq 0 \qquad (2.89)$$

Using the transformation in (2.55), the aperiodicity constraints for the discrete case become

$$F(1) \geq 0 \qquad F(0) \geq 0 \qquad (2.90)$$

and
$$|\Delta_{2n-1}| \geq 0 \qquad (2.91)$$

where $|\Delta_{2n-1}|$ is obtained from (2.34) with the entries obtained from (2.56) and (2.57).

† For discussion of relative stability (discrete case), see E. I. Jury and D. D. Šiljak, "A note on relative damping in linear discrete systems," *Proc. ETAN Conference, 1965.* Also, *Teoretski Prilog Automatika*, Zagreb, Yugoslavia, pp. 67–70, 1970.

d. Stability Constraints (Discrete Case) [13, 18]

The characteristic equation of the linear discrete system can be described as follows:

$$F(z) = a_n z^n + a_{n-1} z^{n-1} + \cdots + a_1 z + a_0 \quad a_n > 0 \quad (2.92)$$

One can readily ascertain that $F(1) \geq 0$ and $(-1)^n F(-1) \geq 0$ constitute two of the critical constraints, because

$$F(1) = a_n \prod_{i}^{1 \cdots n} (1 - z_i) \quad (2.93)$$

and

$$(-1)^n F(-1) = a_n \prod_{i}^{1 \cdots n} (1 + z_i) \quad (2.94)$$

where the z_i are the roots of $F(z) = 0$.

It is seen from (2.93) and (2.94) that both $F(1) = 0$ and $F(-1) = 0$ constitute the critical conditions when the real roots move outside the unit circle at $z = 1$ and $z = -1$.

As proven elsewhere [18], the other critical constraint that constitutes the complex roots moving out of the unit circle is given as

$$|\Delta_{n-1}^-| = a_n^{n-1} \prod_{i<k}^{1 \cdots n} (1 - z_i z_k) \quad (2.95)$$

where $|\Delta_{n-1}^-|$ is obtained from (2.42).

The preceding identity can be considered as the corresponding Orlando formula for the discrete case. If $z_i = \bar{z}_k$(conjugate), then $z_i z_k = |z_i|^2$, and $|\Delta_{n-1}^-|$ changes sign whenever a pair (or odd pairs) of complex roots moves outside the unit circle. Furthermore, for the case of complex roots with even multiplicity, $|\Delta_{n-1}^-| = 0$ is also critical (although it does not change sign). In all cases, the system becomes unstable whenever $|\Delta_{n-1}^-| > 0$.

Similarly, for stability within the unity-shifted unit circle in Section 2.4a, the critical constraints are obtained from (2.67), (2.68), and (2.71) as follows:

$$F(0) \geq 0 \quad F(-2) \leq 0 \quad (2.96)$$

$$|\Delta_{n-1}^-| \geq 0 \quad (2.97)$$

e. Aperiodicity Constraints for Discrete Systems

Using the nonlinear transformation indicated earlier, which transforms the sector $(0, 1)$ into the periphery of the unit circle, the condition for aperiodicity becomes the same as that of stability within the unit circle.

The critical stability constraints are obtained from $F_n^*(W)$ of (2.60). They can be stated as

$$F_n^*(1) = 0 \qquad F_n^*(-1) = 0 \tag{2.98}$$

and

$$|\Delta_{n-1}^-| \geq 0 \tag{2.99}$$

Here $|\Delta_{n-1}^-|$ is obtained from the coefficients of $F_n^*(W)$ similar to the $|\Delta_{n-1}^-|$ of (2.42) obtained from (2.38), with $a_n > 0$.

We will present three examples [18] to illustrate the case of the critical constraints for the design of control systems (continuous and discrete) for stability margin or aperiodicity limits.

Example 2.6.   Assume a continuous feedback system, shown in Figure 2.4, whose open-loop and feedback-transfer function are given as

$$G(s) = \frac{K(s + 1)}{6(s + 4)(s + 2)(s + 1/3)} \tag{2.100}$$

$$H(s) = \frac{354}{(s + 3)(s + 1/2)} \tag{2.101}$$

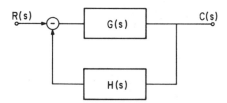

Figure 2.4.   A feedback system for Example 2.6.

In order for the system to become unstable, it is necessary to obtain the gain limit K_{max}. It is readily ascertained that for $0 \leq K \leq \varepsilon$, where ε is a small variation of the gain, the feedback system is stable [since $G(s)H(s)$ represents a stable system]. This fact indicates that the feedback system is initially stable, and one needs to find the value of K_{max} for it to become unstable.

The system characteristic equation is

$$1 + GH(s) = 6s^5 + 59s^4 + 202s^3 + 283s^2$$
$$+ (146 + 354K)s + (24 + 354K) = 0 \tag{2.102}$$

One next obtains $|\Delta_{n-1}| = |\Delta_4|$ from (2.7) as follows:

$$\Delta_4 = \begin{bmatrix} 59 & 283 & 24 + 354K & 0 \\ 0 & 59 & 283 & 24 + 354K \\ 0 & 6 & 202 & 146 + 354K \\ 6 & 202 & 146 + 354K & 0 \end{bmatrix} \qquad (2.103)$$

From (2.103),

$$|\Delta_4| = 53[-\mu^2 + 295\mu + 295240] > 0 \qquad (2.104)$$

where $\mu = 160 + 254K$. It is of interest to note that the preceding inequality is only of second order. For positive gain, the inequality is satisfied when

$$0 < K < 0.893 \qquad (2.105)$$

The other critical constraint, $b_0 > 0$, corresponds to $24 + 354K \geq 0$, which is satisfied for any positive gain. Therefore the maximum gain is

$$K_{\max} \leq 0.893 \qquad (2.106)$$

Example 2.7. Assume a sampled data system, shown in Figure 2.5, with open-loop transfer function

$$G(s) = \frac{1 - e^{-Ts}}{s} e^{-\delta' sT} \frac{K}{s(s + 1)} \qquad (2.107)$$

where $T = 1$ (sampling period) and $\delta' = 1.25$.

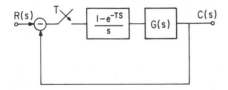

Figure 2.5. A sampled data feedback system for Example 2.7.

The z-transform of $G(s)$ is

$$G_1(z) = 0.2223K \frac{(z + 0.03)(z + 1.755)}{z^2(z - 1)(z - 0.368)} \qquad (2.108)$$

and the characteristic equation is

$$1 + G_1(z) = z^4 - 1.368z^3 + (0.368 + 0.223K)z^2 + 0.3974Kz + 0.0123K = 0$$

For $0 < K < \varepsilon$, the system is stable, as one notices by a rough sketch of the root locus. The critical constraint in this case is $|\Delta_{n-1}^-| = |\Delta_3^-|$, which yields $K_{\max} \le 0.72$. The other critical constraints [i.e., $F(1) = 0$ and $F(-1) = 0$] are satisfied by this gain.

Example 2.8. Consider a unity feedback system, as shown in Figure 2.6, with open-loop transfer function

$$G(s) = \frac{K(s + 3)}{s(s + 1)(s + 4)} \tag{2.109}$$

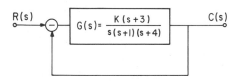

Figure 2.6. A unity feedback system for Example 2.8.

The characteristic equation is

$$1 + G(s) = s(s + 1)(s + 4) + K(s + 3) = 0 \tag{2.110}$$

or

$$s^3 + 5s^2 + (4 + K)s + 3K = 0 \tag{2.111}$$

For $K = 0$ (and $0 \le K < \varepsilon$), the system is aperiodic, as seen from the second-order equation. The critical constraint, in this case, is $|\Delta_{2n-1}| = |\Delta_5|$ from (2.34), which yields

$$36 - 103K + K^2 \ge 0 \tag{2.112}$$

The foregoing inequality is satisfied only when

$$K \le 0.35 \tag{2.113}$$

This is the critical value of gain for the system to be aperiodic because the other critical constraint in equation (2.87) is satisfied.

2.6 CONDITIONS ON ROOT DISTRIBUTION OF POLYNOMIALS

In this section we will present certain theorems on the root distribution of polynomials (complex and real). These theorems will yield information on the number of roots in a certain region in the complex plane. Besides the mathematical interest of this problem, we will show in later chapters the physical and engineering applications. In particular, the problem of positivity and nonnegativity of polynomials, as will be discussed in Chapter 5, can be represented as a special root-distribution problem.

$$
\Delta_{2n-1}=
\begin{vmatrix}
1 & -b_{n-1} & -a_{n-2} & b_{n-3} & a_{n-4} & -b_{n-5} & -a_{n-6} & \cdots & 0 & 0 \\
0 & 1 & -a_{n-1} & -b_{n-2} & a_{n-3} & b_{n-4} & -a_{n-5} & \cdots & 0 & 0 \\
0 & 0 & 1 & -b_{n-1} & -a_{n-2} & b_{n-3} & a_{n-4} & \cdots & 0 & 0 \\
0 & 0 & 0 & 1 & -a_{n-1} & -b_{n-2} & a_{n-3} & \cdots & * & 0 \\
0 & 0 & 0 & 0 & 1 & -b_{n-1} & -a_{n-2} & \cdots & ** & 0 \\
0 & 0 & 0 & 0 & 0 & a_{n-1} & -b_{n-2} & \cdots & 0 & 0 \\
\vdots & \vdots & \vdots & \vdots & \vdots & \vdots & \vdots & & \vdots & \vdots \\
0 & a_{n-1} & -b_{n-2} & -a_{n-3} & b_{n-4} & a_{n-5} & -b_{n-6} & \cdots & 0 & 0 \\
a_{n-1} & -b_{n-2} & -a_{n-3} & b_{n-4} & a_{n-5} & -b_{n-6} & -a_{n-7} & \cdots & 0 & 0
\end{vmatrix}
\tag{2.118}
$$

(The nested principal minors are labelled $\Delta_1,\ \Delta_3,\ \Delta_5,\ \Delta_7$.)

$*\ \ (-1)^{(n+1)/2}b_0$ when n is odd, $\quad (-1)^{n/2}a_0$ when n is even

$**\ \ (-1)^{(n-1)/2}a_0$ when n is odd, $\quad (-1)^{n/2}b_0$ when n is even

Furthermore, it is of interest to point out that the root-distribution problem can be tackled using the inners approach. In this case, the sign variation of the inners determinants plays an important role in obtaining information on the number of roots in a certain region in the plane. In this chapter we will only make a passing remark on the critical cases, and in Chapter 7 we will present the details. However, as the case requires, we will mention the pertinent references. Many excellent mathematical texts exist that present a through discussion of the root-distribution problems. The various root-distribution regions will be given in several theorems.

THEOREM 2.4 [1]. *Given the polynomial having no pure imaginary zeros*

$$F(s) = s^n + (a_{n-1} + jb_{n-1})s^{n-1} + (a_{n-1} + jb_{n-2})s^{n-2} + \cdots$$
$$+ (a_1 + jb_1)s + (a_0 + jb_0) \tag{2.114}$$

where the a_i, b_i are real.

If the $|\Delta_k|$'s of the innerwise matrix Δ_{2n-1} in (2.118) are nonzero for $k = 1, 3, \ldots, 2n - 1$,

$$p = \text{Var}\,[1, |\Delta_1|, |\Delta_3|, \ldots, |\Delta_{2n-1}|] \tag{2.115}$$

$$q = \text{Var}\,[1, |\Delta_1|, |\Delta_3|, \ldots, (-1)^{2n-1}\,|\Delta_{2n-1}|] \tag{2.116}$$

then where Var *means variation of sign and p and q are the number of zeros of $F(s) = 0$ in the half-planes.*

$$\text{Re}\,[s] > 0 \qquad and \qquad \text{Re}\,[s] < 0 \tag{2.117}$$

respectively.

The critical cases of the preceding theorem are discussed in the literature [1,20,21]. The necessary and sufficient condition for $F(s)$ to have all its zeros (roots) in the open left half-plane is that the matrix Δ_{2n-1} be positive inner-wise (pi). The above has been already indicated in Section 2.1.

Remark. When the polynomial is real (i.e., b_i's are zero), then the same conditions hold provided that the various entries of the b_i's = 0 in the inner-wise matrix are inserted. It has been shown by Marden [1] that this matrix reduces to the Hurwitz matrix. For effective computation of the inners determinant, the matrix Δ_{2n-1} for this case can be rearranged by exchanging columns. This point will be discussed in Chapter 7.

THEOREM 2.5 [1]. *Given the real polynomial*

$$F(s) = a_n s^n + a_{n-1}s^{n-1} + a_{n-2}s^{n-2} + \cdots + a_1 s + a_0$$
$$a_n > 0 \tag{2.119}$$

let us write

$$F(s) = G(s^2) + sH(s^2)$$
$$G(u) = a_0 + a_2 u + a_4 u^2 + \cdots \tag{2.120}$$
$$H(u) = a_1 + a_3 u + a_5 u^2 + \cdots$$

Let the nonnull innerwise matrix Δ_n be given below.

Set

$$V_o = \text{Var}\,[1, |\Delta_1|, |\Delta_3|, \ldots] \tag{2.121}$$
$$V_e = \text{Var}\,[1, |\Delta_2|, |\Delta_4|, \ldots]$$

Then if $G(u) \neq 0$, for $u > 0$,

$$p = \text{number of roots in } \text{Re}\,(s) > 0,$$

is given by

$$p = 2V_e \qquad\qquad\qquad \text{for } n = \text{even} \tag{2.122}$$

$$p = 2V_0 - \tfrac{1}{2}[1 - \text{sgn}\, a_{n-1}] \qquad \text{for } n\text{-odd} \tag{2.123}$$

whereas if $H(u) \neq 0$, for $u > 0$,

$$p = 2V_e - \tfrac{1}{2}[\text{sgn}\, a_{n-1} - \text{sgn}\,(a_1 a_0)]$$
$$\text{for } n\text{-even} \tag{2.124}$$

$$p = 2V_o - \tfrac{1}{2}[1 - \text{sgn}\,(a_1 a_0)] \qquad\qquad \text{for } n\text{-odd} \tag{2.125}$$

$$\Delta_n|_{n-\text{even}} = \quad (2.126)$$

$$\Delta_n|_{n-\text{odd}} = \quad (2.127)$$

Remark. When all the coefficients of $F(s)$ are positive, the conditions $G(u) \neq 0$, $H(u) \neq 0$, for $u > 0$, are obviously fulfilled, and vice versa; hence Theorem 2.5 also yields the condition for $p = 0$ that is exactly the Liénard–Chipart criterion. Furthermore, in the above theorem, $q = n - p$.

THEOREM 2.6 [11]. *Given the polynomial*

$$F(z) = a_n z^n + a_{n-1} z^{n-1} + \cdots + a_1 + a_0$$

$$\text{with } a_n \neq 0 \quad (2.128)$$

where $a_n, a_{n-1}, \ldots$ *are complex.*
  If the $|\Delta_k|$*'s of the innerwise matrix* Δ_{2n} *given in* (2.39) *are nonzero for* $k = 2, 4, \ldots, 2n$, *then*

$$p = \text{Var} \, [1, |\Delta_2|, |\Delta_4|, |\Delta_6|, \ldots, |\Delta_{2n}|] \quad (2.129)$$

$$q = \text{Var} \, [1, -|\Delta_2|, |\Delta_4|, \ldots, (-1)^n |\Delta_{2n}|] \quad (2.130)$$

where p and q are the numbers of roots outside and inside the unit circle, respectively. Note that Δ_{2n} is (pi) when all the roots are inside the unit circle. Also, when Δ_{2n} is (ni), Theorem 2.6 gives information on the root distributions (i.e., in this case, $p = 1$ and $q = n - 1$).

THEOREM 2.7 [11]. *When the polynomial $F(z)$ given in (2.128) is real (with $a_n > 0$), the root-distribution problem can be simplified by considering the sign variation of two innerwise matrix determinants of half size as that of the Schur–Cohn matrix Δ_{2n}. For instance, when n is even, the number of roots outside the unit circle is*

$$q = \text{Var} \, [|\Delta_{n-1}^-|, |\Delta_{n-3}^-|, \ldots, |\Delta_1^-|, |\Delta_1^+|, \ldots, |\Delta_{n-1}^+|] \quad (2.131)$$

plus one or zero, depending on the sign $F(1) \times F(-1)$.
  When n is odd, the number of roots outside the unit circle is

$$\text{Var} \, [|\Delta_{n-1}^-|, \ldots, |\Delta_2^-|, 1, |\Delta_2^+|, \ldots, |\Delta_{n-1}^+|],$$

plus one or zero, depending on the sign of $F(1) \times F(-1)$. The number of roots inside the unit circle (i.e., q) is n minus the number p.
  The matrices $\Delta_{n-1}^{\pm}$ are obtained from the following relationship:

$$\Delta_{n-1}^{\pm} = X_{n-1} \pm Y_{n-1} \quad (2.132)$$

where X_{n-1}, Y_{n-1} are given in (2.41).

  When the product of some of the inners determinants of $\Delta_{n-1}^{\pm}$ is zero, one can obtain conditions on the roots to lie on the unit circle or to be reciprocal with respect to the unit circle. For instance, when Δ_{n-1}^- is null innerwise, there exist $n/2$ pairs of reciprocal roots (also includes roots on the unit circle) when n is even and $(n - 1)/2$ such roots when n is odd. In general, if

$$|\Delta_{n-1}^- = |\Delta_{n-3}^-| = \cdots = |\Delta_{n-2K+1}^-| = 0 \qquad (2.133)$$

then $F(z)$ of (2.128) has K pairs of reciprocal roots [11]. Other critical cases are discussed elsewhere [21] and will be further explored in Chapter 7.

THEOREM 2.8 [5, 8].   The number of distinct real roots N of

$$F(z) = a_n z^n + a_{n-1} z^{n-1} + \cdots + a_1 z + a_0 = 0$$

$$\text{with } a_n > 0 \quad (2.134)$$

is

$$N = \text{Var} \, [1, -|\Delta_1^1|, |\Delta_3^1|, \ldots, (-1)^n |\Delta_{2n-1}^1|]$$
$$- \text{Var} \, [1, |\Delta_1^1|, |\Delta_3^1|, \ldots, |\Delta_{2n-1}^1|]. \qquad (2.135)$$

where Δ^1 is the following nonnull innerwise matrix:

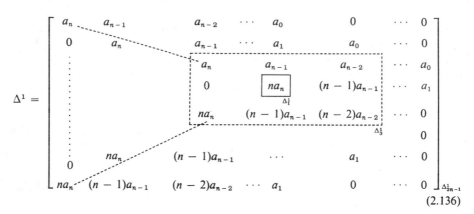

$$(2.136)$$

COROLLARY 2.1.   *The necessary and sufficient condition for all n-distinct roots of (2.134) to be on the real axis is that $\Delta^1 = \Delta_{2n-1}$, given above, be positive innerwise (pi). This can be readily ascertained from (2.135) when we let $N = n$. [Note that the maximum variation of the first term in (2.135) is n.] This corollary presents a proof of Theorem 2.4b.*

THEOREM 2.9 [22].   *The number of distinct positive real roots of $F(z) = 0$ given in (2.134) is*

$$N = \text{Var} \, [1, -|\Delta_2^2|, |\Delta_4^2|, \ldots, (-1)^n |\Delta_{2n}^2|]$$
$$- \text{Var} \, [1, |\Delta_1^1|, |\Delta_3^1|, \ldots, |\Delta_{2n-1}^1|] \qquad (2.137)$$

Note that
$$|\Delta_{2n}^2| = (-1)^n a_0 |\Delta_{2n-1}^1| \qquad (2.138)$$

and
$$|\Delta_2^2| = -a_{n-1} \frac{|\Delta_1^1|}{n} \qquad (2.139)$$

where the nonnull matrix Δ^2 of dimension $2n \times 2n$ is given as

$$
\Delta^2 = \begin{bmatrix}
a_n & \cdots\cdots & a_0 & & 0 & \cdots\cdots & 0 & & 0 & \cdot \\
0 & & & & & & & & & \cdot \\
0 & \;\; a_n & a_{n-1} & & a_{n-2} & & a_{n-3} & \cdots & \cdots & 0 \\
\cdot & \;\; 0 & a_n & & a_{n-1} & & a_{n-2} & \cdots & \cdots & a_0 \\
\cdot & \;\; 0 & na_n & & (n-1)a_{n-1} & & (n-2)a_{n-2} & \cdot & \cdots & \cdot \\
0 & \;\; na_n & (n-1)a_{n-1} & & (n-2)a_{n-2} & & (n-3)a_{n-3} & & \cdots & \cdot \\
0 & & & & & & & & & \cdot \\
na_n & \cdots\cdots & a_1 & & 0 & \cdots\cdots\cdots\cdots\cdots & 0 & & & 0
\end{bmatrix}
$$

$$(2.140)$$

THEOREM 2.10 [5]. The number of distinct negative real roots of $F(z)$ *in (2.134) is*
$$N = \text{Var}\,[1_1, -|\Delta_1^1|, |\Delta_3^1|, \ldots, (-1)^n\,|\Delta_{2n-1}^1|$$
$$- \text{Var}\,[1, -|\Delta_2^2|, |\Delta_4^2|, \ldots, (-1)^n\,|\Delta_{2n}^2|].$$

where Δ^1 and Δ^2 are given in (2.136) and (2.140).

REFERENCES

[1] M. MARDEN, "Geometry of polynomial," *Am. Math. Soc.*, 2nd ed (1966).

[2] R. J. DUFFIN, "Algorithms for classical stability problems," *SIAM Rev.*, **11** (2), 196–213 (April 1963).

[3] A. HURWITZ, "Über die Bedingungen unter welchem eine Gleichung nur Wurzeln mit negativen relen Teilen besitzt," *Math. Ann.*, **46**, 273–284 (1895).

[4] A. LIÉNARD and M. H. CHIPART, "Sur la signe de la partie reelle des racines d'une équation algebrique," *J. Math. Pure Appl.*, **10**, 291–346 (1914).

[5] E. I. JURY and S. M. AHN, "Interchangeability of 'inners' and 'minors,'" *ASME J. Dynamic Systems, Measurement and Control*, 257 (Dec. 1971).

[6] M. R. STOJIĆ and D. D. ŠILJAK, "Generalization of Hurwitz, Nyquist, Mikhailov stability criteria," *IEEE Trans. on Automatic Control*, **AC-10**, 250–254 (July 1965).

[7] E. I. JURY, "Stability, root clustering and 'inners,'" *Proceedings of the Fifth IFAC World Congress, Paris, France, June 12–17, 1972*, Part 4. Distributed by the Instrument Society of America, Paper No. 35.2.

[8] A. T. FULLER, "Conditions for aperiodicity on linear systems," *Brit. J. Appl. Phys.*, **6,** 450 (Dec. 1955).

[9] M. I. ROMANOV, "Algebraic criteria for aperiodicity of linear systems," *Dokl. Akad. Nauk, SSSR*, **12f** (2), 291–294 (1959).

[10] A. COHN, "Über die Anzahl der Wurzeln einer algebraischen Gleichung in einem Kreise," *Math. Z.*, **14–15,** 110–148 (1914).

[11] E. I. JURY, "Inners, approach to some problems of system theory," *IEEE Trans. on Automatic Control*, **AC-16,** 233–240 (June 1971).

[12] B. D. O. ANDERSON and E. I. JURY, "A simplified Schur–Cohn test," *IEEE Trans. on Automatic Control*, **AC-18,** 157–163 (April 1973).

[13] E. I. JURY, *Theory and Application of the Z-Transform Method*. New York: Wiley, 1964, Chap. 3.

[14] E. SZARANIEC, "Stability, instability, and aperiodicity test for linear discrete systems," *Automatica*, **9,** 513–516 (1973).

[15] E. I. JURY and J. TSCHAUNER, "On the inners formulation of stability condition within the shifted unit circle," *Regelungstechnik*, **7,** 299–300 (July 1971).

[16] R. COURANT and D. HILBERT, *Methods of Mathematical Physics*. New York: Interscience, 1962, Vol. II.

[17] E. I. JURY and S. M. AHN, "The theory of inners applied to networks," *IEEE Int. Symp. on Circuit Theory* (1970).

[18] E. I. JURY and T. PAVLIDIS, "Stability and aperiodicity constraints for system design, *IEEE Prof. Group of Circuit Theory*, **CT-10** (1), 137–141 (March 1963).

[19] F. R. GANTMACHER, *The Theory of Matrices*. New York: Chelsea, 1959, Vol. II, pp. 196–197.

[20] S. H. LEHNIGK, *Stability Theorems for Linear Motions, with an Introduction to Lyapunov's Direct Method*. Englewood Cliffs, N.J.: Prentice-Hall, 1966.

[21] B. M. BROWN, "On the distribution of the zeros of a polynomial," *The Quarterly Journal of Mathematics*, **16** (63), 241–256 (Sept. 1965).

[22] E. I. JURY, "The three *p*'s in system theory." Paper presented at the International Symposium in Circuit Theory, Los Angeles, Cal., April 1972. *Symposium Proceedings*, pp. 32–37.

CHAPTER 3

Stability and Positive Definite Symmetric Matrices

In this chapter we will present the stability and root-clustering problems in terms of positive definite (pd) symmetric matrices. Such representation is well known from the early works of Hermite [1], Markov [2], Lyapunov [3], Schur–Cohn [4,5], and others [6]. This representation, similar to the innerwise matrices, also offers a unifying feature—indeed, as will be shown in this chapter, that symmetric and innerwise matrices are related by matrix multiplication. Each of these forms of matrices offers certain advantages in the calculation of the determinants and in proving certain theorems. In order not to duplicate the discussions of the preceding chapter, here, we will present only the proofs of certain stability criteria that were stated before. Furthermore, we will offer some new results that shed light or complement the discussions of Chapter 2. In particular, the discussion of Section 3.6 will generalize the region of the root-clustering problem and present the key to obtaining more enlarged regions.

3.1 STABILITY OF LINEAR TIME-INVARIANT CONTINUOUS SYSTEMS

In this section we will present a proof of the "reduced Hermite criterion" [7,8] and will indicate its relationship to the Liénard–Chipart criterion discussed in Section 2.1. This discussion will also present a proof of the earlier-stated Liénard–Chipart criterion.

THE HERMITE AND THE REDUCED HERMITE CRITERIA [1,8]

Let $F(s)$ be a real polynomial in a variable s of degree n:

$$F(s) = \sum_{k=0}^{n} a_k s^k \qquad a_n > 0 \quad (3.1)$$

51

We define the $n \times n$ Hermite matrix P associated with the polynomial $F(s)$† in (3.1) as

$$p_{ij} = \sum_{k=1}^{i} (-1)^{k+i} a_{n-k+1} a_{n-i-j+k}$$

$$\begin{aligned} & & j \geq i, \quad j + i = \text{even} \\ & = p_{ji} & j < i, \quad j + i = \text{even} \quad (3.2) \\ & = 0 & j + i = \text{odd} \end{aligned}$$

The Hermite criterion [1] states that $F(s)$ in (3.1) is Hurwitz‡ if and only if P is positive definite (i.e., $P > 0$).

In order to present the reduced Hermite criterion [8], we define the matrices C and D as follows:

$$c_{ij} = \sum_{k=1}^{2i-1} (-1)^{k+2i-1} a_{n-k+1} a_{n-2i-2j+k+2} \qquad j \geq i$$

$$= c_{ji} \qquad\qquad\qquad\qquad\qquad\qquad\qquad\quad i < j \quad (3.3)$$

and

$$d_{ij} = \sum_{k=1}^{2i} (-1)^{k+2i} a_{n-k+1} a_{n-2i-2j+k} \qquad j \geq i$$

$$= d_{ji} \qquad\qquad\qquad\qquad\qquad\qquad\qquad\quad i < j \quad (3.4)$$

(*Note:* $a_\ell = 0$ if ℓ is not in the range $0, 1, \ldots, n$.)

If $n = 5$ in (3.1), then C and D from (3.3) and (3.4) give

$$C = \begin{bmatrix} a_5 a_4 & a_5 a_2 & a_5 a_0 \\ a_5 a_2 & a_5 a_0 - a_4 a_1 + a_3 a_2 & a_3 a_0 \\ a_5 a_0 & a_3 a_0 & a_1 a_0 \end{bmatrix}$$

$$D = \begin{bmatrix} -a_5 a_2 + a_4 a_3 & -a_5 a_0 + a_4 a_1 \\ -a_5 a_0 + a_4 a_1 & -a_3 a_0 + a_2 a_1 \end{bmatrix} \qquad (3.5)$$

It is noticed that both C and D are submatrices of the Hermite matrix P in (3.2), obtainable by eliminating certain rows and columns. Accordingly, as n is even or odd, C is $n/2 \times n/2$ or $(n+1)/2 \times (n+1)/2$ and is obtained by deleting even-numbered rows and columns from P. The matrix D is $n/2 \times n/2$ or $(n-1)/2 \times (n-1)/2$ and is obtained by deleting odd-numbered rows and columns from P.

It may be noted that, by reordering of rows and columns, the matrix P becomes the direct sum of C and D. Hence the Hermite criterion requires that both C and D be checked for positive definiteness.

Furthermore, we comment on another property of the matrices C and D.

---

† The Hermite matrix associated with the complex polynomial $F(s)$ is presented later in (3.108) as well as in (6.33) and (6.34).
‡ A Hurwitz Polynomial is defined as one whose zeros all have negative real parts.

The polynomial $s^n F(s^{-1})$ clearly has roots that are the reciprocals of those of $F(s)$; denote the associated C and D matrices by C_r and D_r. Then one can check that with n even, C_r is D with reversal of rows and columns, whereas D_r is C with reversals of rows and columns. If n is odd, then C_r is C with reversal of rows and columns, and D_r is D with reversals of rows and columns. The preceding property is important in proving all the cases of Theorem 3.1.

THEOREM 3.1 (REDUCED-ORDER HERMITE CRITERION [8]). *The polynomial $F(s)$ in* (3.1) *is Hurwitz (i.e., all roots lie in the left half-plane) if and only if any one of the following four sets of inequalities holds:*

$$C > 0, \quad a_0 > 0, \quad a_2 > 0, \quad a_4 > 0, \ldots \tag{3.6}$$

$$C > 0, \quad a_0 > 0, \quad a_1 > 0, \quad a_3 > 0, \ldots \tag{3.7}$$

$$D > 0, \quad a_0 > 0, \quad a_2 > 0, \quad a_4 > 0, \ldots \tag{3.8}$$

$$D > 0, \quad a_0 > 0, \quad a_1 > 0, \quad a_3 > 0, \ldots \tag{3.9}$$

Proof: The necessity of conditions (3.6) to (3.9) is evident from the preceding discussions. In order to prove sufficiency, we will show if for two cases and the other cases can be similarly proven based on the properties of C and D mentioned earlier. We will show that when (3.6) holds, it follows that $D > 0$.

Case 1. Assume (3.6) and that $n = 2m$. Define the matrix

$$A_1 = \begin{bmatrix} -\dfrac{a_{n-2}}{a_n} & -\dfrac{a_{n-4}}{a_n} & -\dfrac{a_{n-6}}{a_n} & \cdots & -\dfrac{a_0}{a_n} \\ 1 & 0 & 0 & \cdots & 0 \\ 0 & 1 & 0 & \cdots & 0 \\ \vdots & \vdots & \vdots & & \vdots \end{bmatrix} \tag{3.10}$$

By direct calculation, we can verify that the (i, j) elements of CA_1 are

$$-c_{i1}\frac{a_{n-2j}}{a_n} + c_{i,j+1} = -a_n a_{n-2i+1}\frac{a_{n-2j}}{a_n}$$

$$+ \sum_{k=1}^{2i-1} (-1)^{k+2i-1} a_{n-k+1} a_{n-2i-2j+k}$$

$$j \geq i - 1$$

$$= \sum_{k=1}^{2i} (-1)^{k+2i-1} a_{n-k+1} a_{n-2i-2j+k}$$

$$j \geq i - 1$$

$$= -d_{ij} \qquad\qquad j \geq i - 1 \tag{3.11}$$

The first equality follows from (3.3) and the last from (3.4). It is similarly checked that (3.11) also holds for $j < i - 1$. Hence, using the symmetry of D in (3.4), we have

$$CA_1 = A_1'C = -D \tag{3.12}$$

Since C is positive definite by assumption, it has a nonsingular square root [9] $C^{1/2}$, so that

$$C^{1/2}A_1C^{-1/2} = -C^{-1/2}DC^{-1/2} \tag{3.13}$$

The preceding equation implies that A_1 is similar to a symmetric matrix and has all real eigenvalues. These eigenvalues are the zeros of the polynomial

$$s^{n/2} + \frac{a_{n-2}}{a_n} s^{(n/2)-1} + \cdots + \frac{a_0}{a_n} \tag{3.14}$$

By assumption (3.6), the above polynomial must be positive for all non-negative s, and so all eigenvalues of A_1 are negative real. Hence, from (3.13), $D > 0$ (positive definite) is proven.

Case 2.   Assume (3.8) and that $n = 2m$. The condition $C > 0$ follows from the result of using (3.6) and the fact that the appropriate matrix associated with $s^n F(s^{-1})$ is D_r. Note that for this case D_r is C with reversals of rows and columns. The same is true for other cases, and when $n = 2m + 1 =$ odd, a minor variation establishes the theorem.

CONNECTION WITH THE LIÉNARD–CHIPART CRITERION [10,11]

There are connections between the leading principal minors of the Hurwitz and the Hermite matrices.† Liénard and Chipart were the first to indicate this connection, which was reestablished by others. These connections show that the odd-order leading principal minors of the Hurwitz matrix [or the equivalent innerwise matrix determinants; see (2.7) and (2.8) for a specific n] are the same (within a constant) as the leading principal minors of the matrix C, whereas the even-order Hurwitz determinants, $|\Delta_2|, |\Delta_4|, \ldots$ [note generalization for any "n" of (2.8)] (or the corresponding innerwise matrix determinants) are the same (within a constant) as the leading principal minors of D. Accordingly, the Liénard–Chipart criterion inequalities are (3.6) to (3.9).

Based on this observation, we can state the following definition of "interchangeability of 'inners' and minors arrays" [11].

DEFINITION 3.1 [11].   *The inners of matrix Q and the leading principal minor arrays of matrix P are said to be interchangeable if the signs of the*

† The leading principal minors of Hermite matrices are also known as the "Bezoutians" [25].

determinants of all the inners are the same as the sign of all the leading principal minors.†

From the definition it follows that the reduced-order Hermite matrix is interchangeable with the innerwise matrix of (2.10). Furthermore, the reduced Hermite matrix being (pd) is equivalent to the Liénard–Chipart innerwise matrix being (pi), and vice versa. In Section 3.4 we will establish the connection between the symmetric matrix and the innerwise matrix associated with the root clustering and the root distribution of a polynomial.

3.2   STABILITY OF LINEAR TIME-INVARIANT DISCRETE SYSTEMS [12]

In complete analogy with the discussion of the preceding section, we will present the symmetric forms of the stability criteria for discrete systems. These criteria will include the well-known Schur–Cohn criterion [4,5], the recently developed symmetrical form of the determinantal stability criterion discussed in Section 2.3, and the reduced Schur–Cohn criterion [12]. The latter is similar to the reduced Hermite criterion discussed earlier.

SCHUR-COHN-FUJIWARA CRITERION [4, 5, 6]

Consider the real polynomial

$$F(z) = \sum_{i=0}^{n} a_i z^i \qquad\qquad a_n > 0 \quad (3.15)$$

and associate with it the $n \times n$ symmetric matric $C = (\gamma_{ij})$ defined by

$$\gamma_{ij} = \sum_{p=1}^{\min(i,j)} (a_{n-i+p} a_{n-j+p} - a_{i-p} a_{j-p}) \qquad (3.16)‡$$

For $n = 3$ one has, for example,

$$C = \begin{bmatrix} a_3^2 - a_0^2 & a_2 a_3 - a_1 a_0 & a_1 a_3 - a_2 a_0 \\ a_2 a_3 - a_1 a_0 & a_3^2 + a_2^2 - a_0^2 - a_1^2 & a_2 a_3 - a_1 a_0 \\ a_1 a_3 - a_2 a_0 & a_2 a_3 - a_1 a_0 & a_3^2 - a_0^2 \end{bmatrix} \qquad (3.17)$$

and for $n = 4$ one has

$$C = \begin{bmatrix} a_4^2 - a_0^2 & a_3 a_4 - a_1 a_0 & a_2 a_4 - a_2 a_0 & a_1 a_4 - a_3 a_0 \\ a_3 a_4 - a_1 a_0 & a_4^2 + a_3^2 - a_1^2 - a_0^2 & a_3 a_4 + a_2 a_3 - a_2 a_1 - a_1 a_0 & a_2 a_4 - a_2 a_0 \\ a_2 a_4 - a_2 a_0 & a_3 a_4 + a_2 a_3 - a_2 a_1 - a_1 a_0 & a_4^2 + a_3^2 - a_1^2 - a_0^2 & a_3 a_4 - a_1 a_0 \\ a_1 a_4 - a_3 a_0 & a_2 a_4 - a_2 a_0 & a_3 a_4 - a_1 a_0 & a_4^2 - a_0^2 \end{bmatrix} \qquad (3.18)$$

† This definition should be distinguished from the one implicit in Section 1.4.
‡ The associated symmetric matrix for the complex polynomial $F(z)$ is given in (3.123).

Remarks

1. From (3.17) and (3.18) one can establish the property of the arbitrary dimension matrix C: not only is there symmetry about the main diagonal, but there is also symmetry about the cross diagonal. This property would enable us, as will be shown later, to simplify the Schur–Cohn test.

2. The matrix C in (3.16) is the counterpart of the matrix (with real coefficients) (2.39) discussed in Section 2.3. In Section 3.4 we will indicate the connection between these equivalent matrix determinants (i.e., interchangeability).

Let us adopt the notation $n_+(X)$, $n_-(X)$ and $n_0(X)$ to denote the number of positive, negative, and zero eigenvalues of a symmetric matrix X. The result of Schur–Cohn, as extended by Fujiwara, is as follows [6].

THEOREM 3.2. *With $F(z)$ and C as defined above, the number of zeros z_i of $F(z)$ for which $|z_i| < 1$ and for which z_i^{-1} is not a zero is $n_+(C)$; the number of zeros z_i for which $|z_i| > 1$ and for which z_i^{-1} is not a zero is $n_-(C)$; the number of zeros z_i for which $|z_i| = 1$ or for which z_i^{-1} is also a zero is $n_0(C)$.*

From the theorem it follows that $F(z)$ is stable (or all roots are inside the unit circle in z-plane) if and only if C is positive definite (pd) or, equivalently written, $C > 0$.

SYMMETRICAL MATRIX FORM OF THE DETERMINANTAL CRITERION

Let $F(z)$ in (3.15) be an even polynomial—that is,

$$F(z) = a_{2m}z^{2m} + a_{2m-1}z^{2m-1} + \cdots + a_0, \quad a_{2m} > 0 \quad (3.19)$$

We associate with $F(z)$ two $m \times m$ symmetric matrices $A = (\alpha_{ij})$ and $B = (\beta_{ij})$ defined by

$$\alpha_{ij} = \sum_{p=1}^{\min(i,j)} (a_{2m+p-i}a_{2m+p-j} - a_{i-p}a_{j-p})$$

$$+ \sum_{p=1}^{i} (a_{2m+p-i}a_{j+p-1} - a_{i-p}a_{2m+1-j-p}) \quad (3.20)$$

and

$$\beta_{ij} = \sum_{p=1}^{\min(i,j)} (a_{2m+p-i}a_{2m+p-j} - a_{i-p}a_{j-p})$$

$$- \sum_{p=1}^{i} (a_{2m+p-i}a_{j+p-1} - a_{i-p}a_{2m+1-j-p}) \quad (3.21)$$

In case $n = 4$, one has, for example,

$$A = \begin{bmatrix} a_4^2 - a_0^2 + a_4a_1 - a_0a_3 & a_4a_3 - a_0a_1 + a_4a_2 - a_0a_2 \\ a_4a_3 - a_0a_1 + a_4a_2 - a_0a_2 & a_4^2 + a_3^2 - a_0^2 - a_1^2 + a_4a_3 + a_3a_2 - a_0a_1 - a_1a_2 \end{bmatrix} \quad (3.22)$$

and

$$B = \begin{bmatrix} a_4^2 - a_0^2 - a_4 a_1 + a_0 a_3 & a_4 a_3 - a_0 a_1 - a_4 a_2 + a_0 a_2 \\ a_4 a_3 - a_0 a_1 - a_4 a_2 + a_0 a_2 & a_4^2 + a_3^2 - a_0^2 - a_1^2 - a_4 a_3 - a_3 a_2 + a_0 a_1 + a_1 a_2 \end{bmatrix} \quad (3.23)$$

Let $F(z)$ in (3.15) be an odd-degree polynomial. That is,

$$F(z) = a_{2m-1} z^{2m-1} + a_{2m-2} z^{2m-2} + \cdots + a_0$$
$$a_{2m-1} > 0 \quad (3.24)$$

With $F(z)$ we associate symmetric matrices $A = (\alpha_{ij})$ and $B = (\beta_{ij})$, of dimensions $m \times m$ and $(m-1) \times (m-1)$, respectively, defined by

$$\alpha_{ij} = \sum_{p=1}^{\min(i,j)} (a_{2m-1-i+p} a_{2m-1-j+p} - a_{i-p} a_{j-p})$$
$$+ \sum_{p=1}^{i} (a_{2m-1-i+p} a_{j+p-1} - a_{i-p} a_{2m-j-p})$$
$$0 \le i, \quad j \le m-1$$
$$= \sum_{p=1}^{\min(i,j)} (a_{2m-1-i+p} a_{2m-1-j+p}) - a_{i-p} a_{j-p}$$
$$i = m, \quad j < m \text{ and } i < m, \quad j = m$$
$$= \frac{1}{2} \sum_{p=1}^{m} (a_{m-1+p}^2 - a_{m-p}^2) \qquad i = j = m \quad (3.25)$$

and

$$\beta_{ij} = \sum_{p=1}^{\min(i,j)} (a_{2m-1-i+p} a_{2m-1-j+p} - a_{i-p} a_{j-p})$$
$$- \sum_{p=1}^{i} (a_{2m-1-i+p} a_{j+p-1} - a_{i-p} a_{2m-j-p}) \qquad (3.26)$$

In case $n = 3$, one has

$$A = \begin{bmatrix} a_3^2 - a_0^2 + a_1 a_3 - a_2 a_0 & a_2 a_3 - a_1 a_0 \\ a_2 a_3 - a_1 a_0 & \frac{1}{2}(a_3^2 + a_2^2 - a_1^2 - a_0^2) \end{bmatrix} \quad (3.37)$$

$$B = [a_3^2 - a_0^2 - a_1 a_3 + a_2 a_0] \qquad (3.28)$$

In case $n = 5$, one has for B

$$B = \begin{bmatrix} a_5^2 - a_0^2 - a_5 a_1 + a_0 a_4 & a_5 a_4 - a_0 a_1 - a_5 a_2 + a_0 a_3 \\ a_5 a_4 - a_0 a_1 - a_5 a_2 + a_0 a_3 & a_4^2 - a_1^2 - a_4 a_2 + a_1 a_3 + a_5^2 - a_0^2 - a_5 a_3 + a_0 a_2 \end{bmatrix} \quad (3.29)$$

RELATION OF THE MATRICES C, A, AND B

Consider first the case of even $n = 2m$. We shall show how A and B can be found via an orthogonal transformation of C. Examination of the definitions of α_{ij}, β_{ij}, and γ_{ij} shows easily that

$$\alpha_{ij} = \gamma_{ij} + \gamma_{i,2m+1-j} \qquad \beta_{ij} = \gamma_{ij} - \gamma_{i,2m+1-j} \qquad (3.30)$$

so that the individual entries of A and B are easily related to individual entries of C. Also, using the double symmetry of C, one can readily verify that

$$\begin{bmatrix} A & 0 \\ 0 & B \end{bmatrix} = \frac{1}{2} \begin{bmatrix} I & I \\ -I & I \end{bmatrix} J_e C J_e \begin{bmatrix} I & -I \\ I & I \end{bmatrix} \qquad (3.31)$$

$$J_e = \begin{bmatrix} I_m & & 0_m & \cdots \\ \hline & 0 & \cdots & 0 & 1 \\ \vdots & 0 & \cdots & 1 & 0 \\ 0_m & \vdots & & \vdots & \vdots \\ \vdots & 1 & \cdots & 0 & 0 \end{bmatrix} \qquad (3.32)$$

The effect of postmultiplying (3.31) by J_e is to reverse the order of the last m columns of C, while the postmultiplication by

$$\frac{1}{\sqrt{2}} \begin{bmatrix} I & -I \\ I & I \end{bmatrix} \qquad (3.33)$$

then serves to form the sum and difference in the formula (3.30) for α_{ij} and β_{ij}.

In the case of odd $n = 2m - 1$, examination of the definitions of α_{ij}, β_{ij}, and γ_{ij} in (3.25), (3.26), and (3.16) shows that

$$\begin{aligned} \alpha_{ij} &= \gamma_{ij} + \gamma_{i,2m-j} & i,j \le m-1 \\ \alpha_{ij} &= \gamma_{ij} \\ & \quad i = m, \;\; j < m \text{ or } i < m, \;\; j = m \quad (3.34) \end{aligned}$$

$$\alpha_{mm} = \tfrac{1}{2}\gamma_{mm}$$

while
$$\beta_{ij} = \gamma_{ij} - \gamma_{i,2m-j} \qquad (3.35)$$

We can also verify that

$$\begin{bmatrix} A & 0 \\ 0 & B \end{bmatrix} = \tfrac{1}{2} \begin{bmatrix} I_{m-1} & 0 & I_{m-1} \\ \hline 0 & 1 & 0 \\ \hline -I_{m-1} & 0 & I_{m-1} \end{bmatrix} J_0 C J_0 \begin{bmatrix} I_{m-1} & 0 & -I_{m-1} \\ \hline 0 & 1 & 0 \\ \hline I_{m-1} & 0 & I_{m-1} \end{bmatrix} \qquad (3.36)$$

$$
J_0 = \begin{bmatrix}
I_m & \vdots & 0_{m \times m-1} & \cdots & \\
\hline
& 0 & \cdots & 0 & 1 \\
0_{m-1 \times m} & 0 & \cdots & 1 & 0 \\
& \vdots & \vdots & & \vdots & \vdots \\
& 1 & \cdots & 0 & 0
\end{bmatrix} \tag{3.37}
$$

With $\dotplus$ denoting direct sum, it is evident for n-even and n-odd from (3.31) and (3.36) that $n_+(A \dotplus B) = n_+(C)$, $n_-(A \dotplus B) = n_-(C)$, and $n_0(A \dotplus B) = n_0(C)$. This implies the following statement:

"If $F(z)$ is stable, the theorem implies $C > 0$. Hence $A > 0$ and $B > 0$. Conversely, if $A > 0$ and $B > 0$, then $C > 0$ and by the theorem $F(s)$ is stable."

Remarks

1. The computational burden in computing $A > 0$, $B > 0$ is much less than the computational burden of $C > 0$.

2. The stability test requiring $A > 0$ and $B > 0$ is equivalent to the determinantal criterion in Section 2.3—that is, (2.40) to (2.42)—and vice versa. In Section 3.4 we will also show that $\Delta_n^{\pm}$ (being (pi)) is equivalent to $A > 0, B > 0$ and vice versa. Hence they are interchangeable.

REDUCED SCHUR–COHN CRITERION [12]

In the following material we will present the reduced Schur–Cohn criterion in a form of a theorem for n-even and n-odd. The proof will be given in Appendix B.

We associate with $F(z)$ in (3.15), and for $n = 2m$, the polynomial $P(s)$, defined by

$$
P(s) = (s - 1)^{2m} F\left(\frac{s + 1}{s - 1}\right) = a_{2m}(s + 1)^{2m} + a_{2m-1}(s + 1)^{2m-1}(s - 1)
$$

$$
+ \cdots + a_0(s - 1)^{2m} \tag{3.38}
$$

Evidently all zeros of $F(z)$ lie in $|z| < 1$ if and only if all zeros of $P(s)$ lie in Re $[s] < 0$. Equation (3.38) represents the biliner transformation from inside the unit circle into the open left half-plane [13].

Suppose that the coefficients b_i are defined by

$$
P(s) = b_{2m}s^{2m} + b_{2m-1}s^{2m-1} + b_{2m-2}s^{2m-2} + \cdots + b_0 \tag{3.39}
$$

then
$$
b_i = \sum_{r=0}^{2m}\left[\sum_j (-1)^{r+i-j} a_r \binom{r}{j}\binom{2m-r}{i-j}\right] \tag{3.40}
$$

The summation over j is governed by $\max(0, 2m - r - i) \leq j \leq$

min (i, r).] Furthermore, a necessary condition for the roots of $P(s)$ to lie in the left half-plane is that $b_{2i} > 0$, $i = 0, 1, \ldots, m$.

THEOREM 3.3. *With definitions as above, the necessary conditions for $F(z)$ to be stable (all roots are inside the unit circle) are that A and B are positive definite and that the b_i are positive.*

Any of the following are sufficient conditions.

1. $A > 0$, $\quad b_{2i} > 0$ $\qquad\qquad\qquad\qquad\qquad i = 0, 1, \ldots, m$ (3.41)

2. $A > 0$, $\quad b_{2m} > 0$, $\quad b_0 > 0$, $\quad b_{2i+1} > 0$ $\quad i = 0, 1, \ldots, m - 1$ (3.42)

3. $B > 0$, $\quad b_{2i} > 0$ $\qquad\qquad\qquad\qquad\qquad i = 0, 1, \ldots, m$ (3.43)

4. $B > 0$, $\quad b_{2m} > 0$, $\quad b_0 > 0$, $\quad b_{2i+1} > 0$ $\quad i = 0, 1, \ldots, m - 1$ (3.44)

Similarly, for odd-degree $F(z)$—that is, $n = 2m - 1$—we have

$$b_i = \sum_{r=0}^{2m-1} \left[\sum_j (-1)^{r+i+1-j} a_r \binom{r}{j} \binom{2m-1-r}{i-j} \right] \qquad (3.45)$$

and Theorem 3.4.

THEOREM 3.4. *The necessary conditions for $F(z)$ to be stable are that $A > 0$, $B > 0$, and $b_i > 0$, $i = 0, 1, \ldots, 2m - 1$. Any of the following are sufficient conditions:*

1. $A > 0$, $\quad b_{2m-1} > 0$, $\quad b_{2i} > 0$ $\qquad\qquad i = 0, 1, \ldots, m - 1$ (3.46)

2. $A > 0$, $\quad b_0 > 0$, $\quad b_{2i+1} > 0$ $\qquad\qquad i = 0, 1, \ldots, m - 1$ (3.47)

3. $B > 0$, $\quad b_{2m-1} > 0$, $\quad b_{2i} > 0$ $\qquad\qquad i = 0, 1, \ldots, m - 1$ (3.48)

4. $B > 0$, $\quad b_0 > 0$, $\quad b_{2i+1} > 0$ $\qquad\qquad i = 0, 1, \ldots, m - 1$ (3.49)

Remarks

1. It is evident that the preceding theorem for n-even and n-odd is equivalent to the simplified determinantal criterion discussed in Section 2.3. It is analogous to the relationship between the reduced Hermite criterion and the Liénard–Chipart criterion for the continuous case.

2. For computational purposes, it may be easier to use (3.48) and (3.49) for the even case. Then the even-degree polynomial conditions can be obtained as a limiting case from the odd-degree condition (see example in Section 3.4).

3.3 RELATIONSHIPS BETWEEN REAL AND COMPLEX HURWITZ POLYNOMIALS [14, 33]

In this section we will prove results relating the zero distribution of a real polynomial and the zero distribution of a complex polynomial of approximately half the degree. The main result is one that guarantees that both polynomials are simultaneously Hurwitz or simultaneously not Hurwitz, provided that approximately half the coefficients of the real polynomial are positive. Moreover, we will consider relationships between real aperiodic polynomials and real and complex Hurwitz polynomials. Doing so will enable us to prove the aperiodicity condition stated in Section 2.2. The importance of this relationship is to enable the reader to obtain a simple proof of the well-known results of the Liénard–Chipart (or reduced Hermite) criterion as well as to exhibit what is, in effect, a reduced Lyapunov equation, as will be discussed in Section 3.5.

THEOREM 3.5. *For real a_i, $i = 0, 1, \ldots, 2n$, consider the three polynomials*

$$F_0(s) = \sum_{i=0}^{2n} a_i s^i \tag{3.50}$$

$$F_1(s) = \sum_{i=0}^{n} [a_{2i}(j)^i + a_{2i+1}(j)^{i+1}]s^i \tag{3.51}$$

$$F_2(s) = \sum_{i=0}^{n} [a_{2i}(-j)^i - a_{2i-1}(-j)^{i+1}]s^i \tag{3.52}$$

[In (3.51) and (3.52) one sets $a_{2n+1} = 0$ and $a_{-1} = 0$.] Suppose that either $a_{2i} > 0$ for $i = 0, 1, \ldots, n$ or $a_0 > 0$, $a_{2n} > 0$, and $a_{2i+1} > 0$ for $i = 0, 1, \ldots, n - 1$. Then if any of the polynomials $F_k(s)$, $k = 0, 1, 2$, has its zeros in Re $[s] < 0$, the other two polynomials have this property. Before proceeding with the proof, we present the corresponding theorem for n-odd.

THEOREM 3.6. *For real a_i, $i = 0, 1, \ldots, 2n + 1$, consider the three polynomials*

$$F_0(s) = \sum_{i=0}^{2n+1} a_i s^i \tag{3.53}$$

$$F_1(s) = \sum_{i=0}^{n} [a_{2i}(j)^i + a_{2i+1}(j)^{i+1}]s^i \tag{3.54}$$

$$F_2(s) = \sum_{i=0}^{n} [a_{2i}(-j)^i - a_{2i-1}(-j)^{i+1}]s^i \tag{3.55}$$

[In (3.55) it is noted that $a_{2n+2} = 0$ and $a_{-1} = 0$.] Suppose that either $a_{2n+1} > 0$ and $a_{2i} > 0$ for $i = 0, 1, \ldots, n$ or $a_0 > 0$ and $a_{2i+1} > 0$ for $i = 0, 1, \ldots, n$. Then if any of the polynomials $F_k(s)$, $k = 0, 1, 2$, has their zeros in Re $[s] < 0$, the other two polynomials have the same property.

In the following material we will present proof of the result for $F_0(s)$ and

$F_1(s)$ in Theorem 3.5, based on the use of the Cauchy index to count the zeros of both polynomials in Re $[s] < 0$. This proof is easily modified to apply to $F_2(s)$ and also to prove the theorem of the odd case by noting the following paragraph.

If one starts with a polynomial $F_0(s) = \sum_{i=0}^{2n+2} a_i s^i$ of degree $2n + 2$ and writes down the associated $F_1(s)$ and $F_2(s)$ defined in Theorem 3.5, and if one then sets $a_{2n+2} = 0$ in the three polynomials, one recovers the three polynomials in (3.53) to (3.55) of Theorem 3.6. This fact can be used as the basis for a proof of Theorem 3.6, given that Theorem 3.5 holds.

Proof of Theorem 3.5: We shall count the zeros of $F_0(s)$ and $F_1(s)$ in Re $[s] < 0$ by observing the change in argument of these polynomials as s moves around a contour comprising the imaginary axis and around a semicircle of arbitrarily large radius extending in the left half-plane.

First, it is easily established that if $F_0(s)$ has a zero $j\omega$, ω real, then $F_1(s)$ has a zero $j\omega^2$, and, conversely, whereas it is impossible for $F_1(s)$ to have a zero $-j\omega^2$, ω real, if $a_{2i} > 0$ for $i = 0, 1, \ldots, n$ or $a_0 > 0$, $a_{2n} > 0$ and $a_{2i+1} > 0$ for $i = 0, 1, \ldots, n - 1$. Hence if *one* of $F_0(s)$ and $F_1(s)$ is Hurwitz, *neither* has a pure imaginary zero. Therefore the change in argument of $F_0(s)$ as s moves from $-j\infty$ to $+j\infty$ is

$$\Delta_{-\infty}^{+\infty} \arctan \frac{\operatorname{Im} F_0(j\omega)}{\operatorname{Re} F_0(j\omega)} \tag{3.56}$$

whereas the change in the argument around a semicircular contour extending into the left half-plane, Re $[s] < 0$ from $+jR$ to $-jR$ for R very large is $2n\pi$. Also, the total change in arg $F_0(s)$ in moving through the semicircular contour and the imaginary axis in the counterclockwise direction is $2\pi N_0$, where N_0 is the number of zeros in Re $[s] < 0$. Accordingly,

$$N_0 = n + \frac{1}{2\pi} \Delta_{-\infty}^{+\infty} \arctan \frac{\operatorname{Im} F_0(j\omega)}{\operatorname{Re} F_0(j\omega)} \tag{3.57}$$

Similarly, with N_1 the number of zeros of $F_1(s)$ in Re $[s] < 0$,

$$N_1 = \frac{n}{2} + \frac{1}{2\pi} \Delta_{-\infty}^{+\infty} \arctan \frac{\operatorname{Im} F_1(j\omega)}{\operatorname{Re} F_1(j\omega)} \tag{3.58}$$

If we adopt the notation

$$F_0(s) = G(s^2) + sH(s^2) \tag{3.59}$$

it follows easily from (3.51) that

$$F_1(s) = G(js) + jH(js) \tag{3.60}$$

and
$$F_2(s) = G(-js) + sH(-js) \tag{3.61}$$

Then
$$F_1(j\omega) = G(-\omega) + jH(-\omega) \tag{3.62}$$

and noting $H(-\omega)$ has lower degree than $G(-\omega)$, we see that

$$\Delta_{-\infty}^{+\infty} \arctan \frac{\operatorname{Im} F_1(j\omega)}{\operatorname{Re} F_1(j\omega)} = \Delta_{-\infty}^{+\infty} \arctan \frac{H(-\omega)}{G(-\omega)} = -\pi I_{-\infty}^{\infty} \frac{H(-\omega)}{G(-\omega)} \tag{3.63}$$

where $I_b^a f(s)$ is the Cauchy index [15] of the function $f(s)$ between the limits b and a. Using standard properties of the Cauchy index, we then have

$$-\Delta_{-\infty}^{+\infty} \arctan \frac{H(-\omega)}{G(-\omega)} = \pi I_0^{\infty} \frac{H(-\omega)}{G(-\omega)} + \pi I_{-\infty}^{0} \frac{H(-\omega)}{G(-\omega)}$$

[using property 3, p. 216, of Reference 15 and $G(0) = a_0 > 0$]. Furthermore,

$$-\Delta_{-\infty}^{+\infty} \arctan \frac{H(-\omega)}{G(-\omega)} = \pi I_0^{\infty} \frac{H(-\omega^2)}{G(-\omega^2)} + \pi I_{\infty}^{0} \frac{H(\omega)}{G(\omega)}$$

(using obvious changes of variable). Also,

$$-\Delta_{-\infty}^{+\infty} \arctan \frac{H(-\omega)}{G(-\omega)} = \pi I_0^{\infty} \omega \frac{H(-\omega^2)}{G(-\omega^2)} + \pi I_{\infty}^{0} \frac{H(\omega)}{G(\omega)} \tag{3.64}$$

(using property 2, p. 216, of Reference 15).

Now consider the second summand in the right side of (3.64). In case $a_{2i} > 0$, then $G(\omega) = \sum_{i=0}^{n} a_{2i}\omega^i$ for $\omega \in (0, \infty)$ is positive, and so $I_{\infty}^0 [H(\omega)/G(\omega)] = 0$. If, alternatively, we have $a_0 > 0$, $a_{2n} > 0$, and $a_{2i+1} > 0$ for $i = 0, 1, \ldots, n - 1$, we obtain

$$I_{\infty}^0 \frac{H(\omega)}{G(\omega)} = I_{\infty}^0 \frac{(\omega + 1)H(\omega)}{G(\omega)}$$

(using property 2, p. 216, of Reference 15). Also,

$$I_{\infty}^0 \frac{H(\omega)}{G(\omega)} = \frac{1}{2} \lim_{\omega \to \infty} \left[\operatorname{sgn} \frac{(\omega + 1)H(\omega)}{G(\omega)} \right] - \frac{1}{2} \lim_{\omega \to 0} \left[\operatorname{sgn} \frac{(\omega + 1)H(\omega)}{G(\omega)} \right]$$

$$- I_{\infty}^0 \frac{G(\omega)}{(\omega + 1)H(\omega)}$$

(using property 5, p. 216, of Reference 15). Finally,

$$I_\infty^0 \frac{H(\omega)}{G(\omega)} = \frac{1}{2} \operatorname{sgn} \frac{a_{2n-1}}{a_{2n}} - \frac{1}{2} \operatorname{sgn} \frac{a_1}{a_0} - I_\infty^0 \frac{G(\omega)}{(\omega+1)H(\omega)} \tag{3.65}$$

[Note that $a_{2i+1} > 0$ implies $(\omega+1)H(\omega) > 0$ for $\omega \in (0, \infty)$, so that $I_\infty^0[G(\omega)/(\omega+1)H(\omega)] = 0$.] Hence,

$$I_\infty^0 \frac{H(\omega)}{G(\omega)} = 0 \tag{3.65a}$$

Returning to (3.64) then, we find that if either $a_{2i} > 0$ for $i = 0, 1, \ldots, n$ or $a_0 > 0$, $a_{2n} > 0$, and $a_{2i+1} > 0$ for $i = 0, 1, \ldots, n-1$, then (noting (3.63))

$$-\Delta_{-\infty}^\infty \arctan \frac{\operatorname{Im} F_1(j\omega)}{\operatorname{Re} F_1(j\omega)} = \pi I_0^\infty \frac{\omega H(-\omega^2)}{G(-\omega^2)} \tag{3.66}$$

Also, we have

$$F_0(j\omega) = G(-\omega^2) + j\omega H(-\omega^2) \tag{3.67}$$

and so

$$-\Delta_{-\infty}^{+\infty} \arctan \frac{\operatorname{Im} F_0(j\omega)}{\operatorname{Re} F_0(j\omega)} = -\Delta_{-\infty}^{+\infty} \arctan \frac{\omega H(-\omega^2)}{G(-\omega^2)}$$

$$= \pi I_{-\infty}^{+\infty} \frac{\omega H(-\omega^2)}{G(-\omega^2)}$$

$$= \pi I_0^\infty \frac{\omega H(-\omega^2)}{G(-\omega^2)} + \pi I_{-\infty}^0 \frac{\omega H(-\omega^2)}{G(-\omega^2)}$$

[using property 3, p. 216, of Reference 15 and $G(0) = a_0 \neq 0$]. Also,

$$-\Delta_{-\infty}^\infty \arctan \frac{\operatorname{Im} F_0(j\omega)}{\operatorname{Re} F_0(j\omega)} = \pi I_0^\infty \frac{\omega H(-\omega^2)}{G(-\omega^2)} - \pi I_\infty^0 \frac{\omega H(-\omega^2)}{G(-\omega^2)}$$

(by change of variable). Furthermore,

$$-\Delta_{-\infty}^\infty \arctan \frac{\operatorname{Im} F_0(j\omega)}{\operatorname{Re} F_0(j\omega)} = 2\pi I_0^\infty \frac{\omega H(-\omega^2)}{G(-\omega^2)} \tag{3.68}$$

(using property 1, p. 216, of Reference 15).

Then (3.57) yields

$$N_0 = n + \frac{1}{2\pi} \Delta_{-\infty}^{\pm \infty} \arctan \frac{\operatorname{Im} F_0(j\omega)}{\operatorname{Re} F_0(j\omega)}$$

$$= n - I_0^\infty \frac{\omega H(-\omega^2)}{G(-\omega^2)} \qquad \text{[by using (3.68)]}$$

$$= n + \frac{1}{\pi} \Delta_{-\infty}^\infty \arctan \frac{\operatorname{Im} F_1(j\omega)}{\operatorname{Re} F_1(j\omega)} \qquad \text{[by using (3.66)]}$$

$$= 2N_1 \qquad \text{[by using (3.58)]}$$

Consequently, under the condition that $a_{2i} > 0$ for $i = 0, 1, \ldots, n$ or $a_0 > 0$, $a_{2n} > 0$, and $a_{2i+1} > 0$ for $i = 0, 1, \ldots, n - 1$, and the condition that one of $F_0(s)$ and $F_1(s)$ has no pure imaginary zero (which implies the same property for the other), the number of zeros of $F_0(s)$ in Re $[s] < 0$ is twice the number of $F_1(s)$ in Re $[s] < 0$. In particular, if one polynomial has all its zeros in Re $[s] < 0$, so does the other.

APERIODIC POLYNOMIALS [14]

A polynomial $f_0(s)$ is termed aperiodic if all its zeros are distinct and negative real. Tests for aperiodicity have been given in Section 2.2. One purpose here is to show how we can prove (2.36) and (2.37) via Theorem 3.5, and, in addition, we will show a new test, which again follows from Theorem 3.5.

THEOREM 3.7. *Suppose that*

$$f_0(s) = \sum_{i=0}^{n} a_i s^i \tag{3.69}$$

and

$$F_0(s) = f_0(s^2) + s\frac{df_0(s^2)}{d(s^2)} \tag{3.70}$$

$$F_1(s) = f_0(js) + j\frac{df_0(js)}{d(js)} \tag{3.71}$$

$$F_2(s) = f_0(-js) + s\frac{df_0(-js)}{d(-js)} \tag{3.72}$$

Suppose also that $a_i > 0$, $i = 0, 1, \ldots, n$. Then aperiodicity of $f_0(s)$ implies and is implied by the Hurwitz nature of $F_0(s)$, $F_1(s)$, or $F_2(s)$.

Before proving this result, we note the following points.

1. $F_0(s)$ is a real $2n$th-degree polynomial, and $F_1(s)$ and $F_2(s)$ are complex nth-degree polynomials.

2.   As shown by the formulas (3.59) to (3.61), $F_1(s)$ and $F_2(s)$ in Theorem 3.7 are defined from $F_0(s)$ via the same procedure as in Theorem 3.5. Positivity of the a_i, in fact, implies that all coefficients of $F_0(s)$ are positive. Accordingly, if any one of $F_0(s)$, $F_1(s)$, and $F_2(s)$ is Hurwitz, so are the other two, by Theorem 3.5.

3.   That aperiodicity of $f_0(s)$ and the Hurwitz property for $F_0(s)$ imply one another is the result, in essence, established by Romanov [16] and Fuller [17].

4.   That aperiodicity of $f_0(s)$ and the Hurwitz property for $F_1(s)$ imply one another is the result of Meerov [18]. Theorem 3.5 therefore provides an immediate link between References 16 and 17 on the one hand and Reference 18 on the other.

5.   That aperiodicity of $f_0(s)$ and the Hurwitz property for $F_2(s)$ imply one another is a new result [14].

Proof of Theorem 3.7:   In view of Theorem 3.5, it is enough to prove the equivalences of the statements that $f_0(s)$ is aperiodic and $F_0(s)$ is Hurwitz. Suppose first that $f_0(s)$ is aperiodic. Then we have

$$n = I_{-\infty}^{0} \frac{f_0'(s)}{f_0(s)} \quad \text{(p. 176, of Ref. 15)}$$

$$= I_{\infty}^{0} \frac{df_0(-\omega^2)/d(-\omega^2)}{f_0(-\omega^2)} \quad \text{(by change of variable)}$$

$$= -I_{0}^{\infty} \frac{df_0(-\omega^2)/d(-\omega^2)}{f_0(-\omega^2)} \quad \text{(by property 1, p. 216, of Ref. 15)}$$

$$= -I_{0}^{\infty} \frac{\omega df_0(-\omega^2)/d(-\omega^2)}{f_0(-\omega^2)} \quad \text{(by property 2, p. 216, of Ref. 15)}$$

$$= \frac{1}{2\pi} \Delta_{-\infty}^{\infty} \arctan \frac{\text{Im } F_0(j\omega)}{\text{Re } F_0(j\omega)} \quad \text{(by using (3.68))} \quad (3.73)$$

Now $F_0(s)$ can have no pure imaginary zero, because Re $F_0(j\omega_0) =$ Im $F_0(j\omega_0)$ $= 0$ implies

$$f_0(-\omega_0^2) = 0 \quad \text{and} \quad \omega_0 \left.\frac{df_0(-\omega^2)}{d\omega^2}\right|_{\omega_0} = 0$$

which implies either that $\omega_0 = 0$ is a zero of $f_0(s)$ or that $f_0(s)$ has a repeated negative real zero. In either case, aperiodicity is contradicted. Accordingly, (3.57) applies.

$$N_0 = n + \frac{1}{2\pi} \Delta_{-\infty}^{+\infty} \arctan \frac{\text{Im } F_0(j\omega)}{\text{Re } F_0(j\omega)} \tag{3.57}$$

and using (3.73), we obtain the number N_0 of zeros of $F_0(s)$ in Re $[s] < 0$ as $2n$; that is, $F_0(s)$ is Hurwitz.

The argument is readily reversible, in order to allow deduction of aperiodicity of $f_0(s)$ from the Hurwitz nature of $F_0(s)$. Thus, the theorem is proved.

3.4   POSITIVE INNERWISE AND POSITIVE DEFINITE SYMMETRIC MATRIX TRANSFORMATIONS [19]

In this section we will somewhat generalize the region for the root clustering of a polynomial. In this generalized region we will present the root-clustering condition in terms of both positive definite and positive innerwise matrices. Also, we will establish the connection between these two types of matrices. In particular, we will indicate that the symmetric and innerwise matrices are related by matrix multiplication. The general region, which will be discussed, will include all the root-clustering regions discussed in Sections 3.1 and 3.2, as well as other regions.

CONDITION ON A POSITIVE DEFINITE HERMITIAN MATRIX

Let $F(z)$ given below

$$F(z) \triangleq \sum_{k=0}^{n} a_k z^k \triangleq a_n \prod_{k=1}^{n} (z - \lambda_k)$$

$$a_n \neq 0, \quad a_k, \lambda_i \in \mathbb{C}, \quad k = 0, 1, 2, \ldots, n, \quad i = 1, 2, \ldots, n \tag{3.74}$$

be a polynomial whose root clustering is to be investigated. Let the region Γ be given as follows:

$$\Gamma \triangleq \{z | |\gamma(z)|^2 - |\delta(z)|^2 > 0\} \tag{3.75}$$

In (3.75), $\gamma(z)$ and $\delta(z)$ are given polynomials and

$$|\gamma(z)|^2 - |\delta(z)|^2 \not\equiv \text{constant} \quad \text{for all } z \in \Gamma \tag{3.76}$$

1. If $\delta(z) = z - z_0$, $\gamma(z) = r^2$, $r \in R$, then Γ in (3.75) is the circle centered at z_0 with radius r. If $z_0 = 0$, $r = 1$, then Γ is the unit circle.

2. If $\gamma(z) = \rho z - \beta$, $\delta(z) = \rho z - \varepsilon$, ρ, β, $\varepsilon \in \mathbb{C}$, then Γ is one of the two half-planes divided by a straight line. In particular, if $\gamma(z) = z - 1$, $\delta(z) = z + 1$, then Γ is the open left half-plane.

3. If $\gamma(z) = z^2 - 1$, $\delta(z) = z^2$, then Γ is the hyperbola.

4. If $\gamma(z) = 2z$, $\delta(z) = z^2 - 1$, then Γ is a certain region (distorted circle) excluding the origin. Similarly, other regions can also be investigated.

Now, consider the two polynomials

$$g(u) \triangleq \prod_{k=1}^{n} [\delta(\lambda_k)u - \gamma(\lambda_k)] \tag{3.77}$$

$$h(u) \triangleq \prod_{k=1}^{n} [\bar{\gamma}(\bar{\lambda}_k)u - \bar{\delta}(\bar{\lambda}_k)] \tag{3.78}$$

where $\bar{x} \triangleq$ complex conjugate of x.

Using (3.77) and (3.78), obtain the following transformation:

$$
\begin{aligned}
(1 &- |u|^2)^{-1}(|g(u)|^2 - |h(u)|^2) \\
&= (1 - |u|^2)^{-1}[|\delta(\lambda_1)u - \gamma(\lambda_1)|^2|\delta(\lambda_2)u - \gamma(\lambda_2)|^2 \cdots |\delta(\lambda_n)u - \gamma(\lambda_n)|^2 \\
&\quad - |\bar{\gamma}(\bar{\lambda}_1)u - \bar{\delta}(\bar{\lambda}_1)|^2|\delta(\lambda_2)u - \gamma(\lambda_2)|^2 \cdots |\delta(\lambda_n)u - \gamma(\lambda_n)|^2] \\
&\quad + (1 - |u|^2)^{-1}[|\bar{\gamma}(\bar{\lambda}_1)u - \bar{\delta}(\bar{\lambda}_1)|^2|\delta(\lambda_2)u - \gamma(\lambda_2)|^2 \cdots |\delta(\lambda_n)u - \gamma(\lambda_n)|^2 \\
&\quad - |\bar{\gamma}(\bar{\lambda}_1)u - \bar{\delta}(\bar{\lambda}_1)|^2|\bar{\gamma}(\bar{\lambda}_2)u - \bar{\delta}(\bar{\lambda}_2)|^2 \cdots |\delta(\lambda_n)u - \gamma(\lambda_n)|^2] + \cdots \\
&\quad + (1 - |u|^2)^{-1}[|\bar{\gamma}(\bar{\lambda}_1)u - \bar{\delta}(\bar{\lambda}_1)|^2|\bar{\gamma}(\bar{\lambda}_2)u - \bar{\delta}(\bar{\lambda}_2)|^2 \cdots |\delta(\lambda_n)u - \gamma(\lambda_n)|^2 \\
&\quad - |\bar{\gamma}(\bar{\lambda}_1)u - \bar{\delta}(\bar{\lambda}_1)|^2|\bar{\gamma}(\bar{\lambda}_2)u - \bar{\delta}(\bar{\lambda}_2)|^2 \cdots |\bar{\gamma}(\bar{\lambda}_n)u - \bar{\delta}(\bar{\lambda}_n)|^2] \tag{3.79}
\end{aligned}
$$

Noting the identity,

$$(1 - |u|^2)^{-1}[|\delta(\lambda_k)u - \gamma(\lambda_k)|^2 - |\bar{\gamma}(\bar{\lambda}_k)u - \bar{\delta}(\bar{\lambda}_k)|^2] = |\gamma(\lambda_k)|^2 - |\delta(\lambda_k)|^2 \tag{3.80}$$

and by using (3.80) in (3.79),

$$
\begin{aligned}
(1 - |u|^2)^{-1}[|g(u)|^2 - |h(u)|^2] &= \sum_{i=1}^{n} \sum_{j=1}^{n} u^{i-1} c_{ij} \bar{u}^{j-1} \\
&= \sum_{k=1}^{n} [|\gamma(\lambda_k)|^2 - |\delta(\lambda_k)|^2]|\xi_k|^2 \tag{3.81}
\end{aligned}
$$

where $$c_{ij} = \bar{c}_{ji} \tag{3.82}$$

and

$$\xi_1 = [\delta(\lambda_2)u - \gamma(\lambda_2)][\delta(\lambda_3)u - \gamma(\lambda_3)]\cdots[\delta(\lambda_n)u - \gamma(\lambda_n)]$$
$$\xi_2 = [\bar{\gamma}(\lambda_1)u - \bar{\delta}(\lambda_1)][\delta(\lambda_3)u - \gamma(\lambda_3)]\cdots[\delta(\lambda_n)u - \gamma(\lambda_n)]$$
$$\vdots \qquad \vdots \qquad \vdots \qquad \vdots \qquad \vdots$$
$$\xi_n = [\bar{\gamma}(\lambda_1)u - \bar{\delta}(\lambda_1)][\bar{\gamma}(\lambda_2)u - \bar{\delta}(\lambda_2)]\cdots[\bar{\gamma}(\lambda_{n-1})u - \bar{\delta}(\lambda_{n-1})]$$

$$(3.83)$$

Therefore it follows from (3.81) that $C = [c_{ij}]$ is at least a semiposive definite Hermitian matrix. If $(\xi_1, \xi_2, \ldots, \xi_n)$ is a linearly independent set of vectors, then C is positive definite (pd), and vice versa, since [from (3.75)]

$$|\gamma(\lambda_k)|^2 - |\delta(\lambda_k)|^2 > 0$$
$$k = 1, 2, \ldots, n \quad (3.84)$$

A SPECIAL FORM OF THE HERMITIAN MATRIX

Let $h(u)$ be presented as follows:

$$h(u) = \prod_{k=1}^{n} [\bar{\gamma}(\bar{\lambda}_k)u - \bar{\delta}(\bar{\lambda}_k)] \triangleq \sum \bar{d}_k u^k \quad (3.85)$$

$$g(u) = \prod_{k=1}^{n} [\delta(\lambda_k)u - \gamma(\lambda_k)] = (-1)^n \sum_{k=0}^{n} d_{n-k} u^k \quad (3.86)$$

The following lemma will be proven.

LEMMA. *Let $u_n \in \mathbb{C}^n$.*

$$u_n^* C u_n = (1 - |u|^2)^{-1}[|g(u)|^2 - |h(u)|^2]$$
$$= \sum_{k=1}^{n} \left(\left| \sum_{i=1}^{n-k+1} \bar{d}_{n+1-i} u^{n+1-k-i} \right|^2 - \left| \sum_{i=1}^{n-k+1} d_{i-1} u^{n+1-k-i} \right|^2 \right) \quad (3.87)$$

where $u^* \triangleq$ complex conjugate transpose of u.

Proof by Induction: By expansion, (3.87) is readily verified for $n = 1$ and 2. Suppose that (3.87) holds for n; then

$$(1 - |u|^2)^{-1}(|d_0 u^{n+1} + d_1 u^n + \cdots + d_{n+1}|^2$$
$$- |\bar{d}_{n+1} u^{n+1} + \bar{d}_n u^n + \cdots + \bar{d}_0|^2 \quad (3.88)$$

Let

$$X = d_0 u^n + d_1 u^{n-1} + \cdots + d_n$$
$$Y = \bar{d}_n u^n + \bar{d}_{n-1} u^{n-1} + \cdots + \bar{d}_0 \quad (3.89)$$

Then (3.88) becomes

$$(1 - |u|^2)^{-1}(|uX + d_{n+1}|^2 - |\bar{d}_{n+1} u^{n+1} + Y|^2)$$
$$= |d_{n+1}|^2 (1 + |u|^2 + \cdots + |u|^{2n}) - |X|^2 + (1 - |u|^2)^{-1}(|X|^2 - |Y|^2)$$
$$+ 2(1 - |u|^2)^{-1}(\operatorname{Re} \bar{d}_{n+1} uX - \operatorname{Re} \bar{d}_{n+1} u^{n+1} \bar{Y}) \quad (3.90)$$

Since $(1 - |u|^2)^{-1}(|X|^2 - |Y|^2)$ is given by (3.87), the right-hand side of (3.90) becomes

$$\sum_{k=1}^{n+1} \left(\left| \sum_{i=1}^{n-k+1} \bar{d}_{n+2-i} u^{n+2-k-i} \right|^2 - \left| \sum_{i=1}^{n-k+2} d_{i-1} u^{n+2-k-i} \right|^2 \right) \qquad (3.91)$$

Therefore (3.87) holds for $n + 1$.

THEOREM 3.8 (GENERALIZATION OF HERMITE RESULTS). *There exists an $n \times n$ Hermitian matrix C such that for all roots of $F(z) = 0$, given by (3.74), to belong to the region Γ expressed in (3.75), it is necessary and sufficient that C be positive definite (pd)—that is, $C > 0$. Furthermore, all elements of C are rational functions of the a_k's.*

Proof: The necessary condition can be shown by contradiction as follows. Construct C by (3.81). The matrix C is semipositive definite. The set $(\xi_1, \xi_2, \ldots, \xi_n)$, given by (3.83), is a linearly dependent set of vectors if and only if $g(u)$ and $h(u)$, given by (3.77) and (3.78), have common roots. Suppose that such is the case. Then there exists k such that

$$\frac{\bar{\delta}(\bar{\lambda}_k)}{\bar{\gamma}(\bar{\lambda}_k)} = \frac{\gamma(\lambda_k)}{\delta(\lambda_k)} \qquad (3.92)$$

or $$|\gamma(\lambda_k)|^2 - |\delta(\lambda_k)|^2 = 0 \qquad (3.93)$$

which contradicts (3.76)

The sufficiency condition can be readily verified. In order to show that all elements of C are rational functions of the a_k's, consider the d_k's in (3.85). It follows that

$$d_n = \prod_{k=1}^{n} \gamma(\lambda_k)$$

$$\frac{d_{n-1}}{d_n} = -\sum_{k=1}^{n} \frac{\delta(\lambda_k)}{\gamma(\lambda_k)}$$

$$\frac{d_{n-2}}{d_n} = \sum_{k=1}^{n} \sum_{j>k}^{n} \frac{\delta(\lambda_k)\delta(\lambda_j)}{\gamma(\lambda_k)\gamma(\lambda_j)} \qquad (3.94)$$

$$\vdots \qquad \qquad \vdots$$

$$d_0 = (-1)^n \prod_{k=1}^{n} \delta(\lambda_k)$$

Hence, from (3.94), it follows that the d_k's, $k = 0, 1, \ldots, n$, are symmetric; that is, they are invariant under the symmetric group of all the permutations of λ_k, $k = 0, 1, \ldots, n$. It is well known [20] that this symmetry ensures that d_k can be expressed by rational functions of a_k. Noting (3.87), the elements of C are rational functions of d_k.

In a related work Kalman [21] had proved, by purely algebraic methods, the equivalence of this theorem for the region Γ which is the same as that defined in this discussion.

THEOREM 3.9. (GENERALIZATION OF ROUTH–HURWITZ, SCHUR– COHN, FULLER–MEEROV, AND OTHERS). *There exists a $2n \times 2n$ innerwise matrix Δ such that all roots of $F(z) = 0$ in (3.74) belong to region Γ if and only if the matrix Δ defined below is positive innerwise (pi). Furthermore, all elements of Δ are rational functions of the a_k's.*

Proof: The matrix Δ can be constructed from (3.87) as follows. Let R_ℓ and S_ℓ be the matrices

$$
R_\ell = \begin{bmatrix}
\bar{d}_n & \bar{d}_{n-1} & \bar{d}_{n-2} & \cdots & \bar{d}_{n-\ell+1} \\
0 & \bar{d}_n & \bar{d}_{n-1} & \cdots & \bar{d}_{n-\ell+2} \\
\cdot & \cdot & \cdot & \cdot & \cdot \\
0 & 0 & 0 & \cdots & \bar{d}_{n-1} \\
0 & 0 & 0 & \cdots & \bar{d}_n
\end{bmatrix}
\tag{3.95}
$$

$$
S_\ell = \begin{bmatrix}
0 & 0 & 0 & \cdots & d_0 \\
0 & 0 & 0 & d_0 & d_1 \\
\vdots & \vdots & & \vdots & \vdots \\
0 & d_0 & d_1 & \cdots & d_{\ell-2} \\
d_0 & d_1 & d_2 & \cdots & d_{\ell-1}
\end{bmatrix} = S'_\ell
\tag{3.96}
$$

Define u_ℓ^*, where $u_\ell \in \mathbb{C}$, $\ell = 1, 2, \ldots, n$, as

$$
u_\ell^* = (\bar{u}^{\ell-1}, \bar{u}^{\ell-2}, \ldots, \bar{u}, 1)
\tag{3.97}
$$

Then
$$
u_\ell^* [R_\ell^* : -S_\ell^*] \begin{bmatrix} R_\ell \\ S_\ell \end{bmatrix} u_\ell
\tag{3.98}
$$

is equal to
$$
u_\ell^* (R_\ell^* R_\ell - S_\ell^* S_\ell) u =
$$
$$
= \sum_{k=1}^{\ell} \left(\left| \sum_{i=1}^{\ell-k+1} \bar{d}_{\ell+1-i} u^{\ell+1-k-i} \right|^2 - \left| \sum_{i=1}^{\ell-k+1} d_{i-1} u^{\ell+1-k-i} \right|^2 \right)
\tag{3.98a}
$$

The right-hand side of the (3.98a) is similar to (3.87). Thus if one denotes C_l the ℓth principal minor of C in (3.87), it follows that

$$
\begin{vmatrix} R_l & \vdots & \bar{S}_l \\ \cdots & & \cdots \\ S_l & \vdots & \bar{R}_l \end{vmatrix} = (d_n)^{-l} \begin{vmatrix} C_l & \vdots & 0 \\ \cdots & & \cdots \\ S_l & \vdots & \bar{R}_l \end{vmatrix} = C_l
\tag{3.99}
$$

If we let
$$
\Delta = \begin{bmatrix} R_n & \vdots & \bar{S}_n \\ \cdots & & \cdots \\ S_n & \vdots & \bar{R}_n \end{bmatrix}
\tag{3.100}
$$

then C is positive definite if Δ is positive innerwise, and vice versa.

Following the same arguments as in Theorem 3.8, it can readily be shown that the elements of Δ are rational functions of the d_k's (or a_k's), $k = 0, 1, \ldots,$ n.

Therefore Δ is positive innerwise (pi) if and only if $\lambda_k \in \Gamma$, $k = 1, 2, \ldots, n$. As a consequence of the preceding results, we can state the following theorem.

THEOREM 3.10. The positive definite matrix C and the positive innerwise matrix Δ are related by the following.

$$C = R_n^* R_n - S_n^* S_n \qquad (3.101)$$

It is important to note that one can always construct the Hermite-type, positive definite matrix from the positive innerwise matrix. This fact will be clearly indicated in a few examples. Thus (3.101) represents a matrix multiplication transformation that connects the innerwise matrix to the Hermitian symmetric matrix.

Cases

a. *Left half-plane:* As pointed out earlier, the left half-plane is obtained when $\gamma(z) = z - 1$, $\delta(z) = z + 1$ is substituted in (3.75). Furthermore, in (3.74) let

$$a_k = a_k' + ja_k''$$
$$a_k', a_k'' \in R, \quad k = 0, 1, 2, \ldots, n - 1 \qquad (3.102)$$

and
$$a_n \equiv 1 \qquad (3.103)$$

By normalizing (i.e., letting $a_n \equiv 1$), the innerwise matrix in (3.100) reduces to dimension $2n - 1 \times 2n - 1$. It is of interest that such reduction is not possible for all regions of Γ. From (3.100), one obtains Δ. This innerwise matrix has been presented in a form discussed by Marden [22] and is given in (2.2).

The premultiplying matrix in Theorem 3.10, Equation (3.101), is as follows:†

$$\begin{bmatrix} 0 & 0 & 0 & 0 & \cdots & 0 & 0 & 0 & 1 \\ -a_{n-2}'' & -a_{n-3}' & a_{n-4}'' & a_{n-5}'' & \cdots & -a_{n-4}'' & -a_{n-3}' & a_{n-2}' & 0 \\ -a_{n-3}' & a_{n-4} & a_{n-5}' & \cdots & \cdots & \cdots & -a_{n-4}'' & -a_{n-5}' & 0 \\ \cdot & \cdot & \cdot & \cdot & \cdot & \cdot & \cdot & \cdot & \cdot \\ * & 0 & 0 & 0 & \cdots & \cdots & \cdots & {}^*_* & 0 \end{bmatrix}$$

$$(3.104)$$

$$* \ (-1)^{n/2} a_0'' \text{ when } n \text{ is even}, \quad (-1)^{(n-1)/2} a_0' \text{ when } n \text{ is odd} \qquad (3.105)$$

$${}^*_* \ (-1)^{(n/2)+1} a_0' \text{ when } n \text{ is even}, \quad (-1)^{(n-1)/2} a_0'' \text{ when } n \text{ is odd} \qquad (3.106)$$

† Note that C = equation (3.104) $\times$ Δ.

From Theorem 3.10, after noting

$$m = 2n \qquad b_m = 1,$$
$$b_{m-1} = a'_{n-1}, \quad b_{m-2} = -a''_{n-1}, \quad b_{m-3} = -a''_{n-2}, \quad b_{m-4} = -a'_{n-2},$$
$$b_{m-5} = -a'_{n-3}, \quad b_{m-6} = a''_{n-3}, \quad b_{m-7} = a''_{n-4}, \dots$$

(sign changes every four elements) (3.107)

the symmetric matrix C becomes

$$C = (c_{ij})$$

$$c_{ij} = \sum_{k=1}^{2i-1} (-1)^{k+2i-1} b_{m-k+1} b_{m-2i+k+2} \qquad j \geq i$$

$$c_{ji} = c_{ij} \qquad\qquad\qquad j < i, \quad i,j = 1, 2, \dots, n$$

(3.108)

It is of interest to note that when $a''_k = 0$, and $a'_k = a_k$—that is, $F(z)$ is a real polynomial—(3.108) reduces to that given in (3.2).

The special and important case of stability in linear continuous systems arises when all the a'_ks are real and positive. In this situation the dimension of C can be reduced from $n \times n$ to $n/2 \times n/2$ (n-even) [(3.3)] or to $(n-1)/2 \times (n-1)/2$ (n-odd) [(3.4)], and the dimension of Δ from $2n - 1 \times 2n - 1$ to $n - 1 \times n - 1$ [(2.10)]. This, in turn, yields the reduced-Hermite and the Liénard–Chipart criteria discussed in Sections 3.1 and 2.1, respectively. The premultiplying matrices, in this case, become the ones below.

For n-even,

$$\begin{bmatrix} 0 & 0 & 0 & \cdots & 0 & 0 & 0 & a_n \\ a_{n-3} & a_{n-5} & a_{n-7} & \cdots & -a_{n-8} & -a_{n-6} & -a_{n-4} & 0 \\ a_{n-5} & a_{n-7} & a_{n-9} & \cdots & -a_{n-10} & -a_{n-8} & -a_{n-6} & 0 \\ \cdot & \cdot & \cdot & \cdots & \cdot & \cdot & \cdot & \cdot \\ a_3 & a_1 & 0 & \cdots & 0 & -a_0 & -a_2 & 0 \\ a_1 & 0 & 0 & \cdots & 0 & 0 & -a_0 & 0 \end{bmatrix}$$

(3.109)

For n-odd,

$$\begin{bmatrix} a_{n-2} & a_{n-4} & a_{n-6} & \cdots & -a_{n-7} & -a_{n-5} & -a_{n-3} \\ a_{n-4} & a_{n-6} & a_{n-8} & \cdots & -a_{n-9} & -a_{n-7} & -a_{n-5} \\ \cdot & \cdot & \cdot & \cdots & \cdot & \cdot & \cdot \\ a_3 & a_1 & 0 & \cdots & 0 & -a_0 & -a_2 \\ a_1 & 0 & 0 & \cdots & 0 & 0 & -a_0 \end{bmatrix}$$

(3.110)

Example 3.1. Let

$$F(s) = \sum_{k=0}^{6} a_k s^k \qquad\qquad a_6 > 0 \quad (3.111)$$

For the roots of $F(s) = 0$ to lie in the open left half-plane, it is necessary and sufficient that

$$a_k > 0 \qquad k = 0, 1, 2, \ldots, 6 \quad (3.112)$$

and, from (3.3), the following 3×3 symmetric matrix must be positive definite (pd).

$$C = \begin{bmatrix} a_6 a_5 & a_6 a_3 & a_6 a_1 \\ a_6 a_3 & a_6 a_1 - a_5 a_2 + a_4 a_3 & -a_5 a_0 + a_4 a_1 \\ a_6 a_1 & -a_5 a_0 + a_4 a_1 & -a_3 a_0 + a_2 a_1 \end{bmatrix} \quad (3.113)$$

Using (3.101), the preceding symmetric matrix is generated by the matrix multiplication of the premultiplying matrix in (3.109) and the Liénard–Chipart (or Hurwitz) matrix written in innerwise form as in (2.10):

$$\begin{bmatrix} 0 & 0 & 0 & 0 & a_6 \\ a_3 & a_1 & -a_0 & -a_2 & 0 \\ a_1 & 0 & 0 & -a_0 & 0 \end{bmatrix} \begin{bmatrix} a_6 & a_4 & a_2 & a_0 & 0 \\ 0 & a_6 & a_4 & a_2 & a_0 \\ 0 & 0 & a_5 & a_3 & a_1 \\ 0 & a_5 & a_3 & a_1 & 0 \\ a_5 & a_3 & a_1 & 0 & 0 \end{bmatrix} \equiv [C : 0] \quad (3.114)$$

Example 3.2. Let

$$F(s) = \sum_{k=1}^{5} a_k s^k \qquad\qquad a_5 > 0 \quad (3.115)$$

For the roots of $F(s) = 0$ to lie in the open left half-plane, it is necessary and sufficient that

$$a_k > 0 \qquad k = 0, 1, 2, \ldots, 5 \quad (3.116)$$

and, from (3.4), that the following 2×2 symmetric matrix be (pd).

$$D = \begin{bmatrix} a_4 a_3 - a_5 a_2 & -a_5 a_0 + a_4 a_1 \\ -a_5 a_0 + a_4 a_1 & a_2 a_1 - a_3 a_0 \end{bmatrix} \quad (3.117)$$

From (3.110), the preceding matrix can be generated as follows.

$$\begin{bmatrix} a_3 & a_1 & -a_0 & -a_2 \\ a_1 & 0 & 0 & -a_0 \end{bmatrix} \begin{bmatrix} a_4 & a_2 & a_0 & 0 \\ 0 & a_4 & a_2 & a_0 \\ 0 & a_5 & a_3 & a_1 \\ a_5 & a_3 & a_1 & 0 \end{bmatrix} = [D:0] \quad (3.118)$$

The foregoing discussions illustrate the relationship between the reduced-Hermite and the Liénard–Chipart criteria. This relationship has also been noted by Liénard–Chipart [10] and later by many authors [11,23–25], following this early work.

b. *Unit circle [11]:* If we let $\gamma(z) = 1$, $\delta(z) = z$ in (3.75), then, as pointed out earlier, Γ becomes

$$\Gamma = \{z\,|\,1 - |z|^2 > 0\} \tag{3.119}$$

which is the region inside the unit circle. If we write $h(u)$ and $g(u)$ in (3.85) and (3.86) as

$$h(u) = \sum_{k=0}^{n} \bar{a}_k u^k \tag{3.120}$$

$$g(u) = (-1)^n \sum_{k=0}^{n} a_{n-k} u^k \tag{3.121}$$

the innerwise matrix Δ from Theorem 3.9 becomes

$$\Delta = \begin{bmatrix}
\bar{a}_n & \bar{a}_{n-1} & \cdots & a_3 & \bar{a}_2 & \bar{a}_1 & 0 & 0 & 0 & \cdots & \bar{a}_0 \\
& & & \bar{a}_n & \bar{a}_{n-1} & \bar{a}_{n-2} & 0 & 0 & \bar{a}_0 & \cdots & \bar{a}_{n-3} \\
& \cdots & & 0 & \bar{a}_n & \bar{a}_{n-1} & 0 & \bar{a}_0 & \bar{a}_1 & \cdots & \bar{a}_{n-2} \\
& \cdots & & 0 & 0 & \bar{a}_n & \bar{a}_0 & \bar{a}_1 & \bar{a}_2 & \cdots & \bar{a}_{n-1} \\
& \cdots & & 0 & 0 & a_0 & a_n & a_{n-1} & a_{n-2} & \cdots & a_1 \\
& \cdots & & 0 & a_0 & a_1 & 0 & a_n & a_{n-1} & \cdots & a_2 \\
& \cdots & & a_0 & a_1 & a_2 & 0 & 0 & a_n & \cdots & a_3 \\
& & & & & & & & & \cdots & \vdots \\
a_0 & a_1 & \cdots & a_{n-1} & a_{n-2} & a_{n-1} & 0 & 0 & 0 & \cdots & a_n
\end{bmatrix} \tag{3.122}$$

It may be noted that this matrix is of dimension $2n \times 2n$, and it cannot be reduced to $2n - 1 \times 2n - 1$ as for the case in the left half-plane.

The premultiplying matrix in Theorem 3.9 (3.101) is

$$\begin{bmatrix}
a_n & 0 & 0 & 0 & \cdots & 0 & 0 & 0 & -\bar{a}_0 \\
a_{n-1} & a_n & 0 & 0 & \cdots & 0 & 0 & -\bar{a}_0 & -\bar{a}_1 \\
a_{n-2} & a_{n-1} & a_n & 0 & \cdots & 0 & -\bar{a}_0 & -\bar{a}_1 & -\bar{a}_2 \\
\vdots & \vdots & & & & & & & \vdots \\
a_1 & a_2 & a_3 & a_4 & \cdots & -\bar{a}_{n-4} & -\bar{a}_{n-3} & -\bar{a}_{n-2} & \bar{a}_{n-1}
\end{bmatrix} \tag{3.123}$$

The elements of C in Theorem 3.8 are

$$
\begin{aligned}
c_{ij} &= \sum_{k=1}^{i} a_{n+k-i}\bar{a}_{n+k-j} - \bar{a}_{i-k}a_{j-k} & i \le j \\
c_{ji} &= \bar{c}_{ij} & i > j, \quad i,j = 1,2,\ldots,n
\end{aligned}
$$

(3.123a)

It may be noted that when all the a_i's are real, this equation is identical to the γ_{ij} given in (3.16). Equation (3.123a) represents the Schur–Cohn symmetric matrix as discussed in Section 3.2.

The special and important case of stability of linear discrete systems occurs when all the a_k's are real. In this case, the stability criterion can be considerably simplified, as indicated in Section 3.2 and also presented in Section 2.3. The transformation between the symmetric and innerwise matrices for this case can also be obtained, similar to the continuous case, by using the premultiplying matrix relationship as follows.

1. For even-degree $F(z)$, let $n = 2m$ in (3.19). The premultiplying matrix P is given as

$$
P = \begin{bmatrix}
0 & & -a_0 & a_n & 0 \\
 & -a_0 & -a_1 & a_{n-1} & a_n \\
 & \cdot & & \cdot & \cdot & \cdot \\
 & \cdot & & & \cdot & \cdot & \cdot \\
 & \cdot & & & & \cdot & \cdot & \cdot \\
-a_0 & & & \cdots & & & a_n
\end{bmatrix}
$$

(3.124)

Here P is an $n/2 \times n$ matrix such that

$$
P\Delta_n^+ = [0:A]
$$

(3.125)

where $A = (\alpha_{ij})$ is given in (3.20) and $\Delta_n^+ = X_n + Y_n$ can be generated as in (2.41):

$$
X_n = \begin{bmatrix}
a_n & a_{n-1} & \cdots & a_1 \\
0 & a_n & \cdots & a_2 \\
\vdots & & & \vdots \\
\cdot & \cdot \cdot \cdot \cdot \cdot \cdot & a_n
\end{bmatrix}
\qquad
Y_n = \begin{bmatrix}
0 & & \cdots & & a_0 \\
\vdots & & & & \vdots \\
0 & a_0 & \cdots & a_{n-2} \\
a_0 & a_1 & \cdots & a_{n-1}
\end{bmatrix}
$$

(3.126)

Example 3.3. Let $n = 4$. We obtain from (3.124) and (3.126)

$$
P = \begin{bmatrix}
0 & -a_0 & a_4 & 0 \\
-a_0 & -a_1 & a_3 & a_4
\end{bmatrix}
$$

(3.127)

$$\Delta_4^+ = \begin{bmatrix} a_4 & a_3 & a_2 & a_0 + a_1 \\ 0 & a_4 & a_3 + a_0 & a_1 + a_2 \\ 0 & a_0 & a_4 + a_1 & a_2 + a_3 \\ a_0 & a_1 & a_2 & a_3 + a_4 \end{bmatrix} \tag{3.128}$$

$$P\Delta_4^+ = \begin{bmatrix} 0 & 0 & \vdots & a_4(a_4 + a_1) - a_0(a_3 + a_0) & a_4(a_3 + a_2) - a_0(a_1 + a_2) \\ 0 & 0 & \vdots & a_4(a_3 + a_2) - a_0(a_1 + a_2) & a_4^2 + a_3^2 - a_1^2 + a_4 a_3 + a_3 a_2 - a_0^2 - a_0 a_1 - a_1 a_2 \end{bmatrix} \tag{3.129}$$

Hence $$P\Delta_4^+ = [0:A] \tag{3.130}$$

where A is the same as that given in (3.22).

2.   For odd-degree $F(z)$, let $n = 2m - 1$ in (3.24). The premultiplying matrix Q to obtain (3.26)—that is, $B = (\beta_{ij})$ is given as

$$Q = \begin{bmatrix} \cdot & \cdot & \cdot & a_0 & a_n & \cdot & \cdot & \cdot & 0 \\ 0 & a_0 & a_1 & a_{n-1} & a_n & & & \vdots \\ \vdots & \cdot & \cdot & \vdots & \vdots & \vdots & & \vdots & \cdot & \cdot & \vdots \\ a_0 & \cdot & & \cdot & & \cdot & & \cdot & a_n \end{bmatrix} \tag{3.131}$$

where Q is an $(n-1)/2 \times n - 1$ matrix such that

$$Q\Delta_{n-1}^- = [0:B] \tag{3.132}$$

Example 3.4.   Let $n = 5$. We obtain from (3.131) and (2.41)

$$Q = \begin{bmatrix} 0 & a_0 & a_5 & 0 \\ a_0 & a_1 & a_4 & a_5 \end{bmatrix} \tag{3.133}$$

$$\Delta_4^- = X_4 - Y_4 = \begin{bmatrix} a_5 & a_4 & a_3 & a_2 - a_0 \\ 0 & a_5 & a_4 - a_0 & a_3 - a_1 \\ 0 & -a_0 & a_5 - a_1 & a_4 - a_2 \\ -a_0 & -a_1 & -a_2 & a_5 - a_3 \end{bmatrix} \tag{3.134}$$

From (3.132), we have

$$Q\Delta_4^- = \begin{bmatrix} 0 & 0 & \vdots & a_5(a_5 - a_1) + a_0(a_4 - a_0) & a_5(a_4 - a_2) + a_0(a_3 - a_1) \\ 0 & 0 & \vdots & a_5(a_4 - a_2) + a_0(a_3 - a_1) & a_5(a_5 - a_3) + a_4(a_4 - a_2) + a_1(a_3 - a_1) + a_0(a_2 - a_0) \end{bmatrix} \tag{3.135}$$

Hence $$Q\Delta_4^- = [0:B] \tag{3.136}$$

where B is the same matrix given in (3.29).

It is of interest to note that we can obtain the matrix B for the even case from the odd one by effecting the following limiting process:

$$a_n \rightarrow a_{n-1}$$
$$\vdots \quad \vdots \qquad\qquad\qquad n\text{-odd} \quad (3.137)$$
$$a_i \rightarrow a_{i-1}$$
$$a_0 \rightarrow 0$$

Based on the preceding limiting process, the new symmetric matrix for the even case to be obtained from the odd is then $\tilde{B} = (\tilde{\beta}_{ij})$:

$$\tilde{\beta}_{ij} = \sum_{p=1}^{\min(i,j)} (a_{2m-i+p}a_{2m-j+p} - a_{i-p-1}a_{j-p-1})$$
$$- \sum_{p=1}^{i} (a_{2m-i+p}a_{j+p-2} - a_{i-p-1}a_{2m-j-p+1})$$
$$0 \le i, \ j \le m-1 \quad (3.138)$$

Example 3.5. Let $n = 4$. Thus

$$g(z) = a_4 z^4 + a_3 z^3 + a_2 z^2 + a_1 z + a_0 = 0$$
$$a_4 > 0 \quad (3.139)$$

Let us define

$$F(z) = zg(z) + b_0 \qquad\qquad b_0 = 0 \quad (3.140)$$
$$F(z) = a_4 z^5 + a_3 z^4 + a_2 z^3 + a_1 z^2 + a_0 z + b_0 = 0 \quad (3.141)$$

or $\qquad F(z) = b_5 z^5 + b_4 z^4 + b_3 z^3 + b_2 z^2 + b_1 z + b_0 = 0 \quad (3.142)$

where we have set

$$a_4 = b_5, \quad a_3 = b_4, \quad a_2 = b_3, \quad a_1 = b_2, \quad a_0 = b_1 \quad (3.143)$$

From (3.26), we have for $F(z)$

$$B = \begin{bmatrix} b_5(b_5 - b_1) + b_0(b_4 - b_0) & b_5(b_4 - b_2) + b_0(b_3 - b_1) \\ b_5(b_4 - b_2) + b_0(b_3 - b_1) & b_5(b_5 - b_3) + b_4(b_4 - b_4) + b_1(b_3 - b_1) + b_0(b_2 - b_0) \end{bmatrix}$$
$$(3.144)$$

and from (2.41),

$$\Delta_4^- = X_4 - Y_4 = \begin{bmatrix} b_5 & b_0 & b_3 & b_2 - b_0 \\ 0 & b_5 & b_4 - b_0 & b_3 - b_1 \\ 0 & -b_0 & b_5 - b_1 & b_4 - b_2 \\ -b_0 & -b_1 & -b_2 & b_5 - b_3 \end{bmatrix} \quad (3.145)$$

Now we introduce the following limiting process (change of notation):

$$b_5 \rightarrow a_4, \quad b_4 \rightarrow a_3, \quad b_3 \rightarrow a_2, \quad b_2 \rightarrow a_1, \quad b_1 \rightarrow a_0, \quad b_0 \rightarrow 0 \quad (3.146)$$

From (3.144) and (3.146), we obtain

$$\tilde{B} = \begin{bmatrix} a_4(a_4 - a_0) & a_4(a_3 - a_1) \\ a_4(a_3 - a_1) & a_4(a_4 - a_2) + a_3(a_3 - a_1) + a_0(a_2 - a_0) \end{bmatrix}$$

(3.147)

The preceding matrix can be directly obtained from (3.138) for $n = 5$. Also, using the change of notation in (3.145), we obtain

$$\tilde{\Delta}_4^- = \begin{bmatrix} a_4 & a_3 & a_2 & a_1 \\ 0 & a_4 & a_3 & a_2 - a_0 \\ 0 & 0 & a_4 - a_0 & a_3 - a_1 \\ 0 & -a_0 & -a_1 & a_4 - a_2 \end{bmatrix}$$

(3.148)

One can easily verify that

$$\tilde{B} \text{ (pd)} \Leftrightarrow \tilde{\Delta}_4^- \text{ (pi)}$$

(3.149)

Since $a_4 > 0$, $\tilde{\Delta}_4^-$ (pi) $\Leftrightarrow \Delta_3^-$ (pi.) This result can easily be verified by actual expansion of $\tilde{\Delta}_4^-$ and Δ_3^-, Δ_2^- and Δ_1^-. Therefore, for stability, we can utilize the following relationship for $n = 2m$ (even).

$$\tilde{Q}\Delta_{n-1}^- = [0 : \tilde{B}]$$

(3.150)

where

$$\tilde{Q} = \begin{bmatrix} & \cdots & a_n & & \cdots & \\ 0 & a_0 & a_{n-1} & & a_n & 0 \\ a_0 & a_1 & a_{n-2} & & a_{n-1} & a_n \\ a_0 & a_1 & \cdots & a_{n-2} & a_{n-1} & a_n \end{bmatrix}$$

(3.151)

In (3.151), $\tilde{Q}$ is of dimension $n/2 \times n - 1$. In a similar manner, we can obtain the *premultiplying matrix P_B* for $n = 2m$ related to the symmetric matrix B given in (3.21):

$$P_B = \begin{bmatrix} & \cdots & a_0 & a_n & & \cdots & \\ & 0 & & & & & \\ & & a_0 & a_1 & a_{n-1} & a_n & \\ & & & & & 0 & \\ & a_0 & a_1 & a_{n-2} & a_{n-1} & a_n & \\ a_0 & \cdots & & & & & a_n \end{bmatrix}$$

(3.152)

where P_B is an $(n/2) \times n$ matrix such that

$$\tilde{P}_B \Delta_n^- = [0:B] \tag{3.153}$$

and it follows that

$$B_{(pd)} \Leftrightarrow \Delta_{n(pi)}^- \tag{3.154}$$

Similarly, we can obtain additional premultiplying matrices that connect other symmetric forms with the innerwise matrices.

Therefore (3.132) for $n = 2m - 1$ and (3.150) for $n = 2m$ are the simplest form for stability. Each yields the simplified determinantal stability criterion given in Section 2.3, plus the linear inequalities. Also, the above example establishes the transformation between the reduced Schur–Cohn criterion and the simplified determinantal stability criterion.

c. *Negative real axis.* The problem when all the n distinct roots of (3.74) are on the negative real axis is of importance, as mentioned earlier, in obtaining the aperiodicity condition in linear continuous systems.

The negative real axis is not the region that can be represented by (3.75). However, by imposing the condition that the coefficients in (3.74) be real and positive (a necessary condition for aperiodicity), one can circumvent this critical case by using Theorem 3.9. By invoking any one of the conditions in (3.70) to (3.72), we can check the Hurwitz condition of any of the $F_k(s)$ in these equations. Thus the problem of aperiodicity reduces to that of the open left half-plane roots discussed in part (a). The innerwise matrix, Δ, in Theorem 3.9 is changed as shown on p. 81.

The premultiplying matrix in Theorem 3.10 is

$$\begin{bmatrix} 0 & 0 & 0 & 0 & \cdots & 0 & 0 & 0 & a_n \\ (n-1)a_{n-1} & (n-2)a_{n-2} & (n-3)a_{n-3} & (n-4)a_{n-4} & \cdots & -a_{n-4} & -a_{n-3} & -a_{n-2} & 0 \\ (n-2)a_{n-2} & (n-3)a_{n-3} & (n-4)a_{n-4} & (n-5)a_{n-5} & \cdots & -a_{n-5} & -a_{n-4} & -a_{n-3} & 0 \\ \cdot & \cdot & \cdot & \cdot & \cdots & \cdot & \cdot & \cdot & \cdot \\ a_1 & 0 & 0 & 0 & \cdots & 0 & 0 & -a_0 & 0 \end{bmatrix} \tag{3.156}$$

The element of C in Theorem 3.8 is

$$c_{ij} = \sum_{k=1}^{n-j+1} (n - j + 2 - k)a_{n-j+2-k}a_{n-i+k}$$

$$- (n - i + 1 + k)a_{n-i+1+k}a_{n-j-k+1} \qquad j \geq i \tag{3.157}$$

$$c_{ji} = c_{ij} \qquad\qquad\qquad j < i$$

$$a_k = 0 \qquad\qquad \text{when } k > n, \quad i, j = 1, 2, \ldots, n$$

The condition that $F(z)$ in (3.74) has all its n distinct roots on the real axis

$$\Delta = \begin{vmatrix}
a_n & a_3 & a_2 & a_1 & a_0 & 0 & \cdots & 0 \\
0 & a_n & a_{n-1} & a_{n-2} & a_{n-3} & a_{n-4} & \cdots & 0 \\
0 & 0 & a_n & a_{n-1} & a_{n-2} & a_{n-3} & \cdots & a_0 \\
0 & 0 & 0 & \boxed{na_n} & (n-1)a_{n-1} & (n-2)a_{n-2} & \cdots & a_1 \\
0 & 0 & na_n & (n-1)a_{n-1} & (n-2)a_{n-2} & (n-3)a_{n-3} & \cdots & 0 \\
0 & na_n & (n-1)a_{n-1} & (n-2)a_{n-2} & (n-3)a_{n-3} & (n-4)a_{n-4} & \cdots & 0 \\
\vdots & & & & & & & \vdots \\
na_n & 3a_3 & 2a_2 & a_1 & 0 & 0 & \cdots & 0
\end{vmatrix} \tag{3.155}$$

81

can also be obtained from (3.79). In this case, (3.79) becomes, following Fujiwara [6],

$$(x - y)^{-1}\left\{ F(x)\left[\frac{dF(y)}{dy}\right] - F(y)\left[\frac{dF(x)}{dx}\right]\right\} \tag{3.158}$$

The condition stated above is represented either by C in (3.157) being (pd) or Δ in (3.155) being (pi).

Example 3.6. Let $F(s)$ be as follows [11]:

$$F(s) = \sum_{k=0}^{4} a_k s^k \qquad a_4 > 0 \quad (3.159)$$

For all the distinct roots of $F(s) = 0$ to lie on the negative real axis, it is necessary and sufficient that

1. $a_k > 0$, $\quad k = 0, 1, 2, 3, 4$
2. the following symmetric matrix obtained from (3.157) be positive definite (pd):

$$C = \begin{bmatrix} 4a_4^2 & 3a_3a_4 & 2a_2a_4 & a_1a_4 \\ 3a_3a_4 & 3a_3^2 - 2a_2a_4 & 2a_2a_3 - 3a_1a_4 & a_1a_3 - 4a_0a_4 \\ 2a_2a_4 & 2a_2a_3 - 3a_1a_4 & 2a_2^2 - 4a_0a_4 - 2a_1a_3 & a_1a_2 - 3a_0a_2 \\ a_1a_4 & a_1a_3 - 4a_0a_4 & a_1a_2 - 3a_0a_3 & a_1^2 - 2a_0a_2 \end{bmatrix} \tag{3.160}$$

or, equivalently, the following 7×7 matrix obtained from (3.155), be positive innerwise (pi).

$$\Delta = \begin{bmatrix} a_4 & a_3 & a_2 & a_1 & a_0 & 0 & 0 \\ 0 & a_4 & a_3 & a_2 & a_1 & a_0 & 0 \\ 0 & 0 & a_4 & a_3 & a_2 & a_1 & a_0 \\ 0 & 0 & 0 & 4a_4 & 3a_3 & 2a_2 & a_1 \\ 0 & 0 & 4a_4 & 3a_3 & 2a_2 & a_1 & 0 \\ 0 & 4a_4 & 3a_3 & 2a_2 & a_1 & 0 & 0 \\ 4a_4 & 3a_3 & 2a_2 & a_1 & 0 & 0 & 0 \end{bmatrix} \tag{3.161}$$

The positive innerwise (pi) and the positive definite (pd) transformation, noting (3.156), is as follows:

$$\begin{bmatrix} 0 & 0 & 0 & 0 & 0 & 0 & a_4 \\ 3a_3 & 2a_2 & a_1 & -a_0 & -a_1 & -a_2 & 0 \\ 2a_2 & a_1 & 0 & 0 & -a_0 & -a_1 & 0 \\ a_1 & 0 & 0 & 0 & 0 & -a_0 & 0 \end{bmatrix} \Delta = [(c_{ij}):0] \tag{3.162}$$

HANKEL MATRICES [15]

Consider the following from (3.85) and (3.86):

$$\frac{h(u)}{g(u)} = p_{-1} + \frac{p_0}{u} + \frac{p_1}{u^2} + \frac{p_2}{u^3} + \cdots \tag{3.163}$$

Then

$$d_0 p_{-1} = \bar{d}_n \tag{3.164}$$

$$d_1 p_{-1} + d_0 p_0 = \bar{d}_{n-1} \tag{3.165}$$

$$d_2 p_{-1} + d_1 p_0 + d_0 p_1 = \bar{d}_{n-1} \tag{3.166}$$

Next, R_n in (3.95) can be written as

$$R_n = \begin{bmatrix} p_{-1} & p_0 & p_1 & p_2 & \cdots \\ 0 & p_{-1} & p_0 & p_1 & \cdots \\ 0 & 0 & p_{-1} & p_0 & \cdots \\ \vdots & \vdots & \vdots & \vdots & \vdots \end{bmatrix} \begin{bmatrix} d_0 & d_1 & d_2 & \cdots \\ 0 & d_0 & d_1 & \cdots \\ 0 & 0 & d_0 & \cdots \\ \vdots & \vdots & \vdots & \vdots \end{bmatrix} \tag{3.167}$$

Let P_1 and $\tilde{S}$ denote the first and the second matrix in (3.155), From (3.101), we have

$$C = \tilde{S}^* P_1^* P_1 \tilde{S} - S_n^* S_n \tag{3.168}$$

where S_n is defined in (3.96).

In the special cases—that is, the left half-plane, the unit circle (using a certain transformation [26]), and the negative real axis—it can be shown that C can be written as $T'PT$, where P is the Hankel matrix [given in (3.171) for the Markov parameters [15]]. Since this matrix plays an important role in the stability theory, we can state the following theorem, which is a generalization of Netto's theorem [27] and which is also discussed by Housholder [28].

THEOREM 3.11. *The positive definite (pd) matrix C and the positive inner-wise (pi) matrix Δ are represented by P_1, S, and S_n as follows:*

$$C = \tilde{S}^* P_1^* P_1 \tilde{S} - S_n^* S_n \tag{3.168a}$$

$$\Delta = \left[\begin{array}{c|c} P\tilde{S} & \bar{S}_n \\ \hline S_n & \tilde{\tilde{S}}\bar{P} \end{array} \right] \tag{3.169}$$

Example 3.7 Left Half-Plane. From Theorem 3.11 we obtain the Hankel matrix. In this case, the matrix can replace the Δ or C because S_n and $\tilde{S}$ in (3.168) and (3.169) can be eliminated by redefining P in (3.168) and (3.169) as follows:

$$\frac{b_{n-2}u^{n-1} + b_{n-4}u^{n-3} + \cdots}{b_n u^n + b_{n-1}u^{n-2} + b^{n-3}u^{n-2} + \cdots} = \frac{p_0}{u} + \frac{p_1}{u^2} + \cdots \qquad (3.170)$$

where the b_k's are defined by (3.107)

Therefore, from Theorem 3.11, the matrix C is positive definite and it implies that Δ is positive innerwise, and vice versa. Furthermore, also implied is that the following matrix of dimension $n \times n$ has all its principal minors of positive values.

$$P = \begin{bmatrix} p_0 & p_1 & p_2 & \cdots \\ p_1 & p_2 & p_3 & \cdots \\ \vdots & \vdots & \vdots & \vdots \end{bmatrix} \qquad (3.171)$$

and vice versa.

For the special case when all the a_k's are real and positive in (3.74), the above matrix can be halved similar to the Liénard–Chipart criterion [7]. Similar conclusions can be reached for the unit circle.

3.5 GENERALIZED LYAPUNOV THEOREM [19, 29]

In this section we will obtain a more general form of the Lyapunov matrix equation such that the matrix A, defined below, has all its eigenvalues belonging to the region Γ defined in (3.75). By obtaining such a general form, we can readily deduce the well-established Lyapunov matrix equation for the region in the left half-plane, the inside of the unit circle, and the negative real axis. Also, in this section we will discuss the solution of the Lyapunov matrix equation for the preceding specific region. In particular, we will comment on the solution when A is in companion form as well as in general form. And when A is in companion form, we will establish the connection between the Lyapunov matrix equation and the reduced Hermite theorem and the reduced Schur–Cohn criterion, respectively.

PRELIMINARIES

Let A, given below,

$$A \triangleq (a_{ij}) \qquad\qquad A \in \mathbb{C}^{n \times n} \quad (3.172)$$

be a matrix whose eigenvalues $\alpha_k \in \mathbb{C}, k = 1, 2, \ldots, n$, are under investigation. Let the region Γ be given as in (3.75).

Also, let

$$x \in \mathbb{C}^n \qquad\qquad A, B \in \mathbb{C}^{n \times n} \quad (3.173)$$

$$\bar{x} \triangleq \text{complex conjugate of } x \qquad (3.174)$$

$$x^* = \text{complex conjugate transpose of } x \qquad (3.175)$$

$$A' \triangleq \text{transpose of } A \text{ (as defined in Chapter 1)} \tag{3.176}$$

$$A^* \triangleq \text{complex conjugate transpose of } A \tag{3.177}$$

$$|B| \triangleq \text{determinant value of } B \text{ (as defined in Chapter 1)} \tag{3.178}$$

$$|B|_1^2 = BB^* \text{ (this is not to be taken as } |B|^2) \tag{3.179}$$

THEOREM 3.12 (GENERALIZATION OF LYAPUNOV-TYPE EQUATION [19]. *The matrix A has all its eigenvalues belonging to the region Γ of (3.75) if and only if, for any positive definite $n \times n$ matrix Q, there exists a positive definite $n \times n$ matrix H such that*

$$\bar{\gamma}(A^*)H\gamma(A) - \bar{\delta}(A^*)H\delta(A) = Q \tag{3.180}$$

Proof: In order to prove the sufficiency condition, let $X_k \in \mathbb{C}^n$ be the eigenvector corresponding to the eigenvalue α_k. Then from (3.180),

$$[|\gamma(\alpha_k)|^2 - |\delta(\alpha_k)|^2]X_k^* H X_k = X_k^* Q X_k \tag{3.181}$$

Hence
$$|\gamma(\alpha_k)|^2 - |\delta(\alpha_k)|^2 > 0 \tag{3.182}$$

which satisfies (3.76).

In order to prove the necessary condition from (3.79), let

$$(1 - |u|_1^2)^{-1}(|g(u)|_1^2 - |h(u)|_1^2)_{u=A} = \tilde{H}_1 \qquad |B|_{1,Q}^2 \triangleq BQB^* \tag{3.183}$$

$$(1 - |u|_1^2)^{-1}[|g(u)|_1^2 - |h(u)|_1^2]|_{Q,u=A} = \tilde{H} \tag{3.184}$$

$$\bar{\gamma}(A^*)\tilde{H}_1\gamma(A) - \bar{\delta}(A^*)\tilde{H}_1\delta(A) = \tilde{Q} \triangleq \tilde{Q}_1\tilde{Q}_1^* \tag{3.185}$$

To show that $\tilde{H}_1$ and $\tilde{H}$ are positive definite, consider $\tilde{H}_1$ as defined by (3.81) and (3.83). That is,

$$\tilde{H}_1 = \sum_{k=1}^{n} [|\gamma(\lambda_k)|^2 - |\delta(\lambda_k)|^2]\xi_k\xi_k^* \tag{3.186}$$

where

$$\begin{aligned}\xi_1 &= [\delta(\lambda_2)A - \gamma(\lambda_2)][\delta(\lambda_3)A - \gamma(\lambda_3)]\cdots[\delta(\lambda_n)A - \gamma(\lambda_n)]\\ \xi_2 &= [\bar{\gamma}(\lambda_1)A - \bar{\delta}(\lambda_1)][\delta(\lambda_3)A - \gamma(\lambda_3)]\cdots[\delta(\lambda_n)A - \gamma(\lambda_n)]\\ &\vdots \qquad\qquad\qquad\qquad\qquad \vdots\end{aligned} \tag{3.187}$$

Let PAP^{-1} be the Jordan canonical form of A. That is,

$$PAP^{-1} = \Omega + U \tag{3.188}$$

where Ω is diagonal with the eigenvalues of A and U is the matrix with the elements of 0 and 1, with all nonzero elements above the main diagonal. Let

$$D = \text{diag}\{1, \delta^{-1}, \delta^{-2}, \ldots, \delta^{-(n-1)}\} \tag{3.189}$$

and consider (3.186) and (3.187), where A is replaced by $\tilde{A} = DPAP^{-1}D^{-1}$.

Then it is clear that $\tilde{H}_1$ is positive definite. A similar argument shows that $\tilde{H}$ is also positive definite. Hence it follows from (3.185) that $\tilde{Q}$ is also positive definite. Thus

$$\bar{\gamma}(A^*)\tilde{H}\gamma(A) - \bar{\delta}(A^*)\tilde{H}\delta(A) = \tilde{Q}_1 Q \tilde{Q}_1^* \qquad (3.190)$$

Therefore $H = \tilde{Q}_1^{-1}\tilde{H}(\tilde{Q}_1^*)^{-1}$ is positive definite and satisfies (3.180).

Special Cases

a. *Left-half plane.* By letting $\gamma(A) = A - I$, $\delta(A) = A + I$ in (3.180), we obtain the well-known Lyapunov matrix equation [30]:

$$A^*H + HA = -Q \qquad (3.191)$$

b. *Unit circle.* By letting $\gamma(A) = I$, $\delta(A) = A$ in (3.180), we obtain the well-known Lyapunov matrix equation for the unit circle as follows [31]:

$$A^*HA - H = -Q \qquad (3.192)$$

Similarly, we obtain a relationship of the form of (3.191) for the negative real axis (aperiodicity condition) [11].

CONNECTION BETWEEN LYAPUNOV AND HERMITE (OR REDUCED HERMITE), SCHUR–COHN (OR REDUCED SCHUR–COHN) AND ROMANOV STABILITY CRITERIA

In the analysis of dynamical systems one can represent the motion of the system in terms of the state variables. For instance, the free motion of a dynamical system released from an initial state may be written in matrix form as

$$\dot{x} = Ax \qquad (3.193)$$

where A is a real constant matrix of dimension $n \times n$ and x is the state vector whose components are the state variables. In certain cases, the matrix A is usually represented in a companion form [30]. For such cases, the stability conditions are simple and can be given in any one of the previously discussed stability criteria. The purpose of this section is to show the connection between the Lyapunov matrix equations (3.191) and (3.192) and the other stability criteria. The companion form of the matrix A is given by

$$A = \begin{bmatrix} 0 & 1 & 0 & \cdots & 0 \\ 0 & 0 & 1 & \cdots & 0 \\ \vdots & \vdots & \vdots & & \vdots \\ 0 & 0 & 0 & \cdots & 1 \\ -\dfrac{a_0}{a_n} & -\dfrac{a_1}{a_n} & -\dfrac{a_2}{a_n} & \cdots & -\dfrac{a_{n-1}}{a_n} \end{bmatrix} \qquad a_n \equiv 1 \quad (3.194)$$

It may be noted that if A is given as in (3.194), then $F(s)$, the characteristic equation of the dynamical system, is

$$F(s) = |sI - A| \qquad (3.195)$$

where I is the identity matrix.

It is well known that the asymptotic stability of the system presented in (3.193) is given by the necessary and sufficient condition for the eigenvalues of the matrix A to have negative real part. Thus we can use (3.191) to obtain these conditions. Furthermore, without loss of generality, we can use the symmetric form for the matrices Q and H. Thus the following theorem, due to Lyapunov, can be obtained from (3.191).

THEOREM 3.13 [30.]   *The matrix A of (3.194) is a stability matrix (all eigenvalues in open left half-plane) if and only if, for any real positive definite symmetric matrix Q, the solution for the real symmetric matrix H in (3.191) is positive definite* (pd).

In order to establish the connection between H and the Hermite matrix P defined in (3.2), we define the matrix A and vector **b** by

$$A = \begin{bmatrix} -\dfrac{a_{n-1}}{a_n} & -\dfrac{a_{n-2}}{a_n} & -\dfrac{a_{n-3}}{a_n} & \cdots & -\dfrac{a_0}{a_n} \\ 1 & 0 & 0 & \cdots & 0 \\ 0 & 1 & 0 & \cdots & 0 \\ \vdots & \vdots & \vdots & & \vdots \end{bmatrix} \qquad (3.196)$$

$$\mathbf{b} = \sqrt{2}\begin{bmatrix} a_{n-1} \\ 0 \\ a_{n-3} \\ 0 \\ 1 \\ 1 \\ \vdots \end{bmatrix} \qquad (3.197)$$

Then [32]

$$PA + A'P = -bb' \qquad (3.198)$$

If P is positive definite, it is shown by Parks [32] that $x'Px$ is a Lyapunov function establishing stability (actually, asymptotic stability) of $\dot{x} = Ax$. The converse is also true. Hence the condition that H is (pd) in (3.191) is equivalent to that of the Hermite matrix P obtained from the coefficients of $F(s)$ being (pd), and vice versa. This result establishes the fact that when A

is the companion matrix and Q in (3.191) is appropriately chosen, H in the same equation being (pd) is equivalent to the Hermite matrix P being (pd), and vice versa.

CONNECTION BETWEEN REDUCED HERMITE CRITERION AND LYAPUNOV THEOREM

It has been established in Theorem 3.1 that when the coefficients (or half of them) of $F(s)$ in (3.195) are positive, $P > 0$ is replaced either by $C > 0$ or by $D > 0$—that is, (3.6) to (3.9). Based on the equivalence of $H > 0$ and $P > 0$ discussed above, we can readily establish the fact that when $a_n = 1$ in (3.194), and all the entries (or half of them) in the last row of the companion matrix A in (3.194) are positive, then $H > 0$ can be replaced by either $C > 0$ or $D > 0$ (i.e., by the reduced Hermite criterion).

We can also establish this connection by using the results of Section 3.3 as follows. Let $F_0(s)$ and $F_1(s)$ be given as in (3.50) and (3.51), and let the $a_{2i} > 0$, for $i = 0, 1, \ldots, n$, or $a_0 > 0$, $a_{2n} > 0$, and $a_{2i+1} > 0$, for $i = 0, 1, \ldots, n - 1$. Then we can write the Lyapunov matrix equation corresponding to the complex polynomial $F_1(s)$ as

$$P_1 A_1 + A_1^* P_1 = -b_1 b_1^* \tag{3.199}$$

with the following identifications:

$$P_1 = P_1^*$$
$$= \text{diag} [1, j, -1, -j, 1, j, -1, \ldots] C \, \text{diag} [1, -j, -1, j, 1, -j, 1, \ldots] \tag{3.200}$$

and

$$A_1 = \begin{bmatrix} -\dfrac{a_{2n-1} + ja_{2n-2}}{a_{2n}} & \dfrac{a_{2n-4} + ja_{2n-3}}{a_{2n}} & \dfrac{a_{2n-5} - ja_{2n-6}}{a_{2n}} & \dfrac{a_{2n-1} - ja_{2n-7}}{a_{2n}} & \cdots \\ 1 & 0 & 0 & 0 \\ 0 & 1 & 0 & 0 \\ \vdots & \vdots & \vdots & \vdots \end{bmatrix} \tag{3.201}$$

It can easily be checked that A_1 is the companion matrix associated with the polynomial $F_1(s)$. The symmetric matrix P is the Hermite matrix associated with $F_1(s)$, which can be readily obtained from (3.108) or from other references [33]. This matrix P is identically equivalent to the reduced Hermite matrix C associated with $F_0(s)$—that is, (3.3). Finally,

$$b_1^* = \sqrt{2}[a_{2n-1}, \, -ja_{2n-3}, \, -a_{2n-5}, \, ja_{2n-7}, \, a_{2n-9}, \, -ja_{2n-11} \cdots] \tag{3.202}$$

The stability of the A_1 matrix [or, equivalently, for $F_1(s)$ to be Hurwitz] is equivalent to $P_1 > 0$ [for $F_1(s)$]. The condition $P_1 > 0$ is exactly equivalent to $C > 0$ or $D > 0$ in (3.6) to (3.9). Hence we established the connection between the Lyapunov and the reduced Hermite theorems. Actually, one can define (3.199) as the "reduced Lyapunov equation [14]."

CONNECTION BETWEEN LYAPUNOV AND SCHUR–COHN (OR REDUCED SCHUR–COHN) STABILITY CRITERIA [12]

In this case, the Lyapunov matrix equation is as given in (3.192). It represents the stability of discrete dynamical system given by the vector difference equation

$$\mathbf{x}_n = A x_{n-1} \qquad (3.203)$$

where A is an $n \times n$ real matrix represented in a companion form as in (3.194).

Parks [31] and Kalman [34] have shown that the matrix H in (3.192), with appropriately chosen Q, is identical to the Schur–Cohn matrix given in "C"— (3.16). Hence the condition $H > 0$ is identical to $C > 0$, and vice versa.

Furthermore, it is shown in Section 3.2 that $C > 0$ can be simplified by using the reduced Schur–Cohn criterion. Therefore the Lyapunov matrix equation can be simplified in identical fashion. Note that in this case conditions similar to that on b_i in (3.40) can be imposed on the entries of the preceding A matrix.

CONNECTIONS BETWEEN LYAPUNOV AND ROMANOV CRITERIA [11,16]

This case, which deals with the aperiodicity condition, requires that all the n distinct eigenvalues of A lie on the negative real axis. It is similar to the stability in the left half-plane as shown in Theorem 3.7. Here the matrix A is the companion of $F(s) = a_n s^{2n} + \cdots + a_1 s + a_0$. The Lyapunov matrix equation is similar to (3.191). It is shown in related work [11] that the condition $H > 0$ is equivalent to that of $C > 0$ obtained from (3.157), plus positiveness of the coefficients of $F(s)$, and vice versa. An example of C for $n = 4$ is given in (3.160). The matrix C for the aperiodicity case was first formulated by Romanov [16]. It may be noted that the dimension of C in this case is $n \times n$ while that of H is $2n \times 2n$. Furthermore, if one uses the Romanov characteristic equation (i.e., with complex coefficients), H will be of dimension $n \times n$ and it corresponds to C.

OTHER SYMMETRIC MATRIX FORMS FOR STABILITY

In the preceding discussions we have pointed out that in addition to the Hermite symmetric matrix, the Lyapunov symmetric matrix, as well as the

Hankel symmetric matrix, can be used for checking stability or root clustering. In order to complete the discussion of the symmetric matrix forms, we will mention two other forms that are used or referred to the literature—namely, the Schwarz form and Barnett's approach based on the companion matrix.

Schwarz's Form [35]. Given the system characteristic equation, $F(s)$,

$$F(s) = a_n s^n + a_{n-1} s^{n-1} + \cdots + a_0 \qquad a_n = 1$$

and the companion matrix A,

$$A = \begin{bmatrix} 0 & 1 & 0 & 0 & \cdots \\ 0 & 0 & 1 & 0 & \cdots \\ \vdots & \vdots & \vdots & \vdots & \\ -a_0 & -a_1 & -a_2 & \cdots & -a_{n-1} \end{bmatrix} \qquad (3.204)$$

such that

$$|sI - A| = F(s) \qquad (3.204a)$$

there exists a matrix S (called the Schwarz's form),

$$S = \begin{bmatrix} 0 & 1 & 0 & & \cdots & 0 \\ -b_0 & 0 & 1 & 0 & \cdots & 0 \\ 0 & -b_1 & \cdot & \cdot & \cdot & \cdot \\ \cdot & \cdot & & -b_{n-3} & 0 & 1 \\ \cdot & \cdot & & 0 & -b_{n-2} & -b_{n-1} \end{bmatrix} \qquad (3.205)$$

such that

$$|sI - S| = F(s) \qquad (3.206)$$

Moreover, a necessary and sufficient condition for $F(s) = 0$ to have roots with negative real part is that

$$b_i > 0 \qquad i = 0, 1, 2, \ldots, n-1 \qquad (3.207)$$

It is indicated by Chen and Chu [36], and further extended by Barnett and Storey [37], that a certain matrix T involving the entries of the second, third, fourth, etc., columns of Routh array for $F(s)$ will transform the matrix A into the matrix S:

$$TAT^{-1} = S \qquad (3.208)$$

It should be noted that in the preceding transformation A should be given (or reduced) in companion form.

Barnett's Approach [38].   Barnett, in a series of articles, has indicated the root-clustering conditions obtainable from the companion matrix A of $F(s)$. The final results are presented in terms of same dimension matrices in analogy to other such symmetric forms discussed here. Computationally, this approach has no special significance or simplification as compared to the other forms. However, it represents a unifying approach that is of importance. Here we discuss the stability aspect of this approach.

THEOREM 3.14.   *Let*

$$F(s) = a_n + a_{n-2}s + a_{n-4}s^2 + \cdots \tag{3.209}$$

$$G(s) = a_{n-1} + a_{n-3}s + a_{n-5}s^2 + \cdots \tag{3.210}$$

The necessary and sufficient condition for

$$A(s) = a_0 s^n + a_1 s^{n-1} + a_2 s^{n-2} + \cdots + a_n \tag{3.211}$$

to have all its roots with negative real parts is

(i) $a_n > 0, a_1 > 0, a_3 > 0, a_5 > 0, \ldots$ \hfill (3.212)

(ii)   *when n is even, the sequence G_k, $k = 2, 3, \ldots, n/2$, should be positive. Here G_k is the kth principal minor of the matrix $G(A)J_n/2$ and is the companion matrix of $F(s)$:*

$$A = \begin{bmatrix} 0 & 1 & 0 & \cdots & 0 \\ 0 & 0 & 1 & \cdots & 0 \\ \vdots & \vdots & \vdots & & \vdots \\ -a_n & -a_{n-2} & \cdots & & -a_1 \end{bmatrix} \qquad a_0 \equiv 1 \quad (3.213)$$

Also, $G(A)$ is

$$G(A) = \begin{bmatrix} r_1 \\ r_1 A \\ r_1 A^2 \\ \vdots \\ r_1 A^{n-1} \end{bmatrix} \tag{3.214}$$

and $\qquad r_1 = [a_n, a_{n-1}, a_{n-2}, \ldots, a_0]$ \hfill (3.125)

The $n \times n$ matrix J_n has the property that when $G(A)$ is post multiplied by it, the order of the columns is reversed.

(iii)   *When n is odd, the sequence $(-1)^k F_k$, for $k = 1, 2, \ldots, \frac{1}{2}(n-1)$, and where F_k is the kth leading principal minor of the matrix $F(B)[J(n-1)]/2$, should be positive. The matrix B that is the companion matrix of $G(s)$ is given* *in the form shown in (3.213). In this case, $F(B)$ is given in the form of (3.214), where*

$$r_1 = [b_n - b_0 a_n, b_{n-1} - b_0 a_{n-1}, \ldots, b_1 - b_0 a_1] \tag{3.215a}$$

The coefficients $b_n, b_{n-1}, \ldots,$ and $a_n, a_{n-1}, \ldots$ are to be identified with the coefficients of $F(s)$ and $G(s)$, respectively.

It is evident from the foregoing stability conditions that they correspond to the reduced Hermite criterion, or to the reduced Lyapunov theorem, or to the Liénard–Chipart criterion. Similar results have been achieved for stability within the unit circle that correspond to the Schur–Cohn criterion [38].

Example 3.8. Let $A(s)$ be given as

$$A(s) = s^5 + 4s^4 + 8s^3 + 9s^2 + 5s + 2 \tag{3.216}$$

From (3.209) and (3.210), we have

$$F(s) = 2 + 9s + 4s^2 \tag{3.217}$$

$$G(s) = 5 + 8s + s^2 \tag{3.218}$$

Since $n = 5 =$ odd, we use condition (iii). Then

$$B = \begin{bmatrix} 0 & 1 \\ -5 & -8 \end{bmatrix} \tag{3.219}$$

From (3.214),

$$F(B) = \begin{bmatrix} r_1 \\ r_1 B \end{bmatrix} \tag{3.220}$$

where from (3.215a)

$$r_1 = [2 - 4 \times 5, 9 - 4 \times 8] = [-18 \quad -23] \tag{3.221}$$

Thus

$$r_1 B = [-18 \quad -23] \begin{bmatrix} 0 & 1 \\ -5 & -8 \end{bmatrix} \tag{3.222}$$

and

$$F(B) = \begin{bmatrix} -18 & -23 \\ 115 & 166 \end{bmatrix} \tag{3.222a}$$

Now

$$F(B)J_2 = \begin{bmatrix} -18 & -23 \\ 115 & 166 \end{bmatrix} \begin{bmatrix} 0 & 1 \\ 1 & 0 \end{bmatrix} = \begin{bmatrix} -23 & -18 \\ 166 & 155 \end{bmatrix} \tag{3.223}$$

The sequence $-F_1, F_2$ in the theorem is 23, 343, showing that $A(s)$ is Hurwitz, since all the coefficients are positive.

THE SOLUTION OF THE LYAPUNOV MATRIX
EQUATION FOR THE GENERAL "A"
MATRIX

The advantage of using Lyapunov's approach, as compared to other stability tests, lies in the fact that one can deal directly with the system matrix A itself. In the preceding discussions we dealt with Lyapunov's matrix equation when A is given in companion form. In this case, Lyapunov's method did not present any particular advantage. In the following discussion we will deal with the stability of the matrix A when presented in general form.

We begin our discussion by dealing with (3.191) when A is a real matrix and by taking (without loss of generality) Q and H as symmetric matrices. Then (3.191) can be written as [38]

$$(A^T \otimes I_n + I_n \otimes A')h = -q \qquad (3.224)$$

where h and q are column n^2 vectors formed from the rows of H and Q, respectively, taken in order. To clarify the meaning of (3.224), we present the following definition.

If $A = (a_{ij})$ is an $m \times n$ matrix and $B = (b_{ij})$ an $l \times p$ matrix, then the Kronecker (or direct) product of A and B, written $A \otimes B$, is the $ml \times np$ matrix partitioned as follows:

$$\begin{bmatrix} a_{11}B & a_{12}B & \cdots & a_{1n}B \\ a_{21}B & a_{22}B & & a_{2n}B \\ \vdots & & & \vdots \\ a_{m1}B & a_{m2}B & \cdots & a_{mn}B \end{bmatrix} \qquad (3.225)$$

If A is $n \times n$ and X is $n \times k$, then the equation $AX = C$ can be written as $(A \otimes I_k)x = c$, where $I_k = \text{diag}[1, 1, \ldots, 1]$ unit matrix of order k and x is the column nk vector composed of the rows of x taken in order. That is,

$$X = [x_{11}, x_{12}, \ldots, x_{1k}, x_{22}, \ldots, x_{2k}, \ldots, x_{n1k}, \ldots, x_{nk}]' \qquad (3.226)$$

and c is formed similarly from C. The equation $YA = D$ can be written in a similar fashion as $(I_k \otimes A')y = d$, Y and D both being $k \times n$ matrices.

Based on the above definition, it is clear how (3.224) is obtained from the Lyapunov matrix equation—that is, (3.191).

The matrix on the left of (3.228) is of order n^2 and has characteristic roots [38] $\lambda_i + \lambda_j$, $i, j = 1, 2, \ldots, n$. As can be expected from the symmetry of H, redundant equations and variables can be removed from (3.224) to give

$$B\rho = -\phi \qquad (3.227)$$

where B is of order $\frac{1}{2}n(n + 1)$ and ρ and ϕ are column vectors composed of the elements on and above the principal diagonals of H and Q, respectively. For example, $\rho = [p_{11}, p_{12}, p_{12}, p_{13}, p_{13}, \ldots]'$.

Equation 3.227 is discussed and applied in Section 4.1, where it is shown how the matrix B can be generated from the A matrix following a certain simple rule developed by MacFarlane [39].

The order of the matrix that must be inverted in order to solve (3.191) can be reduced from $[n(n + 1)]/2$ to $[n(n - 1)]/2$. The matrix β to be obtained below has been obtained independently by Fuller [40] using the "bialternate product" of A with itself and earlier by Barnett and Storey [41]† using the skew-symmetric condition as follows [41].

Equation (3.191) can be written

$$(HA + \tfrac{1}{2}Q) + (HA + \tfrac{1}{2}Q)' = 0 \tag{3.228}$$

so

$$HA + \tfrac{1}{2}Q = S \tag{3.229}$$

where S is a real skew-symmetric matrix. Hence

$$H = (S - \tfrac{1}{2}Q)A^{-1} \tag{3.230}$$

provided that A is nonsingular, and S is determined from the condition $H' = H$, which reduces to

$$A'S + SA = \tfrac{1}{2}(A'Q - QA) \tag{3.231}$$

Since the principal diagonal of S is identically zero, (3.231) represents $\tfrac{1}{2}n(n - 1)$ equations for the $\tfrac{1}{2}n(n - 1)$ unknown elements of S, which can be written, similar to (3.227), as

$$\beta s = -\gamma \tag{3.232}$$

The matrix β [of order $\tfrac{1}{2}n(n - 1)$] is also obtainable from the matrix B of (3.227) and $S = [s_{12}, s_{13}, s_{23}, s_{14}, s_{24}, \ldots]'$.

In order to avoid solving for $\tfrac{1}{2}n(n + 1)$ or $\tfrac{1}{2}n(n - 1)$ equations using the Lyapunov matrix theorem for stability of the A matrix, Fuller [40] developed different methods for obtaining the stability conditions. These methods involve three alternatives—namely, the first, second, and third equations of root pair sums. Among the three methods, the third alternative seems to be the most promising, for it involves evaluation of determinants of lower order than the other two. In this development Fuller obtained, as mentioned earlier, a different procedure for presenting the entries of the β matrix, which is based on the bialternate sums of A by itself.

In presenting the various criteria for stability of the A matrix, Fuller observed that his methods yield either n^2 or $\tfrac{1}{2}n(n + 1)$ conditions, while the

---

† A similar simplification for the unit circle has been obtained by Barnett: "Simplification of the Lyapunov matrix equation $A^T PA - P = -Q$," *IEEE*, A–C (Aug. 1974), pp. 446–447.

dimension of A is $n \times n$. He showed that, for $n = 3$, these conditions can be reduced to 3. However, for higher n, a certain redundancy exists, and he conjectured that the conditions should be reduced to n.

In the following discussions we will present Fuller's method based on the third equation of root pair sums in detail. We will also present the stability conditions for $n = 2, 3$, and 4 in an inner form. For $n = 4$, we obtain only four conditions for stability as conjectured by Fuller. These conditions are analogous to the Liénard–Chipart criterion discussed in Section 2.1. In order to obtain the stability constraints for $n = 3$ and $n = 4$, we will make use of the critical stability constraints discussed in Section 2.5.

CONDITIONS FOR THE STABILITY OF THE A MATRIX

To obtain the necessary and sufficient condition for the eigenvalues (latent roots) of the A matrix to lie in the open left half-plane, we present the following theorem due to Fuller [40]. The proof of the theorem is presented in Appendix C.

THEOREM 3.15. *Let $A = (a_{ij})$ be a real square matrix of dimension $n > 1$. Let $\tilde{A} = (\tilde{a}_{nq,rs})$ be the bialternate sum of A with itself. That is, let*

$$\tilde{A} = 2A \cdot I_n \qquad (3.233)$$

(where $\cdot$ denotes bialternate product).† Let $\tilde{A}$ be the square matrix of dimension $m = \frac{1}{2}n(n-1)$ with rows $pq(p = 2, 3, \ldots, n; q = 1, 2, \ldots, p - 1)$, columns labeled $rs(r = 2, 3, \ldots, n; s = 1, 2, \ldots, r - 1)$, and elements given by

$$\tilde{a}_{pq,rs} = \begin{cases} -a_{ps} & \text{if } r = q \\ a_{pr} & \text{if } r \neq p \text{ and } s = q \\ a_{pp} + a_{qq} & \text{if } r = p \text{ and } s = q \\ a_{qs} & \text{if } r = p \text{ and } s \neq q \\ -a_{qr} & \text{if } s = p \\ 0 & \text{otherwise} \end{cases} \qquad (3.234)$$

Then for the characteristics roots of A to have all their real parts negative (i.e., stable matrix), it is necessary and sufficient that in $(-1)^n$ times the characteristic polynomial of A, namely,

$$(-1)^n |A - \lambda I_n| \qquad (3.235)$$

and in $(-1)^m$ times the characteristic polynomial of $\tilde{A}$, namely,

$$(-1)^m |\tilde{A} - \mu I_n| \qquad (3.236)$$

† See Appendix C for definition.

the coefficients of λ^i $(i = 0, 1, \ldots, n - 1)$ *and* $\mu^i = (\lambda_i + \lambda_j)^i$ $(i = 0, 1, \ldots, m - 1)$ *should all be positive.*

CRITICAL STABILITY CONSTRAINTS

From (3.235), we have

$$|A| = \lambda_1 \lambda_2 \ldots \lambda_n \tag{3.237}$$

which becomes zero when a real root λ_i becomes zero; and from (3.236) we have the bialternate sum matrix $\tilde{A}$ satisfying

$$|\tilde{A}| = \prod_{j<i}^{n} (\lambda_i + \lambda_j) \tag{3.238}$$

which becomes zero when a complex conjugate pair of the λ_i becomes imaginary. Hence, following the discussion of Section 2.5a and the preceding theorem, only

$$(-1)^n |A| \geq 0 \quad \text{and} \quad (-1)^{(1/2)n(n-1)} |\tilde{A}| \geq 0 \tag{3.239}$$

are the critical constraints.

We may also note from (2.96) that the determinant of the bialternate sum matrix satisfies

$$|\tilde{A}| = (-1)^{(1/2)n(n-1)} |\Delta_{n-1}| \tag{3.240}$$

Remarks

1. The identity of (3,240) is of great significance in obtaining the stability constraints for $n = 3$ and $n = 4$, as will be shown below, and, in connecting the Liénard–Chipart criterion for $n = 3$ and $n = 4$ to Theorem 3.15.

2. The matrix $\tilde{A}$ plays an important role in stability, as well as in the critical constraints conditions mentioned above. It can be easily generated from the elements of the A matrix itself, as noted in (3.234). Furthermore, in Appendix C we will present a simple procedure for generating $\tilde{A}$ from A directly.

3. It is evident from Theorem 3.15 that the stability conditions are $\frac{1}{2}n(n + 1)$. These conditions can be reduced to n as conjectured by Fuller [40] and as shown below for $n = 4$. Such reduction would give the equivalent of the Liénard–Chipart criterion discussed in Section 2.1.

Example 3.9. For the matrix

$$A = \begin{bmatrix} a_{11} & a_{12} \\ a_{21} & a_{22} \end{bmatrix} \tag{3.241}$$

to have only characteristic roots with negative real part, it is necessary and sufficient that the coefficients of λ^0 and λ^1 in (3.235), that is,

$$\begin{vmatrix} a_{11} - \lambda & a_{12} \\ a_{21} & a_{22} - \lambda \end{vmatrix} \tag{3.242}$$

and the coefficient of μ^0 in (3.236), that is,

$$-(a_{22} + a_{11} - \mu) \tag{3.243}$$

should be positive. That is, it is necessary and sufficient that

$$|A| = \begin{vmatrix} a_{11} & a_{12} \\ a_{21} & a_{22} \end{vmatrix} > 0 \tag{3.244}$$

$$a_{22} + a_{11} < 0 \qquad a_{22} + a_{11} < 0$$

Note that the last two conditions in (3.244) coincide; that is, one of them is redundant. Therefore the stability condition

$$|A| = \begin{vmatrix} a_{11} & a_{12} \\ a_{21} & a_{22} \end{vmatrix} > 0 \qquad a_{11} + a_{22} < 0 \tag{3.245}$$

The stability conditions of (3.245) can also be expressed as Δ_3, given below to be negative innerwise (ni).

$$\Delta_3 = \begin{bmatrix} \prod_{\substack{i=1}}^{2}\left[a_{ii} - \sum_{\substack{j=1 \\ i \neq 1}}^{2} a_{ij}\right] & 0 & \cdots & 0 \\ & & & \\ 0 & \boxed{\sum_{i=1}^{2} a_{ii}} & 0 \\ 0 & 0^{\;\Delta_1} & 1 \end{bmatrix} \tag{3.246}$$

Example 3.10. Let $n = 3$. Then

$$A = \begin{bmatrix} a_{11} & a_{12} & a_{13} \\ a_{21} & a_{22} & a_{23} \\ a_{31} & a_{32} & a_{33} \end{bmatrix} \tag{3.247}$$

To obtain the stability conditions for (3.247), we make use of the Liénard–Chipart criterion of Section 2.1 to get the two conditions on the coefficients of the characteristic polynomial associated with (3.247) to be positive [see also (3.6) to (3.9)] and the third condition on the Hurwitz determinant $|\Delta_2|$

to be positive. In terms of the A matrix, the two coefficients are

$$\text{trace } A = \sum_{i=1}^{3} a_{ii} < 0 \tag{3.248}$$

$$(-1)^3 |A| = \prod_{i=1}^{3} (-1)^3 \left[a_{ii} - \sum_{\substack{j=1 \\ i \neq 1}}^{3} a_{ij} \right] \tag{3.249}$$

From (3.249), the condition that Δ_2 is positive is equivalent to

$$(-1)^3 |\tilde{A}| = \prod_{i=1}^{3} (-1)^3 \left[\tilde{a}_{ii} - \sum_{\substack{j=1 \\ i \neq j}}^{3} \tilde{a}_{ij} \right] > 0 \tag{3.250}$$

From (3.234), the arrays $\tilde{a}_{ij}$ are the entries of the following matrix:

$$\tilde{A} = \begin{bmatrix} a_{11} + a_{22} & a_{23} & -a_{13} \\ a_{32} & a_{33} + a_{11} & a_{12} \\ -a_{31} & a_{21} & a_{33} + a_{22} \end{bmatrix} \tag{3.251}$$

The conditions of (3.238) to (3.250) are equivalent to the matrix Δ_5 given below to be negative innerwise (ni):

$$\Delta_5 = \begin{bmatrix} \prod_{i=1}^{3} (-1)^3 \left[\tilde{a}_{ii} - \sum_{\substack{j=1 \\ i \neq j}}^{3} \tilde{a}_{ij} \right] & 0 & \cdots & 0 & 0 & 0 \\ 0 & \prod_{i=1}^{3} (-1)^3 \left[a_{ii} - \sum_{\substack{j=1 \\ i \neq j}}^{3} a_{ij} \right] & 0 & 0 & 0 \\ 0 & 0 & \sum_{i=1}^{3} a_{ii} & 0 & 0 \\ 0 & 0 & 0 & 1 & 0 \\ 0 & 0 & 0 & 0 & 1 \end{bmatrix} \tag{3.251a}$$

The stability conditions (3.248) to (3.250) are the identical conditions Fuller obtained after eliminating the redundancy when applying Theorem 3.15.

Example 3.11. Let $n = 4$. Then

$$A = \begin{bmatrix} a_{11} & a_{12} & a_{13} & a_{14} \\ a_{21} & a_{22} & a_{23} & a_{24} \\ a_{31} & a_{32} & a_{33} & a_{34} \\ a_{41} & a_{42} & a_{43} & a_{44} \end{bmatrix} \tag{3.251b}$$

To obtain the stability condition for (3.251b), we again invoke the Liénard–Chipart criterion of Section 2.1. In this case, we obtain three conditions on the coefficients of the characteristic polynomial associated with (3.251b) to be positive [see (3.6) to (3.9)], and the fourth condition is obtained from the Hurwitz determinant $|\Delta_3|$ to be positive. These four conditions are as follows:

$$\text{trace } A = \sum_{i=1}^{4} a_{ii} < 0 \tag{3.251c}$$

$$|A| = \prod_{i=1}^{4} \left[a_{ii} - \sum_{\substack{j=1 \\ i \neq 1}}^{4} a_{ij} \right] > 0 \tag{3.251d}$$

$$\sum_{\substack{j=1 \\ i \neq j}}^{4} \sum_{i=1}^{4} a_{ii}a_{jj} - \sum_{j=1}^{4} \sum_{\substack{i=1 \\ i \neq j}}^{4} a_{ij}a_{ji} > 0 \tag{3.251e}$$

$$|\tilde{A}| = \prod_{i=1}^{6} \left[\tilde{a}_{ii} - \sum_{\substack{j=1 \\ i \neq j}}^{6} \check{a}_{ij} \right] > 0 \tag{3.251f}$$

The matrix $\tilde{A}$ for $n = 4$ is given, from (3.234), as

$$\tilde{A} = \begin{bmatrix} a_{11} + a_{22} & a_{23} & -a_{13} & a_{24} & -a_{14} & 0 \\ a_{32} & a_{11} + a_{33} & a_{12} & a_{34} & 0 & -a_{14} \\ -a_{31} & a_{21} & a_{22} + a_{33} & 0 & a_{34} & -a_{24} \\ a_{42} & a_{43} & 0 & a_{11} + a_{44} & a_{12} & a_{13} \\ -a_{41} & 0 & a_{43} & a_{21} & a_{22} + a_{44} & a_{23} \\ 0 & -a_{41} & -a_{42} & a_{31} & a_{32} & a_{33} + a_{44} \end{bmatrix} \tag{3.251g}$$

The four stability conditions (3.251c) to (3.251f) can also be written in an innerwise matrix form. Hence the stability condition reduces to the following Δ_7 matrix to be negative innerwise (ni).

$$\Delta_7 = \begin{bmatrix} \sum\limits_{\substack{j=1 \\ i \neq j}}^{4}\sum\limits_{i=1}^{4} a_{ii}a_{jj} & 0 & 0 & 0 & 0 & 0 & \sum\limits_{\substack{j=1 \\ i \neq j}}^{4}\sum\limits_{i=1}^{4} a_{ij}a_{ji} \\[2ex] 0 & \prod\limits_{i=1}^{6}\left[\tilde{a}_{ii} - \sum\limits_{\substack{j=1 \\ i \neq j}} \tilde{a}_{ij}\right] & 0 & 0 & 0 & 0 & 0 \\[2ex] 0 & 0 & \prod\limits_{i=1}^{4}\left[a_{ii} - \sum\limits_{\substack{j=1 \\ i \neq j}}^{4} a_{ij}\right] & 0 & 0 & 0 & 0 \\[2ex] 0 & 0 & 0 & \sum\limits_{i=1}^{4} a_{ii} & 0 & 0 & 0 \\[2ex] 0 & 0 & 0 & 0 & 1 & 0 & 0 \\[1ex] 0 & 0 & 0 & 0 & 0 & 1 & 0 \\[1ex] 1 & 0 & 0 & 0 & 0 & 0 & 1 \end{bmatrix}$$

$$\tag{3.251h}$$

Remarks

1. For higher-order n, the stability conditions are those of the Liénard–Chipart criterion, and they are equivalent to the conditions of Theorem 3.15. Because of the complexity of the expressions, in terms of the entries of the A matrix of the inner determinants of the Liénard–Chipart criterion, it is advisable, for computational purposes, to use Theorem 3.15 even though some redundancy may exist.

2. In Sections 7.1 and 7.4 we obtain the stability constraints for the A matrix for $n = 3$ and $n = 4$ by using the double-triangularization procedure on the A matrix itself. For higher-order n, it may be possible to utilize the double-triangularization procedure on both A and $\tilde{A}$ to obtain the stability constraints. This might be an interesting topic for future investigations.

3. In order to reduce computational errors by using a computer, we can transform the A matrix first to a diagonally dominant form. This step can be

done by noting the following property: The matrix A is said to be diagonally dominant if [42] and only if

$$|a_{ii}| \geq \sum_{\substack{j=1 \\ i \neq j}}^{n} |a_{ij}| \geq \left| \sum_{\substack{j=1 \\ i \neq j}}^{n} a_{ij} \right|$$

$$i = 1, 2, \ldots, n \quad (3.251\text{i})$$

This equation is equivalent to

$$a_{ii} \geq \sum a_{ij}$$
$$\text{if} \quad a_{ii} > 0, \quad \sum a_{ij} > 0 \quad (3.251\text{j})$$

$$a_{ii} \geq \sum a_{ij}$$
$$\text{if} \quad a_{ii} > 0, \quad \sum a_{ij} < 0 \quad (3.251\text{k})$$

$$a_{ii} \leq \sum a_{ij}$$
$$\text{if} \quad a_{ii} < 0, \quad \sum a_{ij} < 0 \quad (3.251\text{l})$$

$$a_{ii} \leq \sum a_{ij}$$
$$\text{if} \quad a_{ii} < 0, \quad \sum a_{ij} > 0 \quad (3.251\text{m})$$

It should be noted that the diagonally dominant condition constitutes a necessary condition for stability of the A matrix.

4.   The condition for the eigenvalues of the A† matrix to be distinct and negative real and to have magnitude less than unity (stability of linear discrete systems) and relative stability are discussed in Appendix C.

5.   If A is real and has negative diagonal elements and a dominant diagonal, then A is stable in virtue of the position of the Gerschgorin circles [43].

6.   If A is a real matrix with $a_{ik} \geq 0$ for $i \neq k$, then it can be tested for stability, the necessary and sufficient conditions being

$$a_{11} < 0, \begin{vmatrix} a_{11} & a_{12} \\ a_{21} & a_{22} \end{vmatrix} > 0, \ldots, (-1)^n |A| > 0$$

This was established by Sevastyanov and Katelyanski, as reported by Gantmacher [15].

7.   There exists a straightforward procedure [44,45]‡ for obtaining the coefficients of λ^i and μ^i from the elements of A and $\tilde{A}$ of (3.235) and (3.236).

3.6   GENERALIZED REGION FOR ROOT
CLUSTERING [29, 46]

In this section we will present on alternate region Ω that includes half-planes, circles, hyperbolas, ellipses, and parabolas. This alternate region

† Conditions on the eigenvalues of the complex matrix A are discussed in the following: B. D. O. Anderson, N. K. Bose, and E. I. Jury, "A simple test for zeros of a complex polynomial in a sector." *IEEE Trans. on Automatic Control* (Aug. 1974), pp. 437–438.
‡ See also reference 49.

differs from the region Γ discussed in Sections 3.4 and 3.5 because it also contains ellipses and parabolas. In this discussion we will show that the root-clustering condition for the ellipses and parabolas can also be described by rational functions of the coefficients and their complex conjugates.

Let the defining equation of the region Ω be described by

$$\phi(\alpha, \beta) = C_{00} + C_{01}\beta + C_{02}\beta^2 + C_{10}\alpha + C_{11}\alpha\beta + C_{20}\alpha^2 \quad (3.252)$$

where
$$C_{11} + \bar{C}_{11} \leq 0 \quad (3.253)$$

and
$$\Omega = \{z \in \mathbb{C} \mid \operatorname{Re} \phi(\bar{z}, z) > 0\} \quad (3.254)$$

$$A \in \mathbb{C}^{n \times n} \quad (3.255)$$

$$\lambda_i \qquad\qquad i = 1, 2, \ldots, n \quad (3.256)$$

are the eigenvalues of A.

THEOREM 3.16. *Let*

$$\phi(\bar{\lambda}_i, \lambda_j) + \bar{\phi}(\lambda_j, \bar{\lambda}_i) \neq 0$$
$$\textit{for all } \lambda_i, \lambda_j \in \Omega, \quad i, j = 1, 2, \ldots, n \quad (3.257)$$

Under the above condition, $\lambda_i \in \Omega$, $i = 1, 2, \ldots, n$, if and only if, for all positive definite Hermitian $Q \in \mathbb{C}^{n \times n}$, there exists a unique positive definite Hermitian $H \in \mathbb{C}^{n \times n}$ such that

$$(\bar{C}_{00} + C_{00})H + (\bar{C}_{01} + C_{10})A^*H + (\bar{C}_{02} + C_{20})(A^*)^2 H$$
$$+ (\bar{C}_{10} + C_{01})HA + (\bar{C}_{11} + C_{11})A^*HA$$
$$+ (\bar{C}_{20} + C_{02})HA^2 = Q \quad (3.258)$$

*where * denotes the complex-conjugate transpose. Furthermore, the elements of H are rational functions of the elements of A and A^*.*
Proof Necessity: Let $x_K \in \mathbb{C}^n$ be the eigenvector corresponding to the eigenvalue λ_k. Then

$$[\phi(\bar{\lambda}_K, \lambda_K) + \bar{\phi}(\lambda_K, \bar{\lambda}_K)]x_k^* H x_K = x_k^* Q x_K \quad (3.259)$$

Sufficiency: In this case (3.258) is equivalent to n^2 linear equations, whose $n^2 \times n^2$ coefficient matrix has

$$\phi(\bar{\lambda}_i, \lambda_j) + \bar{\phi}(\lambda_j, \bar{\lambda}_i)$$
$$i, j = 1, 2, \ldots, n \quad (3.260)$$

as its eigenvalues [47]. By hypothesis, (3.258) has a unique solution H, whose elements are rational functions of the elements of A and A^*. We first prove that there exists a positive definite $H_0 \in \mathbb{C}^{n \times n}$ and a positive definite $Q_0 \in \mathbb{C}^{n \times n}$ that satisfy (3.258).

Let J be the Jordan form of A. Then there exists a nonsingular $P \in \mathbb{C}^{n \times n}$ such that

$$PAP^{-1} = \Lambda + U \qquad (3.261)$$

where Λ is diagonal and all the elements of U are either 0 or 1, with all nonzero elements located on the diagonal above the main diagonal. For some small $\delta > 0$, define the nonsingular matrix D by

$$D = \text{diag}\,[1, \delta^{-1}, \delta^{-2}, \ldots, \delta^{-(n-1)}] \qquad (3.262)$$

Then $$DPAP^{-1}D = \Lambda + \delta U \qquad (3.263)$$

Now define

$$H_0 = \text{diag}\,[\text{Re}\,\phi(\bar{\lambda}_1, \lambda_1),\, \text{Re}\,\phi(\bar{\lambda}_2, \lambda_2), \ldots,\, \text{Re}\,\phi(\bar{\lambda}_n, \lambda_n)] \qquad (3.264)$$

Then Q_0, defined by (3.258), is positive definite by Gershgorin's theorem [47]. For arbitrary positive definite $Q \in \mathbb{C}^{n \times n}$, let

$$Q_t = tQ + (1 - t)Q_0 \qquad (3.265)$$

where $$0 \le t \le 1 \qquad (3.266)$$

Clearly, Q_t is positive definite. Equation (3.258) has a unique solution $H_t \in \mathbb{C}^{n \times n}$ for $Q = Q_t$. The eigenvalues of H_t are real and vary continuously with t. Hence if we prove that H_t never becomes singular, we complete the proof. Let

$$H_t = \left[\begin{array}{c|c} \hat{H}_t & 0 \\ \hline 0 & 0 \end{array}\right] \qquad (3.267)$$

We substitute $H = H_t$ in (3.258). Then every $n \times n$ element of the matrices on the left side of (3.258) is zero except that of $(C_{11} + \bar{C}_{11})A^*H_tA$. Hence we arrive at the contradiction to the fact that Q is positive definite, since $C_{11} + \bar{C}_{11} \le 0$.

Example 3.12 (Ellipse). Let

$$\phi(\alpha, \beta) = -2\alpha\beta - \alpha^2 + 1 \qquad (3.268)$$

$A \in \mathbb{C}^{n \times n}$ has all its eigenvalues inside the ellipse

$$3x^2 + y^2 - 1 < 0 \qquad (3.269)$$

if and only if for all positive definite Hermitian $Q \in \mathbb{C}^{n \times n}$, there exists a unique positive definite Hermitian $H \in \mathbb{C}^{n \times n}$ such that

$$2H - (A^*)^2H - 4A^*HA - HA^2 = Q \qquad (3.270)$$

and the elements of H are rational functions of the elements of A and A^*.
Proof: We need only show that (3.257) holds. That is, all $\lambda_1 \triangleq x_1 + jy_1$ and $\lambda_2 \triangleq x_2 + jy_2$ belonging to the ellipse satisfy

$$-4\bar{\lambda}_1\lambda_2 - \bar{\lambda}_1^2 - \lambda_2^2 + 2 \neq 0 \tag{3.271}$$

Suppose that the equality holds. Then

$$x_1(2y_2 - y_1) = x_2(2y_1 - y_2) \tag{3.272}$$

and $$4x_1x_2 + 4y_1y_2 + x_1^2 - y_1^2 + x_2^2 - y_2^2 - 2 = 0 \tag{3.273}$$

Let $$\frac{x_1}{x_2} = \frac{2y_1 - y_2}{-y_1 + 2y_2} \triangleq k \tag{3.274}$$

Simple calculation shows that k is well defined. Substituting (3.274) into (3.273), we obtain

$$\frac{-5k^2 + 4k + 1}{(2 + k)^2}y_2^2 + \frac{5k^2 - 4k - 1}{k^2 + 4k + 1} < 0 \tag{3.275}$$

and $$\frac{k^2 + 4k - 5}{(2 + k)^2}y_2^2 + \frac{-k^2 - 4k + 5}{k^2 + 4k + 1} < 0 \tag{3.276}$$

In the region $k \leq -5$ and $k \geq -0.5$, (3.275) and (3.276) contradict each other. In the region $-5 < k < -0.5$, $y_2^2 \geq 1$, which is also a contradiction.

Example 3.13 (Parabola). Let

$$\phi(\alpha, \beta) = \alpha^2 + 2\alpha - \alpha\beta \tag{3.277}$$

Then the parabolic region is given by

$$y^2 < x \tag{3.278}$$

and the matrix equation is

$$HA^2 - 2A^*HA + (A^*)^2H + 2HA + 2(A^*)^2H = Q \tag{3.279}$$

A proof similar to the one given in Example 3.12 follows.

From the discussion of this section we have obtained regions for the root clustering other than those mentioned in region Γ in Sections 3.4 and 3.5. However, since the polynomials in (3.75) and (3.254) are different, we cannot say that the region Ω is more general than Γ and, indeed, the question of which is the largest class of regions in the complex plane where the condition for the root clustering can be expressed by the rational function of only the coefficients and their complex conjugates of the given polynomial still remains open [48].

REFERENCES

[1] C. HERMITE, "Sur le nombre des racines d'une équation algebrique comprise entre des limites données," *J. Reine Angew., Math.*, **52**, 39–51 (1854).

[2] A. A. MARKOV, *Collected Works*, Moscow, 1948. See also Gantmacher, *Theory of Matrices*. New York: Chelsea, 1959, Vol. II, Chap. 5.

[3] A. M. LYAPUNOV, *Le Probléme Générale de la Stabilité du Mouvement.* Princeton, N.J.: Princeton University Press, 1949.

[4] I. SCHUR, "Über Potenzreihen, die in Inners des Einheitskreises besehränkt sind," *J. Für Math.*, **147**, 205–232 (1917).

[5] A. COHN, "Über die Anzahl der Wurzeln einer algebraischen Gleichung in einem Kreise," *Math. Z.*, **14–15**, 110–148 (1914).

[6] H. FUJIWARA, "Über die algebraischen Gleichungen, deren Wurzeln einem Kreise oder in eine liegen," *Math. Z.*, **24**, 160–169 (1926).

[7] E. I. JURY and B. D. O. ANDERSON, "Some remarks on simplified stability criteria for continuous linear system," *IEEE Trans. on Automatic Control*, **AC-17** (3), 371–372 (June 1972).

[8] B. D. O. ANDERSON, "The reduced Hermite criterion with applications to proof of Liénard–Chipart criterion," *IEEE Trans. on Automatic Control*, **AC-17** (5), 669–672 (Oct. 1972).

[9] R. E. BELLMAN, *Introduction to Matrix Analysis.* New York: McGraw-Hill, 1960.

[10] A. LIÉNARD and M. H. CHIPART, "Sur la signe de la partie reelle des racines d'une équation algébrique," *J. Math. Pure Appl., Series b*, **10**, 291–346 (1914).

[11] E. I. JURY and S. M. AHN, "Interchangeability of 'inners' and 'minors,'" *Trans. ASME, J. Dynamic Systems, Measurement and Control*, 257–260 (Dec. 1971).

[12] B. D. O. ANDERSON and E. I. JURY, "A simplified Schur–Cohn test," *IEEE Trans. on Automatic Control*, **AC-18** (2), 157–163 (April 1973).

[13] E. I. JURY, *Analysis and Synthesis of Sampled-Data Control Systems.* New York: Wiley, 1958, pp. 34–38.

[14] B. D. O. ANDERSON and E. I. JURY, "Relationships between real and complex Hurwitz polynomials and real aperiodic polynomials" (Oct. 1972).

[15] F. R. GANTMACHER, *The Theory of Matrices.* New York: Chelsea, 1959, Vol. II, Chap. 5.

[16] M. I. ROMANOV, "Algebraic criteria for aperiodicity of linear systems," *Dokl. Akad. Nauk, SSSR*, **124** (2), 291–294 (1959).

[17] A. T. FULLER, "Condition for aperiodicity in linear systems," *Brit. J. Appl. Phys.*, **6**, 195–198 (June 1955).

[18] M. V. MEEROV, "Criteria for aperiodic systems," *Isvest. Akad. Nauk SSSR, OTN* (12), 1165 (1945).

[19] E. I. JURY and S. M. AHN, "Symmetric and innerwise matrices for the root

clustering and root distribution of a polynomial," *J. Franklin Inst.*, **293** (6), 433–450 (June 1972).

[20] L. WEISNER, *Introduction to the Theory of Equations*. New York: Macmillan, 1938, p. 108. Also in Chap X, Theorem 10, Corollary.

[21] R. E. KALMAN, "Algebraic characterization of polynomials whose zeros lie in a certain algebraic domain," *Proc. Nat. Acad. Sci.*, **64** (3), 818–823 (May 1969).

[22] M. MARDEN, "Geometry of polynomials," *Am. Math. Society*, 2nd ed., Eq. (40.1), p. 179, 1966.

[23] A. RALSTON, "A symmetrical formulation of the Routh–Hurwitz stability criterion," *IEEE Trans. on Automatic Control*, **AC-17**, 50–51 (1972).

[24] P. C. PARKS, "Hermité–Hurwitz and Hermité Bilherz links using matrix multiplication," *Electronics Letters*, **5** (3), 55–57 (Feb. 1967).

[25] A. S. HOUSHOLDER, "Bezoutians, elimination and localization," *SIAM Rev.*, **12** (1), 73–78 (Jan. 1970).

[26] H. A. NOUR–ELDIN, "Ein neues stabilitats Kriterium für Abgetstate Regelsysteme," *Regelungstechmic*, **7**, 301–307 (1971).

[27] E. NETTO, *Vorlesung Über Algebra*. Leipzig: Teubner, 1896, Vol. 1.

[28] A. S. HOUSHOLDER, "Bigradients and the problem of Routh–Hurwitz," *SIAM Rev.*, **10**, 56–66 (1968).

[29] J. L. HOWLAND, "Matrix equations and the separation of matrix eigenvalues," *J. Mathematical Analysis and Applications*, **33**, 683–691 (1971).

[30] S. BARNETT and C. STOREY, *Matrix Methods in Stability Theory*. New York: Barnes and Noble, 1970.

[31] P. C. PARKS, "Lyapunov and Schur–Cohn stability criterion," *IEEE Trans. on Automatic Control*, **AC-9,** 121 (1964).

[32] P. C. PARKS, "A new proof of the Hurwitz stability criterion by the second method of Lyapunov," *Proc. Camb. Phil. Soc.*, **58**, 694–702 (1962).

[33] S. H. LEHNIGK, *Stability Theorems for Linear Motions with an Introduction to Lyapunov's Direct Method*. Englewood Cliffs, N.J.: Prentice-Hall, 1966.

[34] R. E. KALMAN, "On the Hermité–Fujiwara theorem in stability theory," *Quart. Appl. Math.*, **23**, 279–282 (1965).

[35] H. R. SCHWARZ, "Ein Verfahren zur Stabilitätsfrage bes Matrizen-Eigenwerte Probleme," *Z. Angew. Math. Phys.*, **7**, 473–500 (1956).

[36] C. F. CHEN and H. CHU, "A matrix for evaluating Schwarz's form," *IEEE Trans. on Automatic Control*, **AC-11,** 303–305 (April 1966).

[37] S. BARNETT and C. STOREY, "The Lyapunov matrix equation and Schwarz's form," *IEEE Trans. on Automatic Control*, **AC-12,** 117–118 (Feb. 1967).

[38] S. BARNETT, *Matrices in Control Theory*. London: Van Nostrand Reinhold, 1971.

[39] A. G. J. MACFARLANE, "The calculations of functions of the time and

frequency response of linear constant coefficient dynamical systems," *Quart. J. Mech. Appl. Math.* (*England*), **16**, 259–271 (1963).

[40] A. T. FULLER, "Conditions for a matrix to have only characteristic roots with negative real parts," *J. Mathematical Analysis and Applications*, **23** (1), 71–98 (July 1968).

[41] S. BARNETT and C. STOREY, "Stability analysis of constant linear systems by Lyapunov's second method," *Electronics Letters*, **2**, 165–166 (1966).

[42] J. H. WILKINSON, *The Algebraic Eigenvalue Problem*. London: Oxford University Press, 1965.

[43] O. TAUSKKY–TODD, "On stable matrices," *Colloques Internationaux C.N.R.S.* No. 165, Editions du C.N.R.S. 15, Quai Anatole France, Paris, 1968.

[44] V. N. FADEEVA, *Computational Methods in Linear Algebra*. New York: Dover, 1959.

[45] J. I. SOLIMAN and A. K. KEVORKIAN, "A simplified procedure for the computation of the coefficients of the characteristic equation from the elements of the *A* matrix," *Int. J. of Control*, **5** (2), 163–169 (1967).

[46] E. I. JURY and S. M. AHN, "Remarks on the root-clustering of a polynomial in a certain region in the complex plane," *Quart. Appl. Math.* July (1974), pp. 203–205.

[47] P. LANCASTER, *Theory of Matrices*. New York: Academic Press, 1969.

[48] R. D. HILL, "Inertia theory for simultaneously triangulable complex matrices," *Linear Algebra and Its Applications*, **2**, 131–142 (1969).

[49] C. Papaconstantinou, "Construction of the characteristic polynomial of a Matrix", *IEEE Trans. on Automatic Control*, Vol. AC-19, No. 2, April, 1974, p. 149–151.

Inners Representation of Complex Integrals

The problem of evaluating the total integral square (or sum) of a signal arises in the analysis and optimization of feedback control systems, both for continuous and discrete systems and for deterministic and stochastic inputs. It also occurs in communication and digital filtering problems. The first approach to this problem for continuous systems was made by James et al. [1] using the solution of a recurrence formula. A few years later Newton et al. [2] presented a method based on the solution of a matrix equation dependent on whether n, the order of the system, was even or odd. In Popov [3], a method due to Katz [4], for obtaining I_n (the complex integral) as a ratio of two determinants (the denominator is Hurwitz), is discussed. About the same time, a similar method was proposed by Mersman [5]. Furthermore, discussion of this method is presented in References 6 to 8. An approach similar to the Routh table for the stability test was presented by Neklony and Benes [9]. Additional elaboration of this method was given by Effertz [10]. In 1970 Åstrom [11] offered an exposition of this method with a computer program obtained for evaluating I_n. An historical note on the evolution of these methods is discussed by Fuller [12].

An approach based on the Lyapunov method was presented by Kalman and Bertram [13], MacFarlane [14], and further developed by Beshara [15]. The analogous problem for the discrete case has been the subject of intensive investigations [16–23]. Similar to the continuous case, the evaluation of I_n can be obtained by using either determinant method or the table form. In Åstrom [11], an enlightening discussion of the table form with computer programs is presented.†

In this chapter we will present the evaluation of I_n for both continuous and discrete cases, using the inners approach as well as the Lyapunov method. In the last section we will connect the evaluation of I_n with the stability test.

We will follow Mersman for the determinantal method for the continuous case, and for the discrete cases we will restate the results in Jury [17]. For the

---

† See also Ref. 20.

Lyapunov method, we will use MacFarlane [14,32] and Man's [24] approaches for the continuous and discrete cases, respectively.

4.1 EVALUATION OF INTEGRAL SQUARE OF SIGNALS, CONTINUOUS CASE

The integral of the square of the signal can be represented in terms of a complex integral, using Parseval's theorem, as

$$\int_0^\infty g^2(t)\,dt = \frac{1}{2\pi j}\int_{-j\infty}^{j\infty} G(s)G(-s)\,ds \tag{4.1}$$

where $G(s)$ is the Laplace transform of $g(t)$.

The proof of this formula can be obtained from the complex convolution theorem [25] as follows. Let

$$G_1(s) = \mathscr{L}[g_1(t)] \tag{4.2}$$

$$G_2(s) = \mathscr{L}[g_2(t)] \tag{4.3}$$

The Laplace transform of $g_1(t)g_2(t)$ is given as

$$G(s) = \mathscr{L}[g_1(t)g_2(t)] = \int_0^\infty g_1(t)g_2(t)e^{-st}\,dt$$
$$\max(\sigma_1, \sigma_2, \sigma_1 + \sigma_2) < \sigma \quad (4.4)\dagger$$

From the inverse Laplace transform formula we have

$$g_2(t) = \frac{1}{2\pi j}\int_{c_2-j\infty}^{c_2+j\infty} G_2(p)e^{tp}\,dp \qquad \sigma_2 < c_2 \quad (4.5)\ddagger$$

Substitute (4.5) into (4.4) to obtain

$$G(s) = \int_0^\infty g_1(t)\frac{1}{2\pi j}\int_{c_2-j\infty}^{c_2+j\infty} G_2(p)e^{tp}\,dp\,e^{-st}\,dt \tag{4.6}$$

Since the functions g_1 and g_2 are Laplace transformable, it is permissible to change the order in which the two integrations in (4.6) are performed,

$$G(s) = \frac{1}{2\pi j}\int_{c_2-j\infty}^{c_2+j\infty} G_2(p)\int_0^\infty g_1(t)e^{-(s-p)t}\,dt\,dp$$
$$\sigma_1 < \sigma - \operatorname{Re}[p] \quad (4.7)$$

---

† Here, σ_1 and σ_2 are the abscissa of absolute convergence of $g_1(t)$ and $g_2(t)$, respectively, and σ is the real part of s.
‡ The value c_2 is the real part of p.

Since the time integral here is $G_1(s - p)$,

$$G(s) = \frac{1}{2\pi j} \int_{c_2 - j\infty}^{c_2 + j\infty} G_2(p)G_1(s - p)\, dp$$

$$\sigma_2 < c_2 < \sigma - \sigma_1, \quad \max(\sigma_1, \sigma_2, \sigma_1 + \sigma_2) < \sigma \quad (4.8)$$

If we assume that $c_2 = 0$, that is, all roots of $G_2(p)$, are in the open left half of the p-plane and $g_1(t) = g_2(t)$ and $s = 0$ and letting the complex variable p be replaced by s, we obtain from (4.4) and (4.8)

$$\int_0^\infty g^2(t)\, dt = \frac{1}{2\pi j} \int_{-j\infty}^{j\infty} G(s)G(-s)\, ds \quad (4.9)$$

which is Parseval's formula.

In order to evaluate the integral in (4.9), we will follow rather closely the method proposed by Mersman [5]. Let

$$I_n = \frac{1}{2\pi j} \int_{-j\infty}^{j\infty} G(s)G(-s)\, ds = \frac{1}{2\pi j} \int_{-j\infty}^{j\infty} \frac{g(s)}{h(s)h(-s)}\, ds \quad (4.10)$$

where

$$g(s) = b_{n-1}s^{2(n-1)} + b_{n-2}s^{2(n-2)} + \cdots + b_0 \quad (4.11)$$

$$h(s) = a_n s^n + a_{n-1}s^{n-1} + \cdots + a_1 s + a_0$$

$$a_n > 0 \quad (4.12)$$

It is assumed that the zeros of $h(s)$ are all distinct (later we will show that, by continuity, the results are valid for multiple zeros) and have negative real part. Then the integration can be performed by means of the residues, and the result is

$$I_n = \sum_{k=1}^{n} A_k \quad (4.13)$$

where A_k is the residue of the integrand in (4.10) at s_k and $h(s_k) = 0$. This expression can be evaluated in terms of the coefficients b_k and a_k by starting with the identity

$$\frac{g(s)}{h(s)h(-s)} = \sum_{k=1}^{n} A_k \left(\frac{1}{s - s_k} - \frac{1}{s + s_k} \right) \quad (4.14)$$

or

$$g(s) = \sum_{k=1}^{n} A_k \left[\frac{h(s)}{s - s_k} h(-s) + \frac{h(s)}{-s - s_k} h(-s) \right] \quad (4.15)$$

Since s_k is a zero of $h(s)$, the quantity $h(s)/(s - s_k)$ is a polynomial; in fact,

$$\frac{h(s)}{s - s_k} = \sum_{j=0}^{n-1} s^{n-1-j} \sum_{i=0}^{j} a_{n-i} s_k^{j-i} \quad (4.16)$$

Substituting (4.16) in (4.15) gives an identity between two polynomials.

Equating coefficients of like powers of s gives a set of simultaneous, linear, algebraic equations for A_k:

$$\sum_{k=1}^{n} \alpha_{\ell k} A_k = (-1)^n \frac{b_{n-\ell}}{2} \qquad \ell = 1, 2, \ldots, n \quad (4.17)$$

where

$$\alpha_{\ell k} = \sum_{i=1}^{n} \xi_{\ell i} s_k^{i-1} \qquad \ell, k = 1, 2, \ldots, n \quad (4.18)$$

$$\xi_{\ell i} = \sum_{j=1}^{n} c_{\ell j} d_{ji} \qquad \ell, i = 1, 2, \ldots, n \quad (4.19)$$

$$c_{\ell j} = a_{n+j-2\ell} \qquad \ell, j = 1, 2, \ldots, n \quad (4.20)$$

$$d_{ij} = (-1)^j a_{n+i-j} \qquad j, i = 1, 2, \ldots, n \quad (4.21)$$

with the convention that $a_k = 0$ if $k < 0$ or $k > n$. With $|\alpha_{\ell k}|$ for the determinant with n rows and n columns having $\alpha_{\ell k}$ in the ℓth row and kth column, the rule for multiplying determinants gives

$$|\alpha_{\ell k}| = |c_{\ell j}| \cdot |d_{ji}| \cdot |s_k^{i-1}| \qquad (4.22)$$

where

$$|d_{ji}| = \begin{vmatrix} -a_n & 0 & 0 & \cdots & 0 \\ a_{n-1} & a_n & & \cdots & 0 \\ -a_{n-2} & -a_{n-1} & & \cdots & 0 \\ \vdots & \vdots & & & \vdots \\ \pm a_1 & \pm a_2 & & \cdots & \pm a_n \end{vmatrix} = (-1)^{(n(n+1))/2} a_n^n \quad (4.23)$$

and

$$|s_k^{i-1}| = \begin{vmatrix} 1 & 1 & \cdots & 1 \\ s_1 & s_2 & \cdots & s_n \\ s_1^2 & s_2^2 & \cdots & s_n^2 \\ \vdots & \vdots & & \vdots \\ s_1^{n-1} & s_2^{n-1} & \cdots & s_n^{n-1} \end{vmatrix} = V_n \qquad (4.24)$$

where V_n is the Vandermonde determinant.

Hence, writing $C_n = |c_{\ell j}|$, we have

$$|\alpha_{\ell k}| = (-1)^{(n(n+1))/2} a_n^n C_n V_n \qquad (4.25)$$

In (4.17), write

$$\beta_\ell = (-1)^n \frac{b_{n-\ell}}{2}$$

for convenience and subtract $\beta_\ell I_n$ from both sides. Recalling (4.13), we can put the resulting system of equations in the form

$$I_n - \sum_{k=1}^{n} A_k = 0 \qquad (4.26)$$

$$\beta_\ell I_n + \sum_{k=1}^{n} (\alpha_{\ell k} - \beta_\ell) A_k = \beta_\ell \qquad 1 \le \ell \le n \quad (4.27)$$

The above is a system of $n + 1$ equations in the $n + 1$ unknowns I_n, A_1, A_2, ..., A_n that can be solved for I_n. First consider the determinant, $|D|$, of the coefficients in the left members of (4.26) and (4.27).

$$|D| = \begin{vmatrix} 1 & -1 & \cdots & -1 \\ \beta_1 & \alpha_{11} - \beta_1 & \cdots & \alpha_{1n} - \beta_1 \\ \beta_2 & \alpha_{21} - \beta_2 & \cdots & \alpha_{2n} - \beta_2 \\ \vdots & \vdots & & \vdots \\ \beta_n & \alpha_{n1} - \beta_n & \cdots & \alpha_{nn} - \beta_n \end{vmatrix} \qquad (4.28)$$

Adding the first column to each of the succeeding columns gives the result

$$|D| = |\alpha_{ij}| = (-1)^{(n(n+1))/2} a_n^n C_n V_n \qquad (4.29)$$

Now $V_n \ne 0$, since all the zeros s_k of $h(s)$ are distinct, and C_n does not vanish, for it is precisely the Hurwitz determinant [26] (except for exchange of rows and columns) of the polynomial $h(s)$, all the roots of which lie in the open left half-plane. Hence $|D| \ne 0$, and the system (4.26) and (4.27) can be solved for I_n directly by Cramer's rule. Thus

$$DI = \begin{bmatrix} 0 & -1 & \cdots & -1 \\ \beta_1 & \alpha_{11} - \beta_1 & \cdots & \alpha_{1n} - \beta_1 \\ \beta_2 & \beta_{21} - \beta_2 & \cdots & \alpha_{2n} - \beta_2 \\ \vdots & \vdots & & \vdots \\ \beta_n & \alpha_{n1} - \beta_n & \cdots & \alpha_{nn} - \beta_n \end{bmatrix} \qquad (4.30)$$

Adding the first column to each succeeding column gives

$$DI = \begin{bmatrix} 0 & -1 & \cdots & -1 \\ \beta_1 & \alpha_{11} & \cdots & \alpha_{1n} \\ \vdots & \vdots & & \vdots \\ \beta_n & \alpha_{n1} & \cdots & \alpha_{nn} \end{bmatrix} \qquad (4.31)$$

By the definition of α_{ij} in (4.18), the above can be factored twice to give

$$
DI = \frac{M}{a_n}
\begin{bmatrix}
0 & 1 & 0 & \cdots & 0 \\
\beta_1 & c_{11} & c_{12} & \cdots & c_{1n} \\
\beta_2 & c_{21} & c_{22} & \cdots & c_{2n} \\
\vdots & \vdots & \vdots & & \vdots \\
\beta_n & c_{n1} & c_{n2} & \cdots & c_{nn}
\end{bmatrix}
\qquad (4.32)
$$

where

$$
M =
\begin{bmatrix}
1 & 0 & \cdots & 0 \\
0 & d_{11} & \cdots & d_{1n} \\
0 & d_{21} & \cdots & d_{2n} \\
\vdots & \vdots & & \vdots \\
0 & d_{n1} & \cdots & d_{nn}
\end{bmatrix}
\begin{bmatrix}
1 & 0 & \cdots & 0 \\
0 & 1 & \cdots & 1 \\
0 & x_1 & \cdots & x_n \\
\vdots & \vdots & & \vdots \\
0 & x_1^{n-1} & \cdots & x_1^{n-1}
\end{bmatrix}
\qquad (4.33)
$$

Thus

$$
(-1)^{(n(n+1))/2} a_n^n C_n V_n I_n = \frac{-1}{a_n}
\begin{vmatrix}
\beta_1 & c_{12} & \cdots & c_{1n} \\
\beta_2 & c_{22} & \cdots & c_{2n} \\
\vdots & \vdots & & \vdots \\
\beta_n & c_{n2} & \cdots & c_{nn}
\end{vmatrix}
(-1)^{(n(n+1))/2} a_n^n V_n
\qquad (4.34)
$$

The relation

$$
\beta_\ell = (-1)^n \frac{b_{n-\ell}}{2}
$$

gives, finally, the desired value of I_n:

$$
I_n = \frac{1}{2\pi j} \int_{-j\infty}^{j\infty} \frac{g(s)\,ds}{h(s)h(-s)} = \frac{(-1)^{n+1}}{2a_n} \frac{G_n}{C_n}
\qquad (4.35)
$$

where

$$
G_n = |g_{ij}| \qquad\qquad 1 \le i, j, n \quad (4.36)
$$

$$
C_n = |c_{ij}| \qquad\qquad 1 \le i, j, n \quad (4.37)
$$

and

$$
c_{ij} = a_{n+j-2i} \qquad\qquad\qquad (4.38)
$$

$$
g_{ij} =
\begin{cases}
b_{n-i} & \text{if } j = 1 \\
c_{ij} & \text{if } j > 1
\end{cases}
\qquad\qquad (4.39)
$$

Since I_n is a continuous function of the coefficients of $h(s)$, and hence of the zeros, (4.35) remains true for multiple zeros.

Remarks

1. The entries in the matrix c_{ij} from (4.37) are recognized as the entries of the Hurwitz matrix [26] except that the rows and columns are exchanged. Hence the determinant value is unchanged.

2. As shown in Chapters 1 and 2, the Hurwitz matrix can be written in inner form. This step can be readily achieved by premultiplying the Hurwitz matrix by the permutation matrix given in Chapter 1. Again, the determinant value is unchanged.

Based on the preceding remarks, we can restate the formula for I_n in inners form as follows:

$$I_n = \frac{(-1)^{n+1}}{2a_n} \frac{|\Delta_n^b|}{|\Delta_n|} \tag{4.40}\dagger$$

For n-odd,

$$\Delta_n = \begin{bmatrix} a_n & \cdots & a_3 & & a_1 & \cdots & & 0 \\ 0 & & & & & & & \\ & & & a_n & a_{n-2} & a_{n-4} & \cdots & a_1 & 0 \\ \vdots & & 0 & \boxed{a_{n-1}} & a_{n-3} & \cdots & a_2 & a_0 \\ & & a_{n-1} & a_{n-3} & a_{n-5} & \cdots & a_0 & 0 \\ 0 & & & & & & & \\ a_{n-1} & \cdots & a_2 & a_0 & 0 & \cdots & & 0 \end{bmatrix} \tag{4.41}$$

For n-even,

$$\Delta_n = \begin{bmatrix} a_{n-1} & \cdots & a_1 & & 0 & \cdots & & 0 \\ 0 & & & & & & & \\ & & & \boxed{a_{n-1}} & a_{n-3} & & 0 & \\ \vdots & & & a_n & a_{n-2} & & & a_0 \\ 0 & & & & & & & \\ a_n & & a_{n-2} & \cdots & a_2 & a_0 & \cdots & 0 \end{bmatrix} \tag{4.42}$$

and Δ_n^b is formed from Δ_n by replacing the last row for n-odd and the first row for n-even by

$$[b_{n-1}, b_{n-2}, \ldots, b_1, b_0] \tag{4.43}$$

$\dagger$ Other techniques for evaluating I_n exist in the literature (see Reference 34).

Example 4.1. Let

$$g(s) = b_2 s^4 + b_1 s^2 + b_0 \tag{4.44}$$

$$h(s) = a_3 s^3 + a_2 s^2 + a_1 s + a_0$$

$$\text{with } a_3 > 0 \tag{4.45}$$

From (4.35) to (4.39), we have

$$I_n = \frac{(-1)^4}{2a_3} \begin{vmatrix} b_2 & a_3 & 0 \\ b_1 & a_1 & a_2 \\ b_0 & 0 & a_0 \\ a_2 & a_3 & 0 \\ a_0 & a_1 & a_2 \\ 0 & 0 & a_0 \end{vmatrix} = \frac{1}{2a_3} \frac{b_2 a_1 a_0 + a_3 a_2 b_0 - b_1 a_0 a_3}{a_2 a_1 a_0 - a_0^2 a_3} \tag{4.46}$$

From (4.40), (4.41), and (4.43), we have

$$I_n = \frac{(-1)^4}{2a_3} \frac{\begin{bmatrix} a_3 & a_1 & 0 \\ 0 & a_2 & a_0 \\ b_2 & b_1 & b_0 \end{bmatrix}}{\begin{bmatrix} a_3 & a_1 & 0 \\ 0 & a_2 & a_0 \\ a_2 & a_0 & 0 \end{bmatrix}} = \frac{1}{2a_3} \frac{b_0 a_2 a_3 + a_0 a_1 b_2 - b_1 a_0 a_3}{a_2 a_1 a_0 - a_0^2 a_3} \tag{4.47}$$

The significance of (4.40) lies in the fact (as will be discussed in the next section) that both the continuous and the discrete cases can be presented in one unified form. Hence the same computational algorithm can be used for evaluating I_n.

It is of interest to note that the inner matrix in the denominator of (4.47) is obtained by premultiplying the Hurwitz matrix by the permutation matrix:

$$\begin{bmatrix} 0 & 1 & 0 \\ 0 & 0 & 1 \\ 1 & 0 & 0 \end{bmatrix} \begin{bmatrix} a_2 & a_0 & 0 \\ a_3 & a_1 & 0 \\ 0 & a_2 & a_3 \end{bmatrix} = \begin{bmatrix} a_3 & a_1 & 0 \\ 0 & a_2 & a_0 \\ a_2 & a_0 & 0 \end{bmatrix} \tag{4.48}$$

LYAPUNOV MATRIX APPROACH

An alternate, but equivalent, method for evaluating I_n can be obtained using Lyapunov matrix equations (discussed in Chapter 3). This approach will also enable us to determine the functionals (or time moments) of system motion. The discussion will mainly follow MacFarlane [14]. The basic idea is the use of pairs of quadratic forms.

Let the free motion of a dynamical system released from an initial state be written in a matrix form as

$$\dot{\mathbf{x}} = A\mathbf{x} \qquad (4.49)$$

where A is a real constant matrix and $\mathbf{x}$ is the state vector whose components are the state variables. Let $\mathbf{x}(0)$ be the vector defining the state from which the vector is released at time $t = 0$.

Let P_1 be any symmetric matrix having real numbers, and denote the quadratic form $x'P_1x$ by $P_1(x, x)$. Define another symmetric matrix, $Q(x, x) \doteq x'Qx$, by

$$Q(x, x) = \frac{d}{dt} P_1(x, x) \qquad (4.50)$$

Then, using (4.49), we obtain

$$A'P_1 + P_1A = Q \qquad (4.51)$$

where the prime indicates the transpose (as indicated in Chapter 1). The preceding matrix equation, due to Lyapunov, is used for stability investigation, as indicated in the preceding chapter. Furthermore,

$$\int_0^T Q(x, x)\, dt = P_1[\mathbf{x}(T), \mathbf{x}(T)] - P_1[\mathbf{x}(0), \mathbf{x}(0)] \qquad (4.52)$$

and, in particular, if the system is asymptotically stable, as assumed (i.e., all the eigenvalues of A have negative real part), then $P_1[\mathbf{x}(\infty), \mathbf{x}(\infty)]$ vanishes and hence

$$\int_0^\infty Q(x, x)\, dt = -P_1[\mathbf{x}(0), \mathbf{x}(0)] \qquad (4.53)$$

This equation represents a functional similar to the earlier-discussed integral I_n. This point will be illustrated later in an example.

In order to obtain formulas for other functionals, let

$$\frac{d}{dt}[tP_1(x, x)] = P_1(x, x) + tQ(x, x) \qquad (4.54)$$

Therefore

$$[tP_1(x, x)]_0^\infty = \int_0^\infty P_1(x, x)\, dt + \int_0^\infty tQ(x, x)\, dt \qquad (4.55)$$

and for an asymptotically stable system,

$$\int_0^\infty tQ(x, x)\, dt = -\int_0^\infty P_1(x, x)\, dt \qquad (4.56)$$

The right-hand side of (4.56) is similar to (4.50) when integrated except that Q is replaced by P_1. Following the integration as in (4.52) and (4.53), and using the asymptotic stability condition, we obtain

$$\int_0^\infty tQ(x, x)\, dt = (-1)^2 P_2[x(0), x(0)] \qquad (4.57)$$

where
$$A'P_2 + P_2A = P_1 \qquad (4.58)$$

Continuing this argument, we obtain

$$\int_0^\infty t^r Q(x, x)\, dt = (-1)^{r+1} r!\, P_{r+1}[x(0), x(0)] \qquad (4.59)$$

where
$$A'P_{(s+1)} + AP_{(s+1)} = P_s$$
$$s = 1, 2, \ldots, r \qquad (4.60)$$

and
$$A'P_1 + P_1A = Q \qquad (4.61)$$

If the nth derivative of x is denoted by $x^{(n)}$, then

$$\frac{d}{dt}[x^{(n)'}P_1 x^{(n)}] = x^{(n)'}[A'P_1 + P_1A]x^{(n)}$$
$$= x^{(n)'}Qx^{(n)} \qquad (4.62)$$

and it follows that for an asymptotically stable system,

$$\int_0^\infty t^r Q[x^{(n)}, x^{(n)}]\, dt = (-1)^{r+1} r!\, P_{(r+1)}[x^{(n)}(0), x^{(n)}(0)]$$
$$= (-1)^{r+1} r!\, P_{(r+1)}[A^n x(0), A^n x(0)] \qquad (4.63)$$

The use of (4.63) requires the solution of the matrix equation (4.61) for P_1, given Q, and the use of (4.59) requires the repeated solution of (4.60) for P_2 given P_1, P_3 given P_2, and so on.

It is indicated in Chapter 3 that a necessary and sufficient condition for asymptotic stability is that $P_1(x, x)$ be positive definite for $Q = -I$ (i.e., negative definite).

SOLUTION OF THE BASIC MATRIX EQUATION (4.61)

Column vectors **p** and **q** are formed from the elements on and below the leading diagonals of symmetric matrices P_1 and Q as shown.

$$\mathbf{p} = \begin{bmatrix} p_{11} \\ p_{12} \\ p_{13} \\ \vdots \\ p_{nn} \end{bmatrix} \qquad \mathbf{q} = \begin{bmatrix} q_{11} \\ q_{12} \\ q_{13} \\ \vdots \\ q_{nn} \end{bmatrix} \tag{4.64}$$

Equation (4.61) can be manipulated to give the simpler equation

$$B\mathbf{p} = \mathbf{q} \tag{4.65}$$

where B is a matrix of order $\frac{1}{2}n(n+1)$ formed in a simple pattern from the matrix A of order n. It is shown [14] that, for asymptotic stability, the inverse of B (i.e., B^{-1}) exists and hence

$$\mathbf{p} = B^{-1}\mathbf{q} \tag{4.66}$$

and for the vectors $\mathbf{p}_1$, $\mathbf{p}_2$, etc. formed from the sequence of matrices P_1, P_2, etc., considered in (4.60),

$$\mathbf{p}_n = B^{-n}\mathbf{q} \qquad\qquad n = 1, 2, \ldots \tag{4.67}$$

After solving (4.66), P may be formed from **p** in a simple way. Multiplying out (4.61) and collecting terms, we find B to be

$$B = \begin{bmatrix} 2a_{11} & 2a_{21} & 0 & 2a_{31} & 0 & 0 & 2a_{41} & 0 & 0 & 0 & \cdots \\ a_{12} & c_{12} & a_{21} & a_{32} & a_{31} & 0 & a_{42} & a_{41} & 0 & 0 & \cdots \\ 0 & 2a_{12} & 2a_{22} & 0 & 2a_{32} & 0 & 0 & 2a_{42} & 0 & 0 & \cdots \\ a_{13} & a_{23} & 0 & c_{13} & a_{21} & a_{31} & a_{43} & 0 & a_{41} & 0 & \cdots \\ 0 & a_{13} & a_{23} & a_{12} & c_{23} & a_{32} & 0 & a_{43} & a_{42} & 0 & \cdots \\ 0 & 0 & 0 & 2a_{13} & 2a_{23} & 2a_{33} & 0 & 0 & 2a_{43} & 0 & \cdots \\ a_{14} & a_{24} & 0 & a_{34} & 0 & 0 & c_{14} & a_{21} & a_{31} & a_{41} & \cdots \\ 0 & a_{14} & a_{24} & 0 & a_{34} & 0 & a_{12} & c_{24} & a_{32} & a_{42} & \cdots \\ 0 & 0 & 0 & a_{14} & a_{24} & a_{34} & a_{13} & a_{23} & c_{34} & a_{43} & \cdots \\ 0 & 0 & 0 & 0 & 0 & 0 & 2a_{14} & 2a_{24} & 2a_{34} & 2a_{44} & \cdots \\ \cdot & \cdot & \cdot & \cdot & \cdot & \cdot & \cdot & \cdot & \cdot & \cdot & \cdots \end{bmatrix}$$

$$\tag{4.68}$$

where c_{ij} is written for $a_{ii} + a_{jj}$.

An investigation of the array for the transpose of B shows that B' may be expressed as the sum of n matrices whose elements are the elements of A in correct relative position, separated by rows and columns of zeros. One column of each of the n matrices is multiplied by 2.

For example, the scheme for $\dot{B}'$ corresponding to an A of order 3 is

$$
B' = \begin{bmatrix}
2a_{11} & a_{12} & 0 & a_{13} & 0 & 0 \\
2a_{21} & c_{12} & 2a_{12} & a_{23} & a_{13} & 0 \\
0 & a_{21} & 2a_{22} & 0 & a_{23} & 0 \\
2a_{31} & a_{32} & 0 & c_{13} & a_{12} & 2a_{13} \\
0 & a_{31} & 2a_{32} & a_{21} & c_{23} & 2a_{23} \\
0 & 0 & 0 & a_{31} & a_{32} & 2a_{33}
\end{bmatrix}
$$

$$
= \begin{bmatrix}
2a_{11} & a_{12} & 0 & a_{13} & 0 & 0 \\
2a_{21} & a_{22} & 0 & a_{23} & 0 & 0 \\
0 & 0 & 0 & 0 & 0 & 0 \\
2a_{31} & a_{32} & 0 & a_{33} & 0 & 0 \\
0 & 0 & 0 & 0 & 0 & 0 \\
0 & 0 & 0 & 0 & 0 & 0
\end{bmatrix}
$$

$$
+ \begin{bmatrix}
0 & 0 & 0 & 0 & 0 & 0 \\
0 & a_{11} & 2a_{12} & 0 & a_{13} & 0 \\
0 & a_{21} & 2a_{22} & 0 & a_{23} & 0 \\
0 & 0 & 0 & 0 & 0 & 0 \\
0 & a_{31} & 2a_{32} & 0 & a_{33} & 0 \\
0 & 0 & 0 & 0 & 0 & 0
\end{bmatrix}
$$

$$
+ \begin{bmatrix}
0 & 0 & 0 & 0 & 0 & 0 \\
0 & 0 & 0 & 0 & 0 & 0 \\
0 & 0 & 0 & 0 & 0 & 0 \\
0 & 0 & 0 & a_{11} & a_{12} & 2a_{13} \\
0 & 0 & 0 & a_{21} & a_{22} & 2a_{23} \\
0 & 0 & 0 & a_{31} & a_{32} & 2a_{33}
\end{bmatrix}
\tag{4.69}
$$

The formation of B' corresponding to an A of any order as a sum of n such matrices follows a certain rule based on the triangle array of integers. Form the triangular array of integers

$$
\begin{array}{ccccc}
1 \\
2 & 3 \\
4 & 5 & 6 \\
7 & 8 & 9 & 10 \\
11 & 12 & 13 & 14 & 15 \\
\cdots\cdots\cdots\cdots\cdots\cdots\cdots \\
\cdots\cdots\cdots\cdots\cdots\cdots\tfrac{1}{2}n(n+1)
\end{array}
\qquad (4.69a)
$$

Complete this to form the symmetrical square array

$$
\begin{array}{cccccc}
1 & 2 & 4 & 7 & 11 & \cdots \quad \cdots \\
2 & 3 & 5 & 8 & 12 & \cdots \quad \cdots \\
4 & 5 & 6 & 9 & 13 & \cdots \quad \cdots \\
7 & 8 & 9 & 10 & 14 & \cdots \quad \cdots \\
11 & 12 & 13 & 14 & 15 & \cdots \quad \cdots \\
\cdots\cdots\cdots\cdots\cdots\cdots\cdots \\
\cdots\cdots\cdots\cdots\cdots \tfrac{1}{2}n(n+1)
\end{array}
\qquad (4.70)
$$

The n matrices that are summed to form B' are obtained as follows. (a) Each of the rows of (4.70) represents one of the n-matrices. For instance, row 1 represents the first matrix, row 2 represents the second matrix, and so on. (b) In examining each of the rows and columns of (4.70), we notice that certain integers are missing. Those missing integers in the first row and column represent the rows and columns of the first matrix that consist wholly of zeros. Similarly, for the second row and column, the missing integers represent the rows and columns of the second matrix that consist wholly of zeros. Continue this process to obtain the zero rows and columns for all the n matrices. (c) Finally, the columns in the n matrices whose positions, counted from the left, correspond to the integers that lie on the main diagonal of (4.70) are to be multiplied by 2.

A similar scheme exists for the formation of B from the elements of A'. The only difference is that the rows of the elements of B are multiplied by 2, instead of the columns. For the recursive solution of (4.67), it is only necessary to find B once.

CALCULATION OF THE QUADRATIC FUNCTIONAL OF THE SYSTEM TIME RESPONSE

Denote by B_{ij} the cofactor of the element at the interaction of the ith row and jth column of B. Then we have for **p**,

$$
\begin{bmatrix} p_{11} \\ p_{12} \\ p_{22} \\ p_{13} \\ p_{23} \\ p_{33} \\ \vdots \\ p_{nn} \end{bmatrix}
= \frac{1}{|B|}
\begin{bmatrix}
B_{11} & B_{21} & B_{n(n+1)/2,1} \\
B_{12} & B_{22} & \cdot \\
B_{13} & B_{23} & \cdot \\
B_{14} & B_{24} & \cdot \\
B_{15} & B_{25} & \cdot \\
B_{16} & B_{26} & \cdot \\
\vdots & \vdots & \vdots \\
B_{1,n(n+1)/2} & B_{2,n(n+1)/2} & B_{n(n+1)/2,n(n+1)/2}
\end{bmatrix}
\begin{bmatrix} q_{11} \\ q_{12} \\ q_{22} \\ q_{13} \\ q_{23} \\ q_{33} \\ \vdots \\ q_{nn} \end{bmatrix}
\tag{4.71}
$$

If the quadratic form $Q(x, x)$ is written explicitly as

$$
Q(x, x) = q_{11}x_1^2 + q_{22}x_2^2 + \cdots + q_{nn}x_n^2 + 2q_{12}x_1x_2 + \cdots + 2q_{n-1,n}x_{n-1}x_n
\tag{4.72}
$$

it is readily seen that

$$
\int_0^\infty x_i^2 \, dt
\tag{4.73}
$$

can be calculated, using (4.72), by putting $q_{ii} = 1$ and all other elements of Q equal to zero. This establishes the connection between the time domain and the frequency domain discussed earlier, for the evaluation of the integral of the square of the signal. Similarly,

$$
P[x(0), x(0)] = p_{11}x_1^2(0) + p_{22}x_2^2(0) + \cdots + 2p_{n-1,n}x_{n-1}(0)x_n(0)
\tag{4.73a}
$$

and it can be seen that $P[x(0), x(0)]$ will reduce to $p_{jj}x_j^2(0)$ if all $x_k(0) = 0$ for $k \neq j$. It then follows, using (4.71), that when an asymptotically stable system is released from the initial condition

$$
\mathbf{x}(0) = \begin{bmatrix} 0 \\ 0 \\ \vdots \\ x_j(0) \\ 0 \\ \vdots \\ b \end{bmatrix}
\tag{4.74}
$$

we have

$$\int_0^\infty x_i^2(t)\, dt = \frac{-B_{i(i+1)/2, j(j+1)/2}}{|B|} x_j^2(0)$$

$$i, j = 1, 2, \dots, n \quad (4.75)$$

Also, using (4.59) and (4.67), and denoting the cofactor of the element at the intersection of the ith row and the jth column of B^m by $B_{ij}^{(m)}$,

$$\int_0^\infty t^r x_i^2(t)\, dt = \frac{(-1)^{r+1} r! \, B_{i(i+1)/2, j(j+1)/2}^{(r+1)}}{|B^{r+1}|} x_j^2(0)$$

$$i, j = 1, 2, \dots, n, \quad r = 1, 2, \dots \quad (4.76)$$

It is of interest to note that (4.75) is similar in form to the evaluation of I_n in (4.40). Although the formulas are obtained in different ways, both yield, as expected, the same result. This fact is illustrated in the following example.

Example 4.2. Consider the feedback system shown in Figure 4.1. Determine

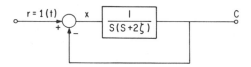

Figure 4.1. Feedback Control System.

the value of the damping ratio $\zeta > 0$ so that when the system is subjected to unit-step input $r(t)$, the following performance index is minimized.

$$I_n = \int_0^\infty x^2(t)\, dt \quad (4.77)$$

Solution: We will solve this problem using the frequency-domain approach [i.e., (4.40)] and the time-domain approach using (4.75).

We first obtain from Figure 4.1 the Laplace transform of $x(t)$ as

$$X(s) = \frac{(s + 2\zeta)}{s^2 + 2\zeta s + 1} \quad (4.78)$$

Using (4.10) and (4.11), we have

$$g(s) = -s^2 + 4\zeta^2$$

$$h(s) = s^2 + 2\zeta s + 1 \qquad n = 2 \quad (4.79)$$

From (4.42) and (4.40), we obtain

$$I_2 = \int_0^\infty x^2(t)\, dt = \frac{(-1)\begin{bmatrix} -1 & 4\zeta^2 \\ 1 & 1 \end{bmatrix}}{2\begin{bmatrix} 2\zeta & 0 \\ 1 & 1 \end{bmatrix}} = \frac{1 + 4\zeta^2}{4\zeta} = \zeta + \frac{1}{4\zeta} \quad (4.80)$$

The optimal value of ζ is obtained when

$$\frac{\partial I_2}{\partial \zeta} = 0 \quad (4.81)$$

or

$$\zeta = \frac{1}{2} \quad (4.82)$$

Using the time-domain approach, we need to minimize [from (4.53)]

$$J = \int_0^\infty x'(t) Q x(t)\, dt \quad (4.83)$$

where

$$\mathbf{x} = \begin{bmatrix} x_1 \\ x_2 \end{bmatrix} = \begin{bmatrix} x \\ \dot{x} \end{bmatrix} \qquad Q = \begin{bmatrix} 1 & 0 \\ 0 & 0 \end{bmatrix} \quad (4.84)$$

The system is assumed to be at rest initially. From Figure 4.1, we obtain

$$\ddot{c} + 2\zeta\dot{c} + c = r \quad (4.85)$$

Noting that $x = r - c$, $r(t) = 1(t)$, we have

$$\ddot{x} + 2\zeta\dot{x} + x = 0 \qquad\qquad t > 0 \quad (4.86)$$

The state-space representation is

$$\begin{bmatrix} \dot{x}_1 \\ \dot{x}_2 \end{bmatrix} = \begin{bmatrix} 0 & 1 \\ -1 & -2\zeta \end{bmatrix}\begin{bmatrix} x_1 \\ x_2 \end{bmatrix} \qquad \begin{bmatrix} x_1(0) \\ x_2(0) \end{bmatrix} = \begin{bmatrix} 1 \\ 0 \end{bmatrix} \quad (4.87)$$

or

$$\dot{x} = Ax \quad (4.88)$$

where A is given in a companion form as

$$A = \begin{bmatrix} 0 & 1 \\ -1 & -2\zeta \end{bmatrix} \quad (4.89)$$

Since A is a stable system (note $\zeta > 0$), the value of I_2 from (4.53) is

$$J = x'(0) P x(0) = \int_0^\infty x_1^2(t)\, dt \qquad x_1(0) = 1 \quad (4.90)$$

To avoid the evaluation of the elements of P, we use (4.75) directly. From (4.68), we have

$$B = \begin{bmatrix} 2a_{11} & 2a_{21} & 0 \\ a_{12} & a_{11} + a_{22} & a_{21} \\ 0 & a_{12} & 2a_{22} \end{bmatrix} = \begin{bmatrix} 0 & -2 & 0 \\ 1 & -2\zeta & -1 \\ 0 & 2 & -4\zeta \end{bmatrix} \tag{4.91}$$

Hence

$$|B| = -8\zeta \tag{4.92}$$

The cofactor $B_{1,1}$ is $8\zeta^2 + 2$. Hence from (4.75),

$$J = \int_0^\infty x_1^2(t)\, dt = \int_0^\infty x^2(t)\, dt = \zeta + \frac{1}{4\zeta} \tag{4.93}$$

which is the same as (4.80) for I_2.

To generate the matrix B, we first obtain B' by using the generating rule based on the triangle arrays, and symmetrical square as indicated in (4.69a) and (4.70).

Triangular array   **Symmetrical square array**

$$\begin{array}{ccc} 1 & & \\ 2 & 3 & \\ 4 & 5 & 6 \end{array} \qquad \begin{array}{ccc} 1 & 2 & 4 \\ 2 & 3 & 5 \\ 4 & 5 & 6 \end{array} \tag{4.94}$$

$$B' = \begin{bmatrix} 0 & 1 & 0 \\ -2 & -2\zeta & 0 \\ 0 & 0 & 0 \end{bmatrix} + \begin{bmatrix} 0 & 0 & 0 \\ 0 & 0 & 2 \\ 0 & -1 & -4\zeta \end{bmatrix} = \begin{bmatrix} 0 & 1 & 0 \\ -2 & -2\zeta & 2 \\ 0 & -1 & -4\zeta \end{bmatrix} \tag{4.95}$$

or

$$B = \begin{bmatrix} 0 & -2 & 0 \\ 1 & -2\zeta & -1 \\ 0 & 2 & -4\zeta \end{bmatrix} \tag{4.96}$$

From the symmetrical square array of (4.94), we notice that the integer 3 is missing from the first row and column, and thus the entries of the third column and third row of the first matrix of B' are zeros. Similarly, the integer 1 is missing from row 2 and column 2 of the array. Hence the first column and first row of the second matrix in B' are zero. The diagonal integer numbers of the square array applicable to the two matrices are 1 and 3. Hence the first column of the first matrix and the third column of the second matrix are multiplied by 2.

It may be noted, in concluding this section, that if A is given in a companion form, then B, and hence (4.75), can be simplified. Moreover, one can also obtain formulas for the time moments functions in the frequency domain. This point has been briefly discussed in the literature [8,27,28]. It appears that for higher-order moments (or functionals), the time-domain approach is more straightforward computationally.

4.2 EVALUATION OF INFINITE SUM OF SQUARE OF SIGNALS, DISCRETE CASE

Similar to the continuous time case, the infinite sum of square of signals can be represented in terms of a complex integral, using Parseval's theorem [16],

$$I_n = \frac{1}{2\pi j} \oint_{\substack{\text{unit} \\ \text{circle}}} F(z)F(z^{-1})z^{-1}\,dz = \sum_{k=0}^{\infty} f^2(kT) \qquad (4.97)$$

where $F(z)$ is the z-transform of $f(nT)$.

The proof of this formula, similar to the continuous case, can be obtained as a special case from the complex convolution formula as follows. Let

$$F(z) = \mathscr{Z}[f(nT)] \qquad G(z) = \mathscr{Z}[g(nT)]$$

where both are analytic for $|z| > R_0$. Then $H(z) = \mathscr{Z}[fg]$ is given as [16]

$$H(z) = \frac{1}{2\pi j} \oint_{|p|=R_1} F(p)G\left(\frac{z}{p}\right)\frac{dp}{p} = \sum_{k=0}^{\infty} f(kT)g(kT)z^{-k} \qquad (4.98)$$

for $|z| > R_0 R_1$, where $R_1 > R_0$. (The path of integration can be chosen as a simple closed curve Γ.)

If we assume that $g(nT) = f(nT)$ and let $z = 1$ in (4.98), we obtain

$$I_n = \sum_{k=0}^{\infty} f^2(kT) = \frac{1}{2\pi j} \oint_{\substack{\text{unit} \\ \text{circle}}} F(z)F(z^{-1})\frac{dz}{z} \qquad (4.99)$$

Here it is assumed that the poles of $F(z)$ are all inside the unit circle.

To evaluate the integral in (4.99), we use the residue method and, later, the state-space representation. The derivation of I_n is based on obtaining [16] the right side of (4.99) as a sum of residues. However, instead of using the explicit expressions for the residues, we obtain a set of $2n + 1$ linear equations, with the sum of the residues as one variable. Because of symmetry, this set reduces to $n + 1$ equations.

Assuming that $F(z)$ is given,

$$F(z) = \frac{B(z)}{A(z)} = \frac{\sum_{i=0}^{n} b_i z^i}{\sum_{i=0}^{n} a_i z^i} = K^2 \frac{\prod_{i=1}^{n}(z - q_i)}{\prod_{i=1}^{n}(z - p_i)} \qquad (4.100)$$

Since the integral of (4.97) is solved by the method of residues, as mentioned above, we are interested in

$$\frac{zA(z)}{z - p_k} = \sum_{i=0}^{n} c_{i,k} z^i \quad k = 0, 1, 2, \ldots, n \quad (4.101)$$

Since we assume p_0 to be the location of the pole at the origin of the z-plane, we have

$$c_{0,k} = 0 \qquad \text{when } k \neq 0 \quad (4.102)$$
$$c_{0,0} = a_0$$

We also have

$$\frac{zA(z-1)}{z - p_k^{-1}} = \frac{zA(z^{-1})p_k}{z(p_k - z^{-1})} = -zp_k \frac{z^{-1}A(z^{-1})}{z^1 - p_k} = -zp_k \sum_{i=0}^{n} c_{i,k} z^{-i}$$

$$(4.103)$$

and

$$\frac{zA(z)A(z^{-1})}{z - p_k} = \sum_{j=0}^{n} z^j \sum_{i=0}^{n} c_{j+1,k} a_i + \sum_{j=1}^{n} z^{-j} \sum_{i=0}^{n} c_{i,k} a_{i-j} \quad (4.104)$$

$$\frac{zA(z)A(z^{-1})}{z - p_k^{-1}} = -zp_k \left[\sum_{j=0}^{n} z^j \sum_{i=0}^{n} c_{i,k} a_{i-j} + \sum_{j=1}^{n} z^{-j} \sum_{i=0}^{n} c_{j+i,k} a_i \right] \quad (4.105)$$

By partial fraction expansion,

$$\frac{B(z)B(z^{-1})}{zA(z)A(z^{-1})} = \sum_{k=0}^{n} \frac{R_{p_k}}{z - p_k} + \sum_{k=1}^{n} \frac{R_{p_k}^{-1}}{z - p_k^{-1}} \quad (4.106)$$

where R_{p_k} equals residue at p_k and $R_{p_k}^{-1}$ equals residue at p_k^{-1}.

Until now we have assumed only simple poles. By the continuity argument, the final results are also valid for multiple poles.

For $k \neq 0$, we have

$$zA(z)A(z^{-1}) = z(z - p_k)(z^{-1} - p_k)[A'(p_k)A'(p_k^{-1})]_{p_k = z}$$
$$= -p_k(z - p_k)(z - p_k^{-1})[A'(p_k)A'(p_k^{-1})]_{p_k = z} \quad (4.107)\dagger$$

and $$R_{p_k} = \frac{B(p_k)B(p_k^{-1})}{-p_k(p_k - p_k^{-1})A'(p_k)A'(p_k^{-1})} = -R_{p_k}^{-1} \quad (4.108)$$

Combining (4.104), (4.105), (4.106), and (4.108), we obtain

$$B(z)B(z^{-1}) = \sum_{k=0}^{n} \sum_{j=0}^{n} \sum_{i=0}^{n} z^j R_{p_k} c_{j+i,k} a_i$$

$$+ \sum_{k=0}^{n} \sum_{j=1}^{n} \sum_{i=0}^{n} z^{-j} R_{p_k} c_{i,k} a_{i-j}$$

$\dagger$ A' denotes the derivative of A with respect to z.

$$+ z \sum_{k=1}^{n} \sum_{j=0}^{n} \sum_{i=0}^{n} z^{j} R_{p_k} p_k c_{i,k} a_{i-j}$$

$$+ z \sum_{k=1}^{n} \sum_{j=1}^{n} \sum_{i=0}^{n} z^{-j} R_{p_k} p_k c_{i+j,k} a_{i} \qquad (4.109)$$

We can also write

$$B(z)B(z^{-1}) = \sum_{j=0}^{n} z^{j} \sum_{i=0}^{n} b_{i+j} b_{i} + \sum_{j=1}^{n} z^{-j} \sum_{i=0}^{n} b_{i-j} b_{i} \qquad (4.110)$$

Let

$$M_i = \sum_{k=0}^{n} c_{i,k} R_{p_k} \qquad (4.111)$$

and

$$Q_i = \sum_{k=1}^{n} c_{i,k} R_{p_k} p_k \qquad (4.112)$$

From (4.101), we have

$$c_{nk} = a_n \quad (k = 0, 1, \ldots, n) \quad \text{and} \quad c_{1k} = -\frac{a_0}{p_k} \quad (k = 1, 2, \ldots, n) \qquad (4.113)$$

Then

$$M_n = a_n \sum_{k=0}^{n} R_{p_k} = I_n a_n \qquad (4.114)$$

In addition,

$$Q_1 = -a_0 \sum_{k=1}^{n} R_{p_k} = -a_0 I_n + a_0 R_0 \qquad (4.115)$$

since

$$c_{0k} = 0 \qquad \text{when } k \neq 0 \quad (4.116)$$

Then

$$Q_0 = 0$$

Combining (4.109) and (4.110), and using the preceding relationship, we have

$$\sum_{j=0}^{n} z^{j} \sum_{i=0}^{n} b_{i+j} b_{i} + \sum_{j=1}^{n} z^{-j} \sum_{i=0}^{n} b_{i-j} b_{i}$$

$$= \sum_{j=0}^{n} z^{j} \sum_{i=0}^{n} M_{j+i} a_{i} + \sum_{j=1}^{n} z^{-j} \sum_{i=0}^{n} M_{i} a_{i-j}$$

$$+ \sum_{j=0}^{n} z^{j+1} \sum_{i=1}^{n} Q_{i} a_{i-j} + \sum_{j=1}^{n} z^{-j+1} \sum_{i=0}^{n} Q_{i+j} a_{i} \qquad (4.117)$$

Equating the coefficients of z^0 in (4.117), we get

$$\sum_{i=0}^{n} b_{i}^{2} = \sum_{i=0}^{n} M_{n-i} a_{n-i} + \sum_{i=0}^{n} a_{n-i} Q_{n-i+1} \qquad (4.118)$$

We notice from (4.117) that the coefficient of z^j is equal to the coefficient of z^{-j}. (Note that $\sum_{i=0}^{n} b_i b_{i+j} = \sum_{i=0}^{n} b_i b_{i-j}$.)

Thus, by summing the equations for j and $-j$,

$$2 \sum_{i=0}^{n} b_i b_{i+j} = \sum_{i=0}^{n} M_{j+n-i} a_{n-i} + \sum_{i=0}^{n} M_{n-i} a_{n-i-j}$$

$$+ \sum_{i=0}^{n-1} Q_{n-i} a_{n-i-j+1} + \sum_{i=0}^{n} Q_{n-i+j+1} a_{n-i}$$

$$= \sum_{i=0}^{n} a_i (M_{j+n-i} + Q_{n-i+j+1})$$

$$+ \sum_{i=0}^{n} a_{i-j} (M_{n-i} + Q_{n-i+1})$$

$$\text{for } j = 1, 2, 3, \ldots, n \quad (4.119)$$

where $a_i = 0$, when $n < i$ and $0 > i$.

It may be noted that (4.119) also yields (4.118) when $j = 0$. Equation (4.119) can also be written as

$$2 \sum_{i=0}^{n} b_i b_{i+j} = \sum_{i=0}^{n} a_{n-i-j} (M_{n+i} + Q_{n-i+1}) + \sum_{i=0}^{n} a_{n-i-j} (M_{n-i} + Q_{n-i+1})$$

$$= \sum_{i=0}^{n} (a_{n-i-j} + a_{n-i+j})(M_{n-i} + Q_{n-i+1})$$

$$\text{for } j = 1, 2, \ldots, n \quad (4.120)$$

Because we assumed that all poles of $F(z)$ are inside the unit circle (hence stability), the $n + 1$ simultaneous linear equations (4.118) and (4.120) can be written as the vector equation

$$\Omega \mathbf{m} = \mathbf{d} \quad (4.121)$$

where Ω is nonsingular and $\mathbf{m}$ and $\mathbf{d}$ are the vectors

$$
\begin{bmatrix}
M_n \\
M_{n-1} + Q_n \\
M_{n-2} + Q_{n-1} \\
\vdots \\
M_0 + Q_1
\end{bmatrix}
\quad \text{and} \quad
\begin{bmatrix}
\sum_{i=0}^{n} b_i^2 \\
2 \sum b_i b_{i+1} \\
2 \sum b_i b_{i+2} \\
\vdots \\
2 b_n b_0
\end{bmatrix}
\quad (4.122)
$$

and Ω is the matrix

$$\Omega = \begin{bmatrix} a_n & a_{n-1} & a_{n-2} & a_{n-3} & \cdots & a_0 \\ a_{n-1} & a_n + a_{n-2} & a_{n-1} + a_{n-3} & a_{n-2} + a_{n-4} & \cdots & a_1 \\ a_{n-2} & a_{n-3} & a_n + a_{n-4} & a_{n-1} + a_{n-5} & \cdots & a_2 \\ \vdots & \vdots & \vdots & & & \vdots \\ a_0 & 0 & 0 & & \cdots & a_n \end{bmatrix}$$

(4.123)

Since, from (4.114),

$$I_n = \frac{M_n}{a_n} \tag{4.124}$$

then [16],

$$I_n = \frac{|\Omega_1|}{a_n |\Omega|} \tag{4.125}$$

where Ω_1 is the matrix formed from Ω by replacing the first column by the vector **d**.

We can reformulate I_n in (4.125) in a simpler form, one related to the stability condition within the unit circle, as will be discussed in the next section. Also, this form is presented directly as an innerwise matrix, as for the continuous case. By so doing, we can use the same computational algorithm for both the continuous and the discrete cases.

To present I_n as a ratio of determinants of two innerwise matrices, we define the following two triangular matrices.

$$X_{n+1} = \begin{bmatrix} a_n & a_{n-1} & \cdots & & a_0 \\ 0 & a_n & a_{n-1} & \cdots & a_1 \\ \vdots & & & & \vdots \\ 0 & \cdots & & & a_n \end{bmatrix}$$

$$Y_{n+1} = \begin{bmatrix} 0 & \cdots & 0 & \cdots & a_0 \\ 0 & \cdots & 0 & \cdots & a_1 \\ \vdots & & \vdots & & \vdots \\ 0 & a_0 & a_1 & \cdots & a_{n-1} \\ a_0 & a_1 & a_2 & \cdots & a_n \end{bmatrix}$$

(4.126)

Let Δ_{n+1}^+ be defined as

$$\Delta_{n+1}^+ = X_{n+1} + Y_{n+1} \tag{4.127}$$

then

$$|\Delta_{n+1}^+| = |X_{n+1} + Y_{n+1}| \tag{4.128}$$

By examining the matrices Ω of (4.123) and $X_{n+1} + Y_{n+1}$ in (4.127), we can readily arrive at the following result:

$$|\Delta_{n+1}^+| = |X_{n+1} + Y_{n+1}| = 2|\Omega| \tag{4.129}$$

Furthermore, the numerator $|\Omega_1|$ of (4.125) is related to $|\Delta_{n+1}^+|$ as follows:

$$|X_{n+1} + Y_{n+1}|_b = 2|\Omega_1| \tag{4.130}$$

where the matrix $[X_{n+1} + Y_{n+1}]_b$ is formed from $X_{n+1} + Y_{n+1}$ by replacing the last row by

$$\left[2b_n b_0, 2\sum b_i b_{i+n-1}, \ldots, 2\sum b_i b_{i+1}, 2\sum_{i=0}^{n} b_i^2 \right] \tag{4.131}$$

In view of the preceding, I_n of (4.125) can be reformulated as

$$I_n = \frac{|X_{n+1} + Y_{n+1}|_b}{a_n |X_{n+1} + Y_{n+1}|} \tag{4.132}\dagger$$

In order to illustrate the relationships (4.129) and (4.130), we present the following example: Let $n = 4$ in (4.100). From (4.125), we have

$$I_n = \frac{|\Omega_1|}{a_n |\Omega|} \tag{4.133}$$

where Ω and Ω_1 are obtained from (4.123) and (4.125) as follows:

$$\Omega = \begin{bmatrix} a_4 & a_3 & a_2 & a_1 & a_0 \\ a_3 & a_4 + a_2 & a_3 + a_1 & a_2 + a_0 & a_1 \\ a_2 & a_1 & a_4 + a_0 & a_3 & a_1 \\ a_1 & a_0 & 0 & a_4 & a_3 \\ a_0 & 0 & 0 & 0 & a_4 \end{bmatrix} \tag{4.134}$$

† As will be shown in Section 4.3, this equation can be alternatively written as

$$I_n = \frac{(-1)^n |X_{n+1} + Y_{n+1}|_b}{2a_n A(1) A(-1) |\Delta_{n-1}^-|}$$

and

$$\Omega_1 = \begin{vmatrix} \sum_{i=0}^{4} b_i^2 & a_3 & a_2 & a_1 & a_0 \\ 2\sum_{i=0}^{4} b_i b_{i+1} & a_4 + a_2 & a_3 + a_1 & a_2 + a_0 & a_1 \\ 2\sum_{i=0}^{4} b_i b_{i+2} & a_1 & a_4 + a_0 & a_3 & a_2 \\ 2\sum_{i=0}^{4} b_i b_{i+3} & a_0 & 0 & a_4 & a_3 \\ 2b_n b_0 & 0 & 0 & 0 & a_4 \end{vmatrix} \qquad (4.135)$$

Multiply the first row of (4.134) by 2 and rotate the 5×5 matrix Ω clockwise by $90°$ (this would not change the sign of the determinant) to obtain

$$2\Omega = \begin{vmatrix} a_0 & a_1 & a_2 & a_3 & 2a_4 \\ 0 & a_0 & a_1 & a_4 + a_2 & 2a_3 \\ 0 & 0 & a_4 + a_0 & a_3 + a_1 & 2a_2 \\ 0 & a_4 & a_3 & a_2 + a_0 & 2a_1 \\ a_4 & a_3 & a_2 & a_1 & 2a_0 \end{vmatrix} \qquad (4.136)$$

Exchanging rows in (4.136) several times, and without changing the sign of the determinant, we obtain

$$2\Omega = \begin{vmatrix} a_4 & a_3 & a_2 & a_1 & 2a_0 \\ 0 & a_4 & a_3 & a_2 + a_0 & 2a_1 \\ 0 & 0 & a_4 + a_0 & a_3 + a_1 & 2a_2 \\ 0 & a_0 & a_1 & a_4 + a_2 & 2a_3 \\ a_0 & a_1 & a_2 & a_3 & 2a_4 \end{vmatrix} \qquad (4.137)$$

Equation (4.137) is immediately recognizable from (4.126) as $\Delta_5^+ = X_5 + Y_5$. Hence

$$|X_5 + Y_5| = 2|\Omega| \qquad (4.138)$$

For other cases of n, rotation by 90°(clockwise) involves a sign change that is canceled in the several row exchanges needed to bring the matrix to the form $X_{n+1} + Y_{n+1}$.

Similarly, in order to obtain the numerator of I_n in (4.133) we trace the first column of (4.135) to the final form of $X_{n+1} + Y_{n+1}$, giving

$$|X_5 + Y_5|_b = 2|\Omega_1| \tag{4.139}$$

Note that the first column (with the first entry multiplied by 2) is traced to the last row in $X_5 + Y_5$. Similar relationships for obtaining n-odd exist.

LYAPUNOV MATRIX APPROACH

As with the continuous case, we can use the discrete Lyapunov matrix equation to determine I_n as well as the functional of the discrete time moments.† The discussion will mainly follow that of Man [24] and is closely related to MacFarlane s work [14].

We are interested in evaluating the time-weighted quadratic performance index of the form

$$J_r = \sum_{k=0}^{n} k^r \mathbf{x}_k' s_k \mathbf{x}_k \tag{4.140}$$

associated with the linear discrete system given by

$$\mathbf{x}_{k+1} = A\mathbf{x}_k \tag{4.141}$$

It is assumed (as in the continuous case) that A is an $n \times n$ constant convergent matrix (i.e., all eigenvalues are inside the unit circle) and s_0 is an $n \times n$ constant positive (or nonnegative) definite symmetric matrix.

With the use of (4.141), one can write (4.140) as

$$J_r = \mathbf{x}_0' \left[\sum_{k=0}^{\infty} k^r A'^k s_0 A^k \right] \mathbf{x}_0 \tag{4.142}$$

When $r = 0$, (4.142) can be immediately evaluated as

$$I_n = J_0 = \mathbf{x}_0' \left[\sum_{k=0}^{\infty} A'^k s_0 A^k \right] \mathbf{x}_0 = \mathbf{x}_0' s_1 \mathbf{x}_0 \tag{4.143}$$

where the $n \times n$ constant positive (nonnegative) definite symmetric matrix s_1 satisfies the matrix equation

$$s_1 - A' s_1 A = s_0 \tag{4.144}$$

† An extension for higher moments based on the complex-integral approach for the discrete case was discussed in the following article: M. H. Hamza and M. E. Ramsy, "Evaluation of time-weighted total square integrals," *Electronics Letters*, **7** (11) (May 3, 1971). See also Ref. 7.

Using the value

$$
s_0 = \begin{bmatrix} 1 & 0 & \cdots & 0 \\ 0 & 0 & \cdots & 0 \\ 0 & & \cdots & 0 \end{bmatrix}
\tag{4.145}
$$

in (4.143), we obtain for the I_n the required sum

$$
I_n = \sum_{k=0}^{\infty} \mathbf{x}_1^2(kT) = \mathbf{x}_0' s_1 \mathbf{x}_0
\tag{4.146}
$$

To evaluate (4.142) for $r \geq 1$, expand the right-hand side and factor out terms in groups of $\sum_{k=0}^{\infty} A'^k S_0 A^k$. Equation (4.142) can be written as

$$
J_r = \mathbf{x}_0' \left[\sum_{i=1}^{\infty} \sum_{k=0}^{\infty} c_i A'^i (A'^k s_0 A^k) A^i \right] \mathbf{x}_0
\tag{4.147}
$$

where the coefficients c_i are given recursively as

$$
c_i = i^r - \sum_{\ell=1}^{i-1} c_\ell
\tag{4.148}
$$

Also, by simple calculation, the above can be written

$$
c_i = i^r - (i-1)^r = \sum_{j=1}^{n} (-1)^{j+1} \binom{r}{j} i^{r-j}
\tag{4.149}
$$

and with the use of (4.143), (4.144), and (4.148), one can simplify (4.147) to

$$
J_r = \mathbf{x}_0' \left[\sum_{i=1}^{\infty} \sum_{j=1}^{r} (-1)^{j+1} \binom{r}{j} i^{r-1} A'^i s_1 A^i \right] \mathbf{x}_0
\tag{4.150}
$$

This equation has the same form as (4.142) except that the powers of the time factors in (4.150) are at least one lower. Hence (4.150) can be evaluated for $r = 1$ in a closed form in the same way as (4.143) and (4.144). Utilizing the development for $r = 1$, one obtains J_r successively.

Example 4.3.   Let $r = 1$. From (4.142) and (4.150), it follows that

$$
J_1 = \mathbf{x}_0' \left[\sum_{k=0}^{\infty} k A'^k s_0 A^k \right] \mathbf{x}_0
$$

$$
= \mathbf{x}_0' \left[\sum_{i=1}^{\infty} A'^i s_1 A^i \right] \mathbf{x}_0
$$

$$
= \mathbf{x}_0' \left[\sum_{i=0}^{\infty} A'^i s_1 A^i - s_1 \right] \mathbf{x}_0
\tag{4.151}
$$

Noting (4.143) and (4.144), the above can be simplified to

$$J_1 = x_0'(s_2 - s_1)x_0 \qquad (4.152)$$

where s_2 is given by

$$s_2 - A's_2A = s_1 \qquad (4.153)$$

Let

$$r = 2$$

From (4.142) and (4.150) it follows that

$$
\begin{aligned}
J_2 &= x_0'\left[\sum_{k=0}^{\infty} k^2 A'^k s_0 A^k\right]x_0 \\
&= x_0'\left[\sum_{i=1}^{\infty} (2i-1)A'^i s_1 A^i\right]x_0 \\
&= x_0'\left[2\sum_{i=0}^{\infty} iA'^i s_1 A^i - \sum_{i=0}^{\infty} A'^i s_1 A^i + s_1\right]x_0 \qquad (4.154)
\end{aligned}
$$

In view of (4.143), (4.144), and (4.151) to (4.153), the above can be simplified to

$$
\begin{aligned}
J_2 &= x_0'[2(s_3 - s_2) - s_2 + s_1]x_0 \\
&= x_0'[2s_3 - 3s + s_1]x_0 \qquad (4.155)
\end{aligned}
$$

where s_3 is given by

$$s_3 - A's_3A = s_2 \qquad (4.156)$$

Table 4.1 Evaluation of J_r in (4.157) for $r = 4, 5, 6, 7, 8, 9, 10$

$J_4 = x_0'[24s_5 - 60s_4 + 50s_3 - 15s_2 + s_1]x_0$

$J_5 = x_0'[120s_6 - 360s_5 + 390s_4 - 180s_3 + 31s_2 - s_1]x_0$

$J_6 = x_0'[720s_7 - 2520s_6 + 3360s_5 - 2100s_4 + 602s_3 - 63s_2 + s_1]x_0$

$J_7 = x_0'[5040s_8 - 20160s_7 + 31920s_6 - 25200s_5 + 10206s_4 - 1932s_3 + 127s_2 - s_1]x_0$

$J_8 = x_0'[40320s_9 - 181440s_8 + 332640s_7 - 317520s_6 + 166824s_5 - 46620s_4$
$\qquad + 6050s_3 - 255s_2 + s_1]x_0$

$J_9 = x_0'[362880s_{10} - 1814400s_9 + 3780000s_8 - 4233600s_7 + 2739240s_6$
$\qquad - 1020600s_5 + 204630s_4 - 18660s_3 + 511s_2 - s_1]x_0$

$J_{10} = x_0'[362880s_{11} - 19958400s_{10} + 46569600s_9 - 59875200s_8 + 46070640s_7$
$\qquad - 21538440s_6 + 5921520s_5 - 877500s_4 + 57002s_3 - 1023s_2 + s_1]x_0$

Proceeding in the same way, similar expressions for J_r for higher values of r† can be derived. It can be concluded that J_r in (4.142) can be written

$$J_r = x_0'\left[\sum_{j=1}^{r+1} b_j s_j\right]x_0 \qquad (4.157)$$

† For $r = 3$, $J_3 = x_0'[6s_4 - 12s_3 + 7s_2 - s_1]x_0$, where $s_4 - A's_4A = s_3$. In Table 4.1, J_r is evaluated for values of r up to 10.

where s_j is given recursively by

$$s_j - A's_jA = s_{j-1} \tag{4.158}$$

The coefficients b_j† can be determined sequentially for $r = 1, 2, 3, \ldots$.

Before concluding this section, it can be remarked that the discrete case for I_n can be also obtained from the continuous case by using the bilinear transformation [19]. However, the direct method introduced in this section is computationally more straightforward and simpler.

4.3 RELATIONSHIP TO STABILITY TESTS [29]

It is remarked in the preceding two sections that the meaningful evaluation of I_n hinges on the stability for both the continuous and the discrete systems. Consequently, the value of I_n is always positive, as it should be for stable systems.

In this section we indicate that stability and evaluation of I_n can be performed simultaneously. If the stability test fails, then the computation of I_n can be terminated.

For the continuous case, the relationship of I_n in (4.40) to stability is rather straightforward. It is given as follows.

1. All the a_i's in (4.12) should be positive.
2. Δ_n should be positive innerwise (pi).

It is recognized that the preceding two conditions are exactly the Liénard–Chipart [30] criterion presented in Section 2.1.

Similarly, for the evaluation of (4.53), we obtain the matrix P_1 from the solution of the Lyapunov matrix equation (4.51) for stability. Also, the matrix B in (4.75) should be tested for stability in the evaluation of the left-hand side of this equation. Indeed, for the example discussed in Section 4.1, the matrix B should be negative innerwise for stability.

For the discrete case, the relationship of I_n in (4.132) to stability is not obtained. Hence in this section we will derive such a relationship. We will state first this relationship and show it for $n = 4$, and indicate later on its derivation [31].

The relationship between the denominator matrix of (4.132) and the stability matrix discussed in Section 2.3, is given by

$$\Delta_{n+1}^+ = X_{n+1} + Y_{n+1} = 2(-1)^n A(1)A(-1)\,\Delta_{n-1}^- \tag{4.159}$$

where $\qquad A(1) = A(z)|_{z=1} \qquad$ and $\qquad A(-1) = A(z)|_{z=-1} \tag{4.160}$

is obtained from (4.100).

† A simple pattern for generating these coefficients exists in the literature [33].

Also, from Chapter 2, we know that $\Delta_{n-1}^- = X_{n-1} - Y_{n-1}$. To show the identity in (4.159), we present the following case. Let $n = 4$. Then from (4.126),

$$X_5 + Y_5 = 2 \begin{bmatrix} a_4 & a_3 & a_2 & a_1 & a_0 \\ 0 & a_4 & a_3 & a_2 + a_0 & a_1 \\ 0 & 0 & a_4 + a_0 & a_3 + a_1 & a_2 \\ 0 & a_0 & a_1 & a_4 + a_2 & a_3 \\ a_0 & a_1 & a_2 & a_3 & a_4 \end{bmatrix} \tag{4.161}$$

Subtract columns 2 and 4 from column 3 and then add columns 1 and 5 to column 3 to obtain

$$X_5 + Y_5 = 2 \begin{bmatrix} a_4 & a_3 & A(-1) & a_1 & a_0 \\ 0 & a_4 & -A(-1) & a_2 + a_0 & a_1 \\ 0 & 0 & A(-1) & a_3 + a_1 & a_2 \\ 0 & a_0 & -A(-1) & a_4 + a_2 & a_3 \\ a_0 & a_1 & A(-1) & a_3 & a_4 \end{bmatrix} \tag{4.162}$$

Add row 3 to rows 2 and 4 and subtract it from rows 1 and 5 to obtain, after factoring out $A(-1)$,

$$X_5 + Y_5 = 2A(-1) \begin{bmatrix} a_4 & a_3 & -a_3 & a_0 - a_2 \\ 0 & a_4 & a_3 + a_2 + a_1 + a_0 & a_1 + a_2 \\ 0 & a_0 & a_4 + a_3 + a_2 + a_1 & a_2 + a_3 \\ a_0 & a_1 & -a_1 & a_4 - a_2 \end{bmatrix} \tag{4.163}$$

Subtract column 1 from 2 and column 4 from 3, to obtain

$$X_5 + Y_5 = 2A(-1) \begin{bmatrix} a_4 & a_3 - a_4 & -a_3 - a_0 + a_2 & a_0 - a_2 \\ 0 & a_4 & a_3 + a_0 & a_1 & a_2 \\ 0 & a_0 & a_4 + a_1 & a_2 + a_3 \\ a_0 & a_1 - a_0 & -a_1 - a_4 + a_2 & a_4 - a_2 \end{bmatrix} \tag{4.164}$$

Add row 2 to row 1 and row 3 to row 4 to obtain

$$X_5 + Y_5 = 2A(-1) \begin{bmatrix} a_4 & a_3 & a_2 & a_1 + a_0 \\ 0 & a_4 & a_3 + a_0 & a_2 + a_1 \\ 0 & a_0 & a_1 + a_4 & a_2 + a_3 \\ a_0 & a_1 & a_2 & a_3 + a_3 \end{bmatrix}$$

$$= 2A(-1) \left\{ \begin{bmatrix} a_4 & a_3 & a_2 & a_1 \\ 0 & a_4 & a_3 & a_2 \\ 0 & 0 & a_4 & a_3 \\ 0 & 0 & 0 & a_4 \end{bmatrix} + \begin{bmatrix} 0 & 0 & 0 & a_0 \\ 0 & 0 & a_0 & a_1 \\ 0 & a_0 & a_1 & a_2 \\ a_0 & a_1 & a_2 & a_3 \end{bmatrix} \right\} \tag{4.165}$$

The matrices in the right-hand side of (4.165) are recognized as $X_4 + Y_4 = \Delta_4^+$. Hence

$$X_5 + Y_5 = 2A(-1)(X_4 + Y_4) \tag{4.166}$$

It is proven elsewhere [16] that the following identity holds in general:

$$X_n + Y_n = A(1)(X_{n-1} - Y_{n-1}) = A(1)\,\Delta_{n-1}^- \tag{4.167}$$

In the particular case we have

$$X_4 + Y_4 = A(1)(X_3 - Y_3) = A(1)\,\Delta_3^- \tag{4.168}$$

Using (4.168) in (4.166), we obtain

$$\Delta_5^+ = X_5 + Y_5 = 2A(-1)A(1)(X_3 - Y_3) = 2A(-1)A(1)\,\Delta_3^- \tag{4.169}$$

Now when n is odd we follow the same procedure as for $n = 4$ and obtain the general relationship [31] indicated in (4.159).

Based on the preceding discussion and on Section 2.3, we can state the connection to stability as follows. The system that has $A(z)$ as its characteristic equation is stable if and only if

1. Δ_{n-1} is positive innerwise (pi).
2. For n-odd, $n \triangleq 2m - 1$,

either

$$\alpha_{2i} > 0 \qquad \alpha_{2m-1} > 0 \qquad \text{or} \qquad \alpha_{2i+1} > 0 \qquad \alpha_0 > 0$$
$$i = 0, 1, \ldots, m - 1 \tag{4.170}$$

Note that

$$\alpha_{2m-1} = A(1) \qquad \text{and} \qquad \alpha_0 = -A(-1)$$

where

$$\alpha_i \triangleq \sum_{r=0}^{2m-1} \left[\sum_j (-1)^{r+i-j+1} a_r \binom{r}{j} \binom{2m-1-r}{i-j} \right] \tag{4.171}$$

For n-even, $n \triangleq 2m$, either

$$\alpha_{2i} > 0 \qquad\qquad\qquad i = 0, 1, \ldots, m$$

or

$$\alpha_{2i+1} > 0 \qquad \alpha_0 = {>}0 \qquad \alpha_{2m} > 0$$
$$i = 0, 1, \ldots, m - 1$$

where

$$\alpha_i = \sum_{r=0}^{2m} \left[\sum_j (-1)^{r+i-j} a_r \binom{r}{j} \binom{2m-r}{i-j} \right] \tag{4.172}$$

Note that

$$\alpha_0 = A(-1) \qquad \alpha_{2m} = A(1)$$

[The summation over j is governed by $\max(0, 2m - r - i) \leqslant j \leqslant \min(i, r)$]

Similar to the continuous case and by using the Lyapunov matrix equation in (4.144), we can guarantee stability in the evaluation $J_0 = I_n$ in (4.146). The discussion of the solution of the Lyapunov matrix equation for stability is given in Chapter 3.

REFERENCES

[1] H. M. JAMES, N. B. NICHOLS, and R. J. PHILLIPS, *Theory of Servomechanisms*. New York: McGraw-Hill, 1947, Sec. 7.9.

[2] G. C. NEWTON, JR., L. A. GOULD, and J. F. KAISER, *Analytical Design of Linear Feedback Controls*. New York: Wiley, 1957, App. E.

[3] E. P. POPOV, *The Dynamics of Automatic Control Systems*, Reading, Mass.: Addison-Wesley, 1962, p. 367.

[4] A. H. KATZ, "On the question of calculating the quadratic criterion for regulation," *Prikl. Mekh.*, **16**, 362 (1952).

[5] W. A. MERSMAN, "Evaluation of an integral occurring in servomechanism theory," *Pacific J. Math.*, **2**, 627–632 (1952).

[6] E. I. JURY and A. G. DEWEY, "A general formulation of the total square integrals for continuous systems," *IEEE Trans. on Automatic Control*, **AC-10** (1), 119–120 (June 1965).

[7] F. SCHNEIDER, "Geschlossene Formeln zur Berechnung der quadratischen und der zeitbeschwerten quadratischen Regelfläche für kontinuierliche und diskrete Systeme," *Regelungstechnik*, **14**, 159 (1966).

[8] J. LEHOCZKY, "The determination of simple quadratic integrals by Routh coefficients," *Periodica Polytech*, **10**, 153–166 (1966).

[9] J. NEKLONY and J. BENES, "Simultaneous control of stability and quality of adjustment," *Proc. 1960 IFAC Cong.*, London, **2**, 734–744 (1961).

[10] F. H. EFFERTZ, "On two coupled matrix algorithms for the evaluation of the RMS, error criterion of linear system," *Proc. IEEE*, **54**, 897–880 (June 1966).

[11] K. J. ÅSTROM, *Introduction of Stochastic Control Theory*. New York: Academic Press, Vol. 70, pp. 116–142 (1970).

[12] A. T. FULLER, "The replacement of saturation constraints by energy constraints in control optimization theory," *Int. J. Control*, **6** (3), 201–207 (1967).

[13] R. E. KALMAN and J. E. BERTRAM, "Control system analysis and design via the second method of Lyapunov," Part I. *Trans. ASME, Series D.*, **82**, 388 (June 1963).

[14] A. G. J. MACFARLANE, "The calculation of functionals of the time and frequency response of a linear constant coefficient dynamical system," *Quart. J. Mech. Appl. Math.*, **16**, Part 2, 250–271 (1963).

[15] R. J. Beshara, "A new evaluation of the mean square integral," *Trans. ASME, J. Dynamic Systems, Measurement and Control*, 242–246 (Dec. 1971).

[16] E. I. Jury, *Theory and Application of the z-Transform Method*. New York: Wiley, 1964, pp. 136–138, 168–172.

[17] E. I. Jury, "A note on the evaluation of the total square integral," *IEEE Trans. on Automatic Control* (*Correspondence*), AC-10, 110–111 (June 1965).

[18] P. LeFevre, *Optimalisation, Statistique des Systemas Dynamiques*. Paris: Dunod, 1965, pp. 103–115.

[19] C. J. Greaves, G. A. Gagne, and G. N. Bordner, "Evaluation of integrals appearing in minimization problems of discrete data system," *IEEE Trans. on Automatic Control*, AC-11, 145–148 (Jan. 1966).

[20] K. J. Åstrom, E. I. Jury, and R. G. Agniel, "A numerical method for the evaluation of complex integrals," *IEEE Trans. on Automatic Control*, AC-15 (4), 468–471 (Aug. 1970).

[21] W. Gersch, "Computation of the mean-square response of a stationary time-discrete system," *IEEE Trans. on Automatic Control*, AC-16 (2), 277 (April 1971).

[22] S. Gutman, "The 'inners' formulation for the total square integral (continuous and discrete time systems)." *M.S. Project No. 72-Sp-147*. Department of Mechanical Engineering, University of California, Berkeley, June 1972.

[23] V. Strejc, "Lecture Notes, University of Minnesota," Dec. 1970. Visiting from the Institute of Information Theory and Automation in Czechoslovakia.

[24] F. T. Man, "Evaluation of time-weighted quadratic performance indices for linear discrete systems," *IEEE Trans. on Automatic Control*, AC-115, 496–497 (Aug. 1970).

[25] M. F. Gardner and J. L. Barnes, *Transients in Linear Systems*. New York: Wiley, 1942, Vol. 1.

[26] F. R. Gantmacher, *Theory of Matrices*. New York: Chelsea, 1959, Vol. II., Chap. XV.

[27] V. Strejc, *Teorie Lineári Regulace*. Prague: Fakultaelektrotechnika Ĉeske Vysoké uĉeni technické V, 1968, pp. 109–118 (1968).

[28] F. Cŝaki and J. Lehoczky, "Comment on 'A Remark on Routh Array'," *IEEE Trans. on Automatic Control*, AC-12, 462–463 (Aug. 1970).

[29] E. I. Jury and S. Gutman, "Inner formulation for the total square integrals (sums)," *Proc. IEEE*, 395–397 (March 1973).

[30] A. Liénard and M. H. Chipart, "Sur la signe de lar partie réelle des racines d'une équation algébrique," *J. Math. Pure et Appl.*, 10, 291–346 (1914).

[31] E. I. Jury and A. G. Dewey, "Relationship between the total square

integral formula and the stability determinants," *Tech. Rept. M-78, E.R.L.*, University of California, Berkeley, July 1964.

[32] A. G. J. MACFARLANE, "Functional-matrix theory for the general linear electrical network," *IEEE Proc.*, **112** (4), 763 (April 1965).

[33] S. GUTMAN, "A unified formulation for the quadratic integral (sum)". *Memorandum No. ERL-M437*, Electronics Res. Lab., College of Engineering, University of California, Berkeley, May, 1974.

[34] S. O. RICE, "Efficient evaluation of integrals of analytic functions by the trapezoidal rule," *Bell Syst. Tech. J.*, **52** (5), 707–722 (May–June 1973).

CHAPTER

5

Positive and Strictly Positive Real Functions and Matrices

The concept of positive real functions, which was first introduced by Brune [1], has found many applications in the synthesis of electric networks [2,3]. In recent years this concept has become of importance in many diverse areas, such as the absolute stability [4–6], hyperstability, optimality [7–9], and sensitivity [8] of dynamic systems.

In this chapter a complete discussion of the concept for both continuous and discrete systems will be given. For design purposes—that is, when the coefficients of the tested polynomial are given other than numbers (i.e., literals)—the inners concept will be emphasized for low-order systems [10]. In the numerical testing for positive real functions, both the Routh table [7,11] for the continuous case and the analog form of the stability table for the discrete case [12,13] will be studied.

Finally, in the last section of this chapter the necessary and sufficient conditions for stability of nonuniformly distributed parameter systems are presented [14]. These conditions yield a finite algorithm, which can be readily utilized for the synthesis of cascade transmission lines [15].

5.1 POSITIVITY AND NONNEGATIVITY OF EVEN POLYNOMIALS THAT ARISE IN CONTINUOUS SYSTEMS

In this section we will present two tests for positivity and nonnegativity conditions. The first test is based on the inners concept discussed in earlier chapters. This test is useful when the coefficients of the even polynomial are presented in literal form. This form is useful for design purposes. However, the test becomes unwieldy if the order of the polynomial is larger than 4. Hence we will present the positivity condition in general form for $n \le 4$.

The second test is based on a modified Routh table as introduced by Šiljak [5] and is useful when the coefficients of the polynomial are given numerically. Thus, this procedure offers a useful numerical test for positivity and nonnegativity. A computer algorithm can be written for this test [16].

We will show that the two methods—the inners approach and the table form—yield the same general conditions.

In concluding this section, we will indicate the use of the critical constraints discussed in Section 2.5 to the test of positivity of real rational functions.

a. Inner's Approach

The definition of the positive real function is well known and has been presented in alternatively different forms in the literature [17]. In the following, we will present one form as discussed in Weinberg [3].

DEFINITION 5.1 *A real rational function*

$$G(s) = \frac{q(s)}{p(s)} \tag{5.1}$$

with relatively prime polynomials (no common cancelable factors) is called positive real (pr) if and only if
  (i)   *the polynomial*

$$f(s) = p(s) + q(s) \tag{5.2}$$

is Hurwitz (all roots are in the open left half of the s-plane), and

(iia) $\text{Re } G(j\omega) \geq 0$ *for all* ω (5.3)

(iib)   *If in* (iia) $\text{Re } G(j\omega) > 0$, *for all* ω, *then it is strictly positive real. Condition* (iia) *can be written by noting that*

$$p(j\omega) = p_r(\omega) + jp_j(\omega) \tag{5.4}$$

$$q(j\omega) = q_r(\omega) + jq_j(\omega) \tag{5.5}$$

as follows:

$$\text{Re } G(j\omega) = \frac{p_r q_r(\omega) + p_j(\omega)q_j(\omega)}{p_r^2 + p_j^2} \tag{5.6}$$

Since $p_r^2 + p_j^2$ *in* (5.6) *is positive for all* ω, *we obtain an algebraic condition for* (iia),

$$\Pi(\omega^2) \geq 0 \qquad \text{*for all* } \omega \quad (5.7)$$

where $\Pi(\omega^2)$ *is an even polynomial in* (ω^2) *and is obtained from the numerator of* (5.6). *The condition* (5.7) *is referred to in this work as the nonnegativity condition of* $\Pi(\omega^2)$.

  Similarly, condition (iib) *yields*

$$\Pi(\omega^2) > 0 \qquad \text{*for all* } \omega \quad (5.8)$$

The preceding condition is referred to in this work as positivity condition. We may note that when $\Pi(\omega^2)$ is a constant polynomial, we also have from (5.7) and (5.8)

$$\Pi(\omega^2) \equiv \Pi(0) \geq 0 \qquad \Pi(\omega^2) \equiv \Pi(0) > 0 \tag{5.9}$$

Remarks

1.   As pointed out in Chapter 1, the inners approach can be applied to test whether $p(s)$ and $q(s)$ in (5.1) are relatively prime [see (1.5)]. Furthermore, the condition for $f(s)$ in (5.2) to be Hurwitz is expressed in inners form in Section 2.1. It is represented by the coefficients of $f(s)$ to be positive and by the matrix Δ_{n-1} in (2.10) for n-odd and n-even to be positive innerwise.

2.   The condition (iib) for a strictly positive real function is represented in terms of $\Pi(\omega^2) = 0$ in (5.8) having *no positive real roots* [4]. The condition for nonnegativity is represented in terms of $\Pi(\omega^2) = 0$ having *no positive real roots of odd multiplicity* [5]. These conditions are well known and can be readily ascertained [2,3]. It may be noted that the condition of positivity is simpler to check than the nonnegativity. Hence the inners approach can be readily used for the test.

3.   The classical test of positivity and nonnegativity is based on the Sturm theorem [2], and the final results are obtained in terms of Sturm sequences [18]. In the following representation we will first obtain the tests based on the inners and later on a modified Routh table. This is in analogy to stability tests based on the Hurwitz criterion and the Routh table. It is well known that the Routh table, and hence Hurwitz determinants, is obtainable from Sturm sequences.

For continuity, we will again give Theorem 2.9 of Section 2.6. This theorem was proven by Fuller and is based on Sturm sequences.

THEOREM 2.9   *The number of distinct positive real roots of $F(z) = 0$ given in (2.134) is*

$$N = \text{Var}\, [1, -|\Delta_2^2|, |\Delta_4^2|, \ldots, (-1)^n\, |\Delta_{2n}^2|] \\ - \text{Var}\, [1, |\Delta_1^1|, |\Delta_3^1|, \ldots, |\Delta_{2n-1}^1|] \tag{2.137}$$

where Δ^1 and Δ^2 are given by (2.136) and (2.140), respectively. Also, the following restated identities may be noted from (2.138) and (2.139):

$$|\Delta_{2n}^2| = (-1)^n a_0\, |\Delta_{2n-1}^1| \tag{2.138}$$

and
$$|\Delta_2^2| = -a_{n-1} \frac{|\Delta_1^1|}{n} \tag{2.139}$$

COROLLARIES

5.1   *The condition for positivity of real polynomials can be obtained from Theorem 2.9 by letting $N = 0$ in (2.137).*

5.2   *The symmetric matrix corresponding to Δ^1 in (2.136), following (3.157), is given by*

$$d_{ij} = \begin{cases} \begin{aligned} \sum_{k=1}^{n-j+1} & (n-j+2-k)a_{n-j+2-k}a_{n-i+k} \\ & -(n-i+1+k)a_{n-i+1+k}a_{n-j-k+1} \end{aligned} & \begin{aligned} j &\geq i \\ j &< i \end{aligned} \\ d_{ji} = d_{ij} & \\ \\ a_k = 0 & \quad when \ k > n, \quad i, j = 1, 2, \ldots, n \end{cases} \quad (5.10)$$

The symmetric matrix corresponding to Δ^2 can be obtained, following the discussion of Section 3.4, by premultiplying (2.140) by the following matrix:

$$\Delta = \begin{bmatrix} (n-1)a_{n-1} & (n-2)a_{n-2} & \cdots & -a_{n-2} & -a_{n-1} \\ (n-2)a_{n-2} & (n-3)a_{n-3} & \cdots & -a_{n-3} & -a_{n-2} \\ \vdots & \vdots & & \vdots & \vdots \\ a_0 & 0 & \cdots & -a_0 & -a_1 \\ 0 & 0 & \cdots & 0 & -a_0 \end{bmatrix} \quad (5.11)$$

It may be noted that the determinants of the inners in (2.137) are the same (within a positive constant) as those corresponding to the leading principal minors of the corresponding symmetric matrices.

Remarks

1.   From Corollary 5.1 it is evident that there exist many alternatives for satisfying some of the positivity conditions—that is, $N = 0$. The number of these alternatives increases significantly as the degree of the polynomial becomes large. This is the main reason why the positivity condition (which is a special root-distribution problem) is much more difficult than stability or the root-clustering problems discussed earlier.

2.   In view of Corollary 5.2, the two inners of dimensions $2n - 1 \times 2n - 1$ and $2n \times 2n$ in (2.136) and (2.140) can be replaced by the corresponding symmetric matrices of half sizes—that is, $n \times n$ dimension. These symmetric matrices also correspond to Hankel symmetric matrices, as discussed by Anderson [19]. Their use will be illustrated in specific cases later on.

3.   In the counting of N in (2.137), some of the $|\Delta_k|$'s could be zero, but $|\Delta^1_{2n-1}| \neq 0$. For such a situation, a modification indicated by Gantmacher

[18] for the Hurwitz determinants can be readily adopted for this case, too.†
A discussion of this point is presented in reference [5]. It should be noted
that $|\Delta^1_{2n-1}|$ can be zero and positivity holds.

4. In (2.138) and (2.139) certain relationships exist between the even
determinants and the odd ones and the coefficients of the polynomial. Such
relationships are utilized to simplify the positivity or nonnegativity conditions
for low-degree polynomials, as shown later. It may be conjectured that, for
higher-degree polynomials, more of these relationships can exist, thereby
simplifying the test for higher-degree polynomials. This conjecture remains
another open question.

THEOREM 5.1 *A sufficient condition [provided (5.2) is also satisfied] for
a real rational function to be positive real (pr) is equivalent to the following
even equation:*

$$\hat{F}(z) = a_n z^{2n} + a_{n-1} z^{2n-2} + \cdots + a_1 z^2 + a_0 \qquad (5.12)$$

*having all its roots distinct and on the imaginary axis. Such a condition is
presented in Section 2.4 and restated in the following:*

(i)   *All coefficients of $\hat{F}(z)$ to be positive.*

(ii)   *The matrix Δ^1 given by (2.136) be positive innerwise or, equivalently,
the symmetric matrix (d_{ij}) given in (5.10) be positive definite. This theorem
establishes the connection between (pi), (pd), and (pr).*

It may be readily noted that condition (i) *or*

$$a_n > 0 \qquad a_0 > 0 \qquad a_k \geq 0$$
$$k = 1, 2, \ldots, n - 1 \qquad (5.13)$$

also constitutes a sufficient condition for positive real functions.

The proof of the theorem follows readily by noticing that the condition of
the distinct imaginary roots is equivalent to the condition of $F(s)$ in (2.9)
having all its n distinct roots on the negative real axis.

THEOREM 5.2 [20].   *In order that the polynomial*

$$\Pi(\omega^2) = \sum_{k=0}^{n} a_{2k} \omega^{2k} \qquad a_0 > 0 \quad (5.14)$$

*have no positive real zeros, it is necessary that the number of sign changes in the
coefficients of (5.14) (starting with a_{2n}) be even.*

The proof of the preceding necessary condition stems easily from Descarte's
Rule of Sign, which states: "The number of positive real roots of $\Pi(\omega^2) = 0$

† A discussion of the critical cases will be presented in Chapter 7.

is either equal to the number of variations of sign between successive terms in $\Pi(\omega^2)$ when arranged in descending powers of ω^2 or less than that number by an even integer."

Since we require no positive real zeros of $\Pi(\omega^2) = 0$ for the positivity test, a necessary condition is that the number of sign variations should be even.

It may be noted that if the number of sign variation is zero (i.e., all coefficients are positive), then the foregoing theorem constitutes a sufficient condition for positivity, as mentioned in Theorem 5.1. Furthermore, the even-sign condition is equivalent to that of $a_{2n} > 0$ and $a_0 > 0$, and vice versa.

Theorem 5.1 and Theorem 5.2 will be effectively utilized in discussing the positivity condition for low-degree polynomials in the following cases.

Cases. Let $n = 2$. Then from (5.14) we have

$$\Pi(\omega^2) = a_4\omega^4 + a_2\omega^2 + a_0 = 0 \quad \text{with } a_0 > 0 \quad (5.15)$$

From Theorem 5.2, we have for positivity

$$a_4 > 0 \tag{5.16}$$

Let $u = \omega^2$ in (5.15), thus obtaining

$$\Pi(u) = a_4u^2 + a_2u + a_0 = 0 \tag{5.17}$$

It is evident that the positivity condition for (5.15) is the same as that of (5.17).

In order to obtain the condition for positivity first, we apply (2.137) of Theorem 2.9 to (5.17):

$$N = 0 = \text{Var}\,[1, -|\Delta_2^2|, |\Delta_4^2|] - \text{Var}\,[1, |\Delta_1^1|, |\Delta_3^1|] \tag{5.18}$$

In this case, Δ_3^1 and Δ_4^2 can be obtained from (2.136) and (2.140) as follows:

$$\Delta^1 = \begin{bmatrix} a_4 & a_2 & a_0 \\ 0 & \boxed{\begin{matrix} 2a_4 & a_2 \\ \end{matrix}}_{\Delta_1^1} & \\ 2a_4 & a_2 & 0 \end{bmatrix} \qquad \Delta^2 = \begin{bmatrix} a_4 & a_2 & a_0 & 0 \\ 0 & \boxed{\begin{matrix} a_4 & a_2 & a_0 \\ 2a_4 & a_2 & 0 \end{matrix}}_{\Delta_2^2} \\ 0 & & & \\ 2a_4 & a_2 & 0 & 0 \end{bmatrix} \tag{5.19}$$

From (2.138) and (2.139), we have

$$|\Delta_4^2| = a_0\,|\Delta_3^1| \tag{5.20}$$

$$|\Delta_2^2| = \left(-\frac{a_2}{2}\,|\Delta_1^1|\right) \tag{5.21}$$

Inserting (5.20) and (5.21) into (5.18), we obtain

$$N = 0 = \text{Var } [1, a_2 \, |\Delta_1^1|, a_0 \, |\Delta_3^1|] - \text{Var } [1, |\Delta_1^1|, |\Delta_3^1|] \qquad (5.22)$$

Noting that $|\Delta_1^1| = 2a_4 > 0$ and $a_0 > 0$, we can distinguish two cases:

1. $a_2 \geq 0$, $N = 0$. This is known from Theorem 5.1 (sufficiency condition). It may be noted that when $a_2 = 0$, it follows that $|\Delta_3| < 0$. Hence whatever sign one assumes for $a_2 \, |\Delta_1^1| = 0$, as required by this critical case using Gantmacher's [18] modification, N is always zero.

2. $a_2 < 0$, for $N = 0$, we require $|\Delta_3^1| = a_4(a_2^2 - 4a_0a_4) < 0$. This condition requires $a_2^2 - 4a_0a_4 < 0$. A condition well known from the general solution of a second-order equation.

To obtain the condition for nonnegativity, we require that $\Pi(\omega^2) = 0$ have no positive real roots of odd multiplicity. In this case, we also allow for $n = 2$, multiple positive real root. It is evident that such can be accomplished when

1. $$a_2 \geq 0 \qquad a_0 \geq 0 \qquad\qquad\qquad (5.23)$$

2. $$a_2 < 0, \qquad \Delta_3 \leq 0, \qquad \text{or} \qquad a_2^2 - 4a_0a_4 \leq 0 \qquad (5.24)$$

It is evident that when $a_2 < 0$ and $a_2^2 - 4a_0a_4 \leq 0$, the two roots of $\Pi(\omega^2)|_{\omega^2 = u} = 0$, are real multiple positive. This can also be ascertained from the fact that when $|\Delta_3| = 0$, there exist two multiple roots (critical case), and since $a_2 < 0$, these two multiple roots are on the positive real axis. (Note that a necessary condition for the multiple roots to exist on the negative real axis is $a_2 > 0$.) This condition is well known from the general solution of the quadratic equation.

Let $n = 3$; from (5.14) we have

$$\Pi(\omega^2) = a_6\omega^6 + a_4\omega^4 + a_2\omega^2 + a_0 = 0$$
$$\text{with } a_0 > 0 \quad (5.25)$$

From Theorem 5.2, we have for positivity

$$a_6 > 0 \qquad\qquad\qquad\qquad (5.26)$$

Let $u = \omega^2$ in (5.25) to obtain

$$\Pi(u) = a_6u^3 + a_4u^2 + a_2u + a_0 = 0 \qquad (5.27)$$

To obtain first the condition for positivity, we use (2.137) to get

$$N = 0 = \text{Var } [1, -|\Delta_2^2|, |\Delta_4^2|, -|\Delta_6^2|] - \text{Var } [1, |\Delta_1^1|, |\Delta_3^1|, |\Delta_5^1|] \qquad (5.28)$$

From (2.139) and (2.138), we have

$$|\Delta_2^2| = -a_4 \frac{|\Delta_1^1|}{3} \tag{5.29}$$

$$|\Delta_6^2| = -a_0 |\Delta_5^1| \tag{5.30}$$

Therefore

$$N = 0 = \text{Var} \left[1, a_4 |\Delta_1^1|, |\Delta_4^2|, a_0 |\Delta_5^1|\right] - \text{Var} \left[1, |\Delta_1^1|, |\Delta_3^1|, |\Delta_5^1|\right] \tag{5.31}$$

In addition to the above relationships, the following identity exists [21].

$$\left(\frac{|\Delta_3^1|}{a_6}\right) = 2(27a_0a_6^2 + 2a_4^3 - 9a_2a_4a_6)^2 + 54a_0 |\Delta_5^1| \tag{5.32}$$

Equation (5.31) indicates that if $|\Delta_5^1|$ is positive when $a_6 > 0$, then $|\Delta_3^1|$ is *necessarily positive*. This fact can be utilized to show that the even determinant $|\Delta_4^2|$ in (5.28) is not needed for computation. This point is discussed in detail below.

1. $$a_4 > 0 \qquad a_2 < 0 \tag{5.33}$$

(a) If $|\Delta_5^1| < 0$ $N = 0$ in (5.31) irrespective of whether $|\Delta_4^2|$ is positive, negative, or zero. (Note that $|\Delta_1^1| = 3a_6$ is always positive; also, when $|\Delta_4^2| = 0$, we can assign to it either positive or negative sign.)
(b) If $|\Delta_5^1| > 0$, then from (5.32), $|\Delta_3^1| > 0$, and hence the criterion for positivity is not satisfied. The reason is that Δ_5 is pi and hence all three distinct roots are on the real axis (see Section 2.4b); and since $a_2 < 0$, then at least one root is on the positive real axis (aperiodicity violated).

2. $$a_4 < 0 \qquad a_2 < 0 \tag{5.34}$$

and $$a_4 = 0 \qquad a_2 < 0 \tag{5.35}$$

(a) If $|\Delta_5^1| < 0$, the criterion for positivity is satisfied. The reason is that Var $[1, -|\Delta_1^1|, |\Delta_4^2|, |\Delta_5^1|]$ in (5.31) cannot have three variations of sign, for this would indicate from (5.31) that $N = 2$, and since this is a third-degree real polynomial, the third root should also be on the real axis, which requires that $|\Delta_5^1|$ be positive (i.e., $|\Delta_5|$ is positive), thus contradicting the imposed condition on $|\Delta_5^1|$. The same condition holds when $a_4 = 0$ and $a_2 < 0$. In this case, $|\Delta_2^2| = 0$ which can be readily assigned any sign according to Gantmacher's procedure [18].
(b) If $|\Delta_5^1| > 0$, from (5.32), $|\Delta_3^1| > 0$. Hence N cannot be zero and the test for positivity is not satisfied. Note that for $a_4 = 0$ and $a_2 < 0$, the aperiodicity condition is violated and hence $N \neq 0$.

In order to justify that $|\Delta_5| \neq 0$ for positivity and to obtain the non-negativity condition, we mention the following points.

The conditions of nonnegativity of the polynomial in (5.25) are given by

1.
$$a_4 \geq 0, \qquad a_2 \geq 0, \qquad a_0 \geq 0 \tag{5.36}$$

and for all other cases

2.
$$|\Delta_5| \leq 0 \tag{5.37}$$

Condition (2) in (5.37) can be readily ascertained as follows.

1. When $|\Delta_5| = 0$, we know that there exists at least a multiple root (see Gantmacher's discussion of Hurwitz critical cases [18]). If this multiple root is on the negative real axis, the third one should also be on the negative real axis (because $a_0 > 0$), and thus a necessary condition is that the coefficients of $\Pi(u)$ in (5.27) should be positive. This point has been covered by (5.36).

2. When the coefficients are not positive, the multiple root when $|\Delta_5| = 0$ should be on the positive real axis, because the third real root (since $a_0 > 0$) should be on the negative real axis. Thus when $a_4 \geq 0$, $a_2 < 0$, or $a_2 \geq 0$, $a_4 < 0$, the condition for nonnegativity is satisfied when $|\Delta_5| \leq 0$. However, the condition for positivity is not satisfied when $|\Delta_5| = 0$. The case of a triple root on the positive real axis cannot exist because $a_0 > 0$ (a necessary requirement for positivity and nonnegativity).

In summary, the following two alternatives are needed for positivity for the third-degree polynomial in (5.27).

1.
$$a_4 \geq 0 \qquad a_2 \geq 0 \tag{5.38}$$

2. all other cases,
$$|\Delta_5^{\frac{1}{2}}| < 0 \tag{5.39}$$

where

$$\Delta_5^{\frac{1}{2}} = \begin{bmatrix} a_6 & a_4 & a_2 & a_0 & 0 \\ 0 & a_6 & a_4 & a_2 & a_0 \\ 0 & 0 & 3a_6 & 2a_4 & a_2 \\ 0 & 3a_6 & 2a_4 & a_2 & 0 \\ 3a_6 & 2a_4 & a_2 & 0 & 0 \end{bmatrix} \tag{5.40}$$

The symmetric matrix corresponding to $\Delta_5^{\frac{1}{2}}$, using (3.156), gives

$$|\Delta_3|_s = \begin{bmatrix} 0 & 0 & 0 & 0 & a_6 \\ 2a_4 & a_2 & -a_0 & -a_2 & 0 \\ a_2 & 0 & 0 & -a_0 & 0 \end{bmatrix} \begin{bmatrix} a_6 & a_4 & a_2 & a_0 & 0 \\ 0 & a_6 & a_4 & a_2 & a_0 \\ 0 & 0 & 3a_6 & 2a_4 & a_2 \\ 0 & 3a_6 & 2a_4 & a_2 & 0 \\ 3a_6 & 2a_4 & a_2 & 0 & 0 \end{bmatrix}$$

$$
= \begin{bmatrix} 3a_6^2 & 2a_6a_4 & a_6a_2 & \vdots \\ 2a_6a_4 & 2a_4^2 - 2a_6a_2 & a_4a_2 - 3a_6a_0 & \vdots & \bigcirc \\ a_6a_2 & a_4a_2 - 3a_6a_0 & a_2^2 - 2a_4a_0 & \vdots \end{bmatrix} \tag{5.41}
$$

If we let

$$
a = \frac{3a_2 - a_4^2}{3} \qquad b = \frac{2a_4^3 - 9a_2a_4 + 27a_0}{27} \qquad a_6 \equiv 1
$$

in (5.27), then $|\Delta_5|$, the determinant of (5.40), is

$$
|\Delta_5| = -\left(\frac{b^2}{4} + \frac{a^3}{27}\right) \tag{5.42}
$$

The preceding equation is exactly the *discriminant* of the third-degree equation $\Pi(u) = 0$, whose general solution is well known.

Let $n = 4$. Then (5.14) can be written as

$$
\Pi(\omega^2) = \omega^8 + p\omega^6 + q\omega^4 + r\omega^2 + s \qquad s > 0 \quad (5.43)
$$

If we let $u = \omega^2$, then

$$
\Pi(u) = u^4 + pu^3 + qu^2 + ru + s \qquad s > 0 \quad (5.44)
$$

To obtain some of the positivity conditions, we use relationship (2.154) as follows:

$$
0 = N = \text{Var}\,[1, |\Delta_1^{\frac{1}{1}}|, |\Delta_3^{\frac{1}{3}}|, |\Delta_5^{\frac{1}{5}}|, |\Delta_7^{\frac{1}{7}}|] - \text{Var}\,[1, -|\Delta_2^{\frac{1}{2}}|, |\Delta_4^{\frac{2}{4}}|, |\Delta_6^{\frac{1}{6}}|, |\Delta_8^{\frac{2}{8}}|]
$$

Similar to the cases $n = 2$ and $n = 3$, we can obtain certain relationships between the even determinant and odd determinants and the coefficients of (5.44) that simplify the positivity test. However, to avoid a detailed discussion of all the alternate cases,† we present an algorithm for the positivity test obtained from the general solution of the fourth-degree equation [22]. This algorithm, which can also be extracted from (5.45), offers a computational simplification. Furthermore, all cases for positivity covered by the algorithm are presented.

Algorithm. If p, q, r are positive, then following (5.13),

$$
\Pi(\omega^2) > 0 \tag{5.45}
$$

Now compute

$$
a = q - \frac{3p^2}{8} \qquad b = \frac{p^3}{8} - \frac{pq}{2} + r
$$

† See also reference [43].

$$c = -\frac{3p^4}{16^2} + \frac{p^2q}{16} - \frac{pr}{4} + s \tag{5.46}$$

$$f = -\frac{a^2 + 12c}{48} \qquad g = \frac{-a^3 + 36ac - 27b^2/2}{854}$$

and follow the steps:

Step 1: If $f^3/27 + g^2/4$ is positive, go to step 8.
If $f^3/27 + g^2/4$ is negative, go to step 2.
If $f^3/27 + g^2/4$ is zero, go to step 7.

Step 2: If a is negative, go to step 3.
If a is positive, $\Pi(\omega^2) > 0$.
If a is zero, $\Pi(\omega^2) > 0$.

Step 3: If $a^2 - 4c$ is negative, $\Pi(\omega^2) > 0$.
If $a^2 - 4c$ is positive, go to step 4.
If $a^2 - 4c$ is zero, go to step 4.

Step 4: If r is positive, go to step 5.
If r is negative, $\Pi(\omega^2) \leq 0$.
If r is zero, $\Pi(\omega^2) \leq 0$.

Step 5: If q is positive, $\Pi(\omega^2) > 0$.
If q is negative, $\Pi(\omega^2) \leq 0$.
If q is zero, $\Pi(\omega^2) \leq 0$.

Step 6: If $a^2 - 4c$ is positive, go to step 4.
If $a^2 - 4c$ is negative, go to step 8.
If $a^2 - 4c$ is zero, go to step 4.

Step 7: If a is positive, go to step 8.
If a is negative, go to step 6.
If a is zero, go to step 8.

Step 8: If b is positive, $\Pi(\omega^2) > 0$.
If b is negative, $\Pi(\omega^2) \leq 0$.
If b is zero, $\Pi(\omega^2) \leq 0$.

It is of special interest to note that in this algorithm the quantities $a^2 - 4c$ and $f^3/27 + g^2/4$ play an important role in checking positivity.

The cases for positivity, in addition to (5.45), are obtained from the algorithm and summarized below.

1. $\qquad\qquad\qquad\quad \cdot F > 0 \quad \text{and} \quad b > 0 \tag{5.47}$

2. $\qquad\qquad\qquad\quad F < 0 \quad \text{and} \quad a \geq 0 \tag{5.48}$

3. $\qquad\qquad F < 0, \quad a < 0, \quad \text{and} \quad a^2 - 4c < 0 \tag{5.49}$

4. $\qquad F \leq 0, \quad a < 0, \quad a^2 - 4c \geq 0, \quad r > 0, \quad \text{and} \quad q > 0 \tag{5.50}$

5. $$F = 0, \quad a < 0, \quad a^2 - 4c < 0, \quad \text{and} \quad b > 0 \qquad (5.51)$$

6. $$F = 0, \quad a \geq 0 \quad \text{and} \quad b > 0 \qquad (5.52)$$

where $$F = \frac{f^3}{27} + \frac{g^2}{4}$$

Remarks

1. For higher-degree polynomials, one can also obtain the positivity condition when $N = 0$. However, for literal coefficients of the polynomial, the conditions become much more complicated.† Furthermore, the number of alternatives for positivity increases significantly, thus making the positivity conditions only of academic interest. For practical cases, it is not always likely that for literal coefficients one would be interested in extracting positivity for higher-degree polynomials. An exception to this exists if only one or two varying parameters are considered in the higher-degree polynomials. In this situation, either the inners approach or the modified Routh table can be used.

2. In some cases, it is known that for certain parameters the polynomial

$$\Pi(\omega^2) = \sum_{k=0}^{n} a_{2k}\omega^{2k}$$

is positive. In such a situation one would desire to know what is the maximum range of this parameter when the polynomial ceases to be positive. This case is called a critical case in analogy to the critical cases for stability, relative stability, and aperiodicity discussed in Section 2.5. A *sufficient* condition for $\Pi(\omega^2)$ to cease to be positive is when $a_0 = 0$. The proof can be readily ascertained from Theorem 5.2, in which it is shown that $a_0 > 0$ is a necessary condition for positivity. Note that the sign of a_0 remains unchanged if a complex root and its conjugate cross the imaginary axis and enter the right half-plane.

b. Modified Routh Table

When the coefficients of the even polynomial $\Pi(\omega^2)$ in (5.7) are given numerically, a convenient numerical method based on the modified Routh table can be obtained. The application of the Routh table to test positive realness was first mentioned by Fryer [23]. In 1969, Šiljak [4] had integrated a simple and straightforward test for positivity (nonnegatively) that can be programmed on a digital computer.

From (5.14) it is noticed that $\Pi(\omega^2)$ is an even polynomial with real coefficients, and therefore it has $2n$ zeros symmetrically distributed with respect

† See also reference [44].

to both the real and the imaginary axes of the ω-plane. Now let us consider the polynomial

$$\Pi(j\omega) = \sum_{k=0}^{n} (-1)^k a_{2k}\omega^{2k} \qquad a_0 > 0 \quad (5.53)$$

and conclude that the previous symmetry is preserved [(5.53) presents a rotation of 90° in counterclockwise direction] but real zeros of $\Pi(\omega^2)$ (if any) become pure imaginary zeros of $\Pi(j\omega)$. Consequently, if $\Pi(j\omega)$ has n zeros with positive real parts, there are no positive real zeros of $\Pi(\omega^2)$, and hence positivity is satisfied. Therefore in order to determine the number of zeros of $\Pi(j\omega)$ with positive real part, we readily replace ω^2 by ω and form the Routh table as below.

Modified Routh Table [4]:

r						
1	ω^{2n}	$(-1)^n a_{2n}$	$(-1)^{n-1}a_{2n-2}$	$\cdots$	$-a_2$	a_0
2	ω^{2n-1}	$(-1)^n n a_{2n}$	$(-1)^{n-1}(n-1)a_{2n-2}$	$\cdots$	$-a_2$	
$\vdots$	$\vdots$	$\vdots$	$\vdots$		$\vdots$	
$(2n+1)$	ω^0	a_0				

$$(5.54)$$

It may be noticed that, in forming the Routh array, it is necessary to form the second row using the coefficients of the derivative of $\Pi(j\omega)$ and continue the test.

If the first column of Routh array yields exactly n changes of sign and $a_0 > 0$, then positivity of $\Pi(\omega^2)$ is guaranteed.

We can further extend the preceding results to obtain the number n of the positive real zeros of $\Pi(\omega^2)$. This can be ascertained by noting that if $\Pi(\omega^2)$ has π positive real zeros, then $\Pi(j\omega)$ will have π purely imaginary zeros. Because of the symmetry of both $\Pi(\omega^2)$ and $\Pi(j\omega)$, the latter will have $(n - \pi)$ zeros with positive real parts. Thus we can state the following result.

The number π of the positive real zeros of $\Pi(\omega^2)$ is

$$\pi = n - \text{Var}\,[(-1)^n a_{2n}, (-1)^n n a_{2n}, \ldots, a_0] \qquad (5.55)$$

where Var is the number of sign variations in the first column of the modified Routh array in (5.54). Note that Var is the number of zeros of $\Pi(j\omega)$ with positive real parts.

It is well known [11] that the first two rows of the Routh array

$$
\begin{array}{cccc}
a_n & a_{n-2} & a_{n-4} & \cdots \\
a_{n-1} & a_{n-3} & a_{n-5} & \cdots \\
b_0 & b_1 & b_2 & \cdots \\
c_0 & c_1 & c_2 & \cdots \\
\vdots & \vdots & \vdots &
\end{array}
\tag{5.56}
$$

are formed by the coefficients of the polynomial $\sum_{k=0}^{n} a_k \omega^k$ to be tested. The next rows are obtained by the recursive algorithm

$$
b_0 = \frac{a_{n-1}a_{n-2} - a_n a_{n-3}}{a_{n-1}} \qquad b_1 = \frac{a_{n-1}a_{n-4} - a_n a_{n-5}}{a_{n-1}}, \ldots
$$

$$
c_0 = \frac{b_0 a_{n-3} - a_{n-4} b_1}{b_0} \qquad c_1 = \frac{b_0 a_{n-5} - a_{n-1} b_2}{b_0}, \ldots
\tag{5.57}
$$

Two special cases may occur. If any number in the first column of (5.56) becomes zero, the algorithm cannot be continued. To remedy this situation, the original polynomial should be multiplied by $(\omega + c)$, where c is a positive constant. Secondly, an entire row may become a row of zeros. In this situation, replace the row of zeros by coefficients of the first derivative of the polynomial formed by the coefficients of the preceding row as indicated in the second row of (5.54).

We can simplify the calculations using the Routh table by utilizing the necessary condition for positivity as stated in Theorem 5.2. This is given in the following [20]:

THEOREM 5.3 *In order for the polynomial $\prod(\omega^2) = 0$ to have no positive real zeros, it is necessary and sufficient that*
(i)   *the number of sign changes between successive term in $\Pi(\omega^2)$ of (5.14) be even or, equivalently, $a_n > 0$, $a_0 > 0$*
(ii)   *the number of sign changes needed to be tested in the first column of the Routh table for the polynomial $\Pi(j\omega)$ in (5.53) is $(n-1)$.*

To prove the above, we discuss the case when n is even. Suppose that we established from the Routh table $(n-1)$ changes of sign in the first column. This indicates that $\Pi(j\omega)$ has at least $(n-1)$ roots with positive real parts. The distribution of these $(n-1)$ roots (odd in number) is one of two alternatives.

1.   The first alternative is that because complex roots appear in conjugate pairs, the total number of real positive roots of $\Pi(\omega^2) = 0$ is either one or odd. Since, for positivity, we require the sign changes in the coefficients of

$\Pi(\omega^2)$ or $\Pi(j\omega)$ to be even, this can happen only if $\Pi(j\omega)$ has n roots with positive real part.

2.   The second alternative is when the sign changes of $\Pi(j\omega)$ are satisfied in addition to the $(n - 1)$ changes of sign. In this case, there exists a single complex root with positive real part. Since complex roots appear in conjugate pairs, $\Pi(j\omega)$ therefore has n roots with positive real part.

For n-odd, similar reasoning leads to the same conclusion. Note that in this case the sign changes required in the coefficients of $\Pi(j\omega)$, corresponding to those of $\Pi(\omega^2)$, be odd.

Theorem 5.3 constitutes, in certain polynomials tested for positivity, a minimum of entries in the first column of the Routh table of $(n + 1)$ to a maximum $(2n)$. It should be noted that since $\Pi(j\omega)$ is of degree $2n$, the usual number of entries in the first column of the Routh table is $(2n + 1)$.

Example 5.1.   We want to test the positivity condition of the even polynomial

$$\Pi(\omega^2) = \omega^8 - 3\omega^6 + 2\omega^4 + \omega^2 + 1 \tag{5.58}$$

From Theorem 5.2, this polynomial satisfies the necessary condition for positivity. Hence we proceed to obtain the necessary and sufficient condition. Form $\Pi(j\omega)$ as follows:

$$\Pi(j\omega) = \omega^8 + 3\omega^6 + 2\omega^4 - \omega^2 + 1 \tag{5.59}$$

From (5.54), form the following Routh table.

ω^8	1	3	2	-1	1
ω^7	4	9	4	-1	0
ω^6	0.75	1	-0.75	1	0
ω^5	$\frac{22}{3}$	16	$-\frac{38}{3}$	0	
ω^4	$-\frac{20}{11}$	$\frac{6}{11}$	1	0	
ω^3	18.2	-8.63	0		
ω^2	0.317	1			
ω^1	-65.9	0			

$$\tag{5.60}$$

Since $n = 4$, we need for positivity only $(n - 1) = 3$ changes of sign. Therefore positivity exists for the polynomial $\Pi(\omega^2)$ in (5.58). Note that in this case we have $2n = 8$ entries in the first column. The last entry in (5.54) is a_0 or 1 for this example.

NONNEGATIVITY OF EVEN POLYNOMIALS

As mentioned earlier, the condition of nonnegativity of $\Pi(\omega^2)$ requires that there exist no positive real roots of odd multiplicity. Thus positive roots of even multiplicity are allowed. In order to obtain a systematic procedure for testing nonnegativity, in reference [5] the following procedure is proposed.

From (5.54), starting at the top, the rows are numbered $r[r = 1, 2, \ldots, (2n + 1)]$; the row preceding each zero row is distinguished by $r_\nu (\nu = 1, 2, \ldots, m)$, the top row being associated with r_0 and the lowest with r_{m+1}. Define

$$2n_\nu = r_\nu - r_{\nu-1}$$
$$\nu = 1, 2, \ldots, (m + 1) \quad (5.61)$$

Further denote the number of sign variations between two consecutive zero rows r_ν and $r_{\nu-1}$ by V_ν and denote π_ν as the number of real positive zeros of $\Pi(\omega^2) = 0$ with multiplicity ν. Based on the above definition, we introduce Theorem 5.4, proved by Šiljak [5].

THEOREM 5.4 *The numbers π_ν of real distinct positive zeros with multiplicity ν of the polynomial $\Pi(\omega^2)$ in (5.14) are given by*

$$\pi_\nu = (n_\nu - V_\nu) - (n_{\nu+1} - V_{\nu+1})$$
$$\nu = 1, 2, \ldots, m \quad (5.62)$$

and with multiplicity $m + 1$ is given by

$$\pi_{m+1} = (n_{m+1} - V_{m+1}) \quad (5.63)$$

where m is the number of zero rows in the corresponding modified Routh array in (5.54) and n_ν is as defined in (5.61). Hence nonnegativity is obtained when $\pi_\nu = 0$ for odd $\nu = 1, 3, \ldots, odd\ (m + 1)$.

The proof of Theorem 5.4 can be ascertained from the following observations:

1.   If the polynomial $\Pi(\omega^2)$ has real positive roots of multiplicity 1, 2, 3, and up to $m + 1$, then the sum of all the positive real roots with their multiplicities is given, following (5.55), by

$$\begin{aligned}
\pi &= \pi_1 + 2\pi_2 + 3\pi_3 + \cdots + \nu\pi_\nu + \cdots + m\pi_m + (m + 1)\pi_{m+1} \\
&= (n_1 - V_1) + (n_2 - V_2) + (n_3 - V_3) + \cdots + (n_\nu - V_\nu) \\
&\quad + \cdots + (n_m - V_m) + (n_{m+1} - V_{m+1})
\end{aligned} \quad (5.64)$$

2.   When a zero row occurs in the modified Routh array of (5.54), it indicates that the polynomial preceding the all-zero row contains all of the multiple roots of $\Pi(\omega^2) = 0$, but each to one degree lower. This is in view of

the differentiation required in the second row. Therefore from (5.55) we have

$$\pi_2 + 2\pi_3 + \cdots + (m-1)\pi_m + m\pi_{m+1}$$
$$= (n_2 - V_2) + (n_3 - V_3) + \cdots + (n_m - V_m)$$
$$+ (n_{m+1} - V_{m+1}) \quad (5.65)$$

Continuing this process, we obtain for the row preceding the final zero row the following condition.

$$\pi_m + 2\pi_{m+1} = (n_m - V_m) + (n_{m+1} - V_{m+1}) \quad (5.66)$$

Finally, the number π_{m+1} of real distinct positive zeros of $\Pi(\omega^2) = 0$ with multiplicities $m + 1$ is given by

$$\pi_{m+1} = (n_{m+1} - V_{m+1}) \quad (5.66a)$$

Substituting (5.66a) into (5.66), we obtain

$$\pi_m = (n_m - V_m) - (n_{m+1} - V_{m+1}) \quad (5.66b)$$

It is noticed that (5.66) is (5.62) for $\nu = m$. Proceeding in this manner by starting from (5.66a) and substituting in the preceding ones in chronological order, we readily verify (5.62). Equation (5.63) is the same as (5.66a). As an example, assume that $\pi(\omega^2)$ has roots of multiplicity one, two, and three. Hence from (5.64) we have the total number of positive real roots, including its multiplicity, as

$$\pi_1 + 2\pi_2 + 3\pi_3 = n_1 - V_1 + n_2 - V_2 + n_3 - V_3 \quad (5.67)$$

From (5.65) we have

$$\pi_2 + 2\pi_3 = n_2 - V_2 + n_3 - V_3 \quad (5.68)$$

and from (5.66a), we have

$$\pi_3 = n_3 - V_3 \quad (5.69)$$

If we substitute (5.69) into (5.68), we get

$$\pi_2 = (n_2 - V_2) - (n_3 - V_3) \quad (5.70)$$

Now, if we substitute (5.70) and (5.69) in (5.67), we obtain

$$\pi_1 = (n_1 - V_1) - (n_2 - V_2) \quad (5.71)$$

Therefore π_1, π_2, and π_3 satisfy (5.62) and (5.63) as given by the theorem. Note if $\Pi(\omega^2) = 0$ has no multiple roots, and hence no zero rows occur in (5.54), then the number of positive real roots is obtained from (5.64) as follows:

$$\pi = (n_1 - V_1) = n - V \quad (5.72)$$

which is identical to (5.55).

Example 5.2. Consider the nonnegativity condition of the following polynomial.

$$\Pi(\omega^2) = a_4\omega^4 + a_2\omega^2 + a_0 \geq 0$$

$$\text{for all real } \omega \geq 0 \quad (5.73)$$

The corresponding Routh array from (5.54) is

r				
1	ω^4	a_4	$-a_2$	a_0
2	ω^3	$2a_4$	$-a_2$	0
3	ω^2	$\left(-\dfrac{a_2}{2}\right)$	a_0	
4	ω^1	$\dfrac{(a_2^2 - 4a_0a_4)}{-a_2}$		
5	ω^0	a_0		

$$(5.74)$$

The necessary conditions satisfying (5.73) are $a_4 > 0$, $a_0 \geq 0$, and we have the following cases.

Case 1. $a_2 \geq 0$. Then (5.73) is satisfied.

Case 2. If $a_2 < 0$, then $a_2^2 - 4a_0a_4 \leq 0$. Applying the theorem for the equality, by replacing it by $-2a_2$ (the coefficient of the derivative of $2a_0 - a_2\omega^2$), we obtain $\pi_1 = (1 - 0) - (1 - 0) = 0$ and $\pi_2 = 1 - 0 = 1$[†] while for the inequality $\pi_1 = (2 - 2) = 0$. Thus nonnegativity is satisfied. The preceding conditions are exactly those of (5.23) and (5.24) obtained from the inners approach. Similar conditions can be obtained for $n = 3$. It may be noted that Theorem 5.3 can be also utilized in Theorem 5.4 to obtain certain simplifications. Furthermore, in this theorem, if $n_\nu \neq 0$ for any odd ν, then nonnegativity is violated and the test can be terminated.

5.2 POSITIVITY AND NONNEGATIVITY OF RECIPROCAL POLYNOMIALS THAT ARISE IN DISCRETE SYSTEMS [12]

In this section we will present an algebraic criterion for positive realness of real rational functions and matrices with respect to the unit circle in the complex plane—the circle positive realness. We define first the circle positive realness of real rational functions relative to the unit circle in the complex

[†] Note that in this case the polynomial $(a_2 + 2a_4\omega^2) = (2a_0 + a_2\omega^2)$ is a multiple zero of $\Pi(\omega^2) = 0$ in (5.73).

plane. Then the problem of testing this kind of positive realness is reduced to the algebraic problem of determining the distribution of zeros of a real polynomial with respect to and on the unit circle. A table form, as well as the inners approach, can be used for such testing. We will also obtain, similar to Section 5.1, a recursive algorithm for circle positive realness. Since the algorithm is based on singular polynomial sequences, we will use some results obtained for such cases.

The algebraic algorithm for testing the circle positive realness can be used in the quantitative analysis of stability, and exponential stability of nonlinear discrete systems, as well as in the stability of multinonlinear discrete control systems. These applications will be discussed in detail in the next section.

Let us consider a real rational function

$$G(z) = \frac{q(z)}{p(z)} \tag{5.75}$$

with relatively prime polynomials

$$p(z) = \sum_{u=0}^{m} p_u z^u \qquad q(z) = \sum_{v=0}^{\ell} q_v z^v \tag{5.76}$$

where $\qquad\qquad m \geq \ell \quad$ and $\quad p_m \neq 0 \tag{5.77}$

DEFINITION 5.2 [12]. *A real rational function $G(z)$ is circle positive real if and only if*

(i) *$G(z)$ is analytic for all $z : |z| > 1$* $\tag{5.78}$

(iia) *Re $G(z) \geq 0$, for all $z : |z| \geq 1$* $\tag{5.79}$

If (iia) *is replaced by strict inequality*

(iib) *Re $G(z) > 0$, for all $z : |z| \geq 1$* $\tag{5.80}$

the function $G(z)$ is said to be strictly positive real. Similar to the discussions of the preceding section, we will study this case separately from the positive real condition.

In order to reduce the analytic conditions of the above definition [i.e., (i) and (iia,b)] to algebraic conditions, we use the following bilinear transformation, which is well known for the continuous case.

$$S(z) = \frac{G(z) - 1}{G(z) + 1} \tag{5.81}$$

Let us denote by $f(z)$ the denominator of the function $S(z)$,

$$f(z) = p(z) + q(z) \tag{5.82}$$

THEOREM 5.5 [12]. *A real rational function $G(z)$ is circle positive real if and only if*

(i) $f(z) = 0$ *implies* $|z| < 1$ (5.83)

(ii)   Re $G(z) \geq 0$ *for all* $z : |z| = 1$ (5.84)

Proof: To show necessity, we note from the above definition that if $G(z)$ is positive real, so is $G(z) + 1$. If $G(z) + 1 = 0$ for some $z : |z| \geq 1$, then for that z we have $G(z) = -1$ or, equivalently, Re $G(z) = -1$, which contradicts the fact that $G(z)$ is circle positive real [conditions (5.78) and (5.79)]. Thus $f(z)$ has all its zeros inside the unit disk $D = \{z : |z| < 1\}$ and (i) of (5.83) follows. Condition (ii) of (5.84) is automatic. In order to demonstrate sufficiency, we use (5.81) and conclude from (5.84) of Theorem 5.5 that

$$1 - |S(z)| \geq 0 \qquad \text{for all } z : |z| = 1 \quad (5.85)$$

By the Maximum Modulus theorem [3], and noting (5.81) and (5.80), we get

$$1 - |S(z)| \geq 0 \qquad \text{for all } z : |z| \geq 1 \quad (5.86)$$

Finally, (5.86), together with (5.81) implies conditions of the definition in (5.78) and (5.79) and the proof is complete.

To test the function $G(z)$ for strict positive realness, condition (ii) of Theorem 5.5 is replaced by

$$\text{Re } G(z) > 0 \qquad \text{for all } z : |z| = 1 \quad (5.87)$$

Condition (i) of (5.83) can readily be tested by using any of the algebraic methods of stability (all zeros inside the unit disk) of linear discrete systems. Among these methods, the inners approach discussed in Section 2.3 and the table form [13] are useful in performing the test.

Condition (ii) in (5.84) is equivalent to the polynomial inequality [by noting (5.75)] given as

$$\tilde{g}(z, z^{-1}) = \frac{1}{2}[p(z)q(z^{-1}) + p(z^{-1})q(z)] \geq 0$$

$$\text{for all } z : |z| = 1 \quad (5.88)$$

The above polynomial can be written as

$$\tilde{g}(z, z^{-1}) = \sum_{k=0}^{n} b_k(z^k + z^{-k}) \quad (5.89)$$

where the coefficient

$$b_k = \frac{1}{2}\sum_{j=k}^{n}(q_j p_{j-k} + q_{j-k}p_s) \quad (5.90)$$

Also,
$$\tilde{g}(z, z^{-1}) = z^{-n}g(z) \tag{5.91}$$

where the polynomial

$$g(z) = \sum_{k=0}^{n} b_k(z^{n+k} + z^{n-k}) \tag{5.92}$$

has exactly the same number of zeros on the unit circle $C = \{z : |z| = 1\}$ as the polynomial $\tilde{g}(z, z^{-1})$, and, also, the multiplicity of the zeros is the same. The number $2n$ is equal to the largest even integer contained in $m + \ell$. It is of interest to note that the polynomial in (5.92) is a reciprocal (or self-inversive) polynomial. That is, if it has a root inside the unit circle, then a reciprocal of that root exists outside the unit circle. Also, it contains roots on the unit circle with any multiplicity. From (5.88), we have the following result.

LEMMA 5.1   *The polynomial $\tilde{g}(z, z^{-1}) \geq 0$ for all $z : |z| = 1$ if and only if $\tilde{g}(z, z^{-1}) \not\equiv b_0$ has no zeros of odd multiplicity on the unit circle $C = \{z : |z| = 1\}$ and $\tilde{g}(z_1, z_1^{-1}) > 0$ for some $z_1 : |z_1| = 1$; or $g(z)$ in (5.89) is identically $b_0 \geq 0$.*

This lemma is analogous to the nonnegativity condition in (5.7), as discussed in Remark 2 of Section 5.1. The condition $\tilde{g}(z_1, z_1^{-1}) > 0$ for some $z_1 : |z_1| = 1$ can be replaced by either one of the following inequalities:

$$\sum_{k=0}^{n} b_k > 0 \qquad \sum_{k=0}^{n} (-1)^k b_k > 0 \tag{5.93}$$

That is, $\tilde{g}(1, 1) = g(1) > 0$, $\tilde{g}(-1, -1) = (-1)^n g(-1) > 0$ from (5.91) and (5.92).

The condition for strictly positive realness can be expressed, noting the above discussion, as follows.

LEMMA 5.2.   *The polynomial $\tilde{g}(z, z^{-1}) > 0$ for all $z : |z| = 1$ if and only if $g(z) \not\equiv b_0$ has no zeros on the unit circle $C = \{z : |z| = 1\}$ and $\sum_{k=0}^{n} b_k > 0$; or $g(z) \equiv b_0 > 0$.*

We discuss first the case of positivity, which is simpler than the non-negativity and later we formulate the latter test. In the following discussion both tests are based on the table form. Hence we briefly introduce the table for counting the number of zeros inside, outside, and on the unit circle.

TABLE FORM [13]

Let the real polynomial $F(z)$ be given as

$$F(z) = a_0 + a_1 z + a_2 z^2 + \cdots + a_k z^k + \ldots + a_n z^n$$
$$a_n > 0 \tag{5.94}$$

Form the following table.

Row	z^0	z^1	z^2	$\cdots$	z^k	$\cdots$	z^{n-1}	z^n
1	a_0	a_1	a_2	$\cdots$	a_{n-k}	$\cdots$	a_{n-1}	a_n
2	a_n	a_{n-1}	a_{n-2}	$\cdots$	a_k	$\cdots$	a_1	a_0
3	$\delta_1 = b_0$	b_1	b_2	$\cdots$	$\cdots$	$\cdots$	b_{n-1}	
4	b_{n-1}	b_{n-2}	b_{n-3}	$\cdots$	$\cdots$	$\cdots$	b_0	
5	$\delta_2 = c_0$	c_1	c_2	$\cdots$	$\cdots$	c_{n-2}		
6	c_{n-2}	c_{n-3}	c_{n-4}	$\cdots$	$\cdots$	c_0		
	$\vdots$	$\vdots$	$\vdots$	$\vdots$	$\vdots$	$\vdots$		
$2n-3$	$\delta_{n-1} = r_0$	r_1	r_2					
$2n-2$	r_2	r_1	r_0	$\cdots$				
$2n-1$	$\delta_n = \ell_0 = r_0^2 - r_2^2$							

$$(5.95)$$

It should be noted that the elements of row $2k + 2$ consist of the elements of the row $2k + 1$ written in reverse order, $k = 0, 1, 2, \ldots$, where

$$b_k = \begin{vmatrix} a_0 & a_{n-k} \\ a_n & a_k \end{vmatrix}, \quad c_k = \begin{vmatrix} b_0 & b_{n-1-k} \\ b_{n-1} & b_k \end{vmatrix}, \quad \cdots, \quad \ell_0 = \begin{vmatrix} r_0 & r_2 \\ r_2 & r_0 \end{vmatrix}$$

$$(5.96)$$

THEOREM 5.6 *Form* $p_k = \delta_1 \delta_2 \delta_3 \cdots \delta_k$, $k = 1, 2, \ldots, n$. *If none of the* δ_k's *is zero, the number of roots inside the unit circle is equal to the number of negative* p_k *and the number outside the unit circle is n minus this number. For testing condition* (i) *in* (5.83), *we require all the zeros of* $F(z) = 0$ *to be inside the unit circle—that is, the number of negative* p_k, $k = 1, 2, \ldots, n$ *is "n."*

If we apply the preceding table to $g(z) = 0$, we notice that the third row is identically zero. This is because $g(z) = 0$ is a reciprocal (self-inversive) polynomial. To proceed with the table and the test, we replace the polynomial preceding $g_1(z)$, the zero row, by a new polynomial formed as follows:

$$g_{\text{new}}(z) = [g_1'(z)]^*$$

$$(5.97)$$

where $g_1'(z) = dg_1(z)/dz$ and $*$ represents the conjugate polynomial.† By performing the above operation for all the identically zero polynomials, we can ascertain the number of roots inside, outside, and on the unit circle. It should be noted that the number of zeros inside the unit circle of the new polynomial $g_{new}(z)$ is the same as that of the polynomial $g_1(z)$ or $g^*(z)$.

For strict positivity, we require the number of roots of $g(z) = 0$ on the unit circle to be zero. This condition is guaranteed if the number of zeros of $g_{new}(z)$ inside the unit circle is "n." This indicates that $g(z) = 0$ has n zeros inside the unit circle, and being a reciprocal or self-inversive polynomial, it has n zeros outside the unit circle [24]. Since the degree of $g(z) = 0$ is $2n$, positivity follows.

Example 5.3.   Let $g(z)$ be given as

$$g(z) = -0.75 + 0.975z + 0.75z^2 + 0.975z^3 - 0.75z^4 \qquad (5.98)$$

Form the following table.

$$
\begin{array}{cccccc}
g(z) & -0.75 & 0.975 & 0.75 & 0.975 & -0.75 \\
g^*(z) & -0.75 & 0.975 & 0.75 & 0.975 & -0.75 \\
\hline
g_1(z) & \delta_1 = 0 & 0 & 0 & 0 &
\end{array}
\qquad (5.99)
$$

To proceed, form

$$g_{new}(z) = \left[\frac{dg(z)}{dz}\right]^* = [0.975 + 1.5z + 2.925z^3 - 3z^3]^*$$
$$= -3 + 2.925z + 1.5z^2 + 0.975z^3 \qquad (5.100)$$

Now we continue the table for the new polynomial $g_{new}(z)$:

Row				
1	-3	2.925	1.5	0.975
2	0.975	1.5	2.925	-3
3	$\delta_1 = 8.06$	-10.24	-7.4	
4	-7.4	-10.24	8.06	
5	$\delta_2 = 9.24$	-152		
6	-152	9.24		
7	$\delta_3 < 0$			

$$(5.101)$$

† Note that $g^*(z) = z^n g(1/z)$, where n is the degree of $g(z)$.

Using Theorem 5.6, we have

$$p_1 = \delta_1 > 0 \qquad p_2 = \delta_1\delta_2 > 0 \qquad p_3 = \delta_1\delta_2\delta_3 < 0$$

Hence $g_{\text{new}}(z)$ or, equivalently, $g(z)$ has one zero inside the unit circle. Being self-inversive, it also has one zero outside the unit circle. Since $n = 4$, it has two zeros on the unit circle, and therefore $g(z)$ is not positive.

Actually, the violation of strict positivity can be readily determined from (5.93). In this case, $\sum_{k=0}^{2} (-1)^k b_k < 0$, and thus one of the necessary conditions is violated.

In summarizing the foregoing discussion, we can restate the strict positivity condition from (5.83) and (5.87) as follows.

A real rational function $G(z)$ is strictly circle positive real if and only if

1. No. $\quad p_k(f) < 0 = m \qquad \sum_{k=0}^{m} a_k > 0 \qquad (-1)^m \sum_{k=0}^{m} (-1)^k a_k > 0 \quad$ (5.102)

where

$$f(z) = \sum_{k=0}^{m} a_k z^k \qquad\qquad a_m > 0 \quad (5.103)$$

2. No. $\quad p_k(g) < 0 = n \qquad \sum_{k=0}^{n} b_k > 0 \qquad \sum_{k=0}^{n} (-1)^k b_k > 0 \qquad$ (5.104)

where $g(z)$ is given in (5.92).

ALGORITHM FOR TESTING THE CIRCLE POSITIVE REALNESS [12]

In order to obtain a recursive algorithm for testing Lemma 5.1, we need to obtain a systematic method for ascertaining the multiplicity of the roots, if they exist, on the unit circle. This method is obtained following these observations:

1. If a root of multiplicity more than one exists on the unit circle, then we must have more identically zero polynomials in the table for counting the root distribution. We assume that ζ is the total number of identically zero polynomials in the sequence of the table, starting with $g(z)$.

2. The polynomial preceding the second zero polynomial contains roots on the unit circle, but with multiplicity reduced by one, because of the process of differentiation. Similarly, the order of multiplicity is reduced by one each time we arrive at a zero polynomial.

We denote the number of roots on the unit circle of multiplicity one as π_1, of multiplicity two as π_2, and of multiplicity μ as π_μ. Evidently $\mu = 1, 2, \ldots$. We denote the degree of $g(z) = 0$ as $2n$. We denote the number of negative $p_k(g)$ in the table for the sequence between the newly formed polynomial preceding the first zero polynomial until we reach the polynomial preceding the second zero polynomial as η_1. Similarly, we have $\eta_2, \eta_3, \ldots$.

Based on these notations, we have for the total number of roots on the unit circle,

$$\pi_1 + 2\pi_2 + 3\pi_3 + \cdots + \mu\pi_\mu = 2n - 2\eta_1 - 2\eta_2 \cdots - 2\eta_\mu \quad (5.105)$$

The number $2n$ can be composed of $2n = n_\alpha = n_1 + n_2 + n_3 + \cdots + n_\mu$. Now, n_1 is the difference between the degree of $g(z) = 0$, which is $2n$, and the degree of the polynomial preceding the first zero row. The number n_2 is the difference between the degrees of the polynomials preceding the second and third zero rows, and so on.

Therefore we can write for (5.105),

$$\pi_1 + 2\pi_2 + 3\pi_3 + \cdots + \mu\pi_\mu = n_1 - 2\eta_1 + n_2 - 2\eta_2 + \cdots + n_\mu - 2\eta_\mu$$
$$(5.106)$$

Similarly, the number of zeros on the unit circle for the new polynomial replacing the second zero row is given by

$$\pi_2 + 2\pi_3 + \cdots + (\mu - 1)\pi_\mu = n_2 - 2\eta_2 + n_3 - 2\eta_3 + \cdots + n_\mu - 2\eta_\mu$$
$$(5.107)$$

Continuing this process, we obtain, for the final zero row,

$$\pi_\mu = n_\mu - 2\eta_\mu \quad (5.107a)$$

In this equation the degree n_μ is the degree of the polynomial preceding the last zero row. Substituting (5.107a) in the preceding one, and this, in turn, in the preceding one, we obtain similar to (5.62),

$$\pi_\mu = (n_\mu - 2\eta_\mu) - (n_{\mu+1} - 2\eta_{\mu+1})$$
$$\mu = 1, 2, \ldots, \zeta \quad (5.108)$$

Here, when $\mu = \zeta$, $n_{\zeta+1}$ and $\eta_{\zeta+1}$ are equal to zero. To explain the preceding formula, let us assume that $\mu = 1, 2, 3$. From (5.106) and (5.107), we have

$$\pi_1 + 2\pi_2 + 3\pi_3 = n_1 - 2\eta_1 + n_2 - 2\eta_2 - 2\eta_3 \quad (5.109)$$

$$\pi_2 + 2\pi_3 = n_2 - 2\eta_2 + n_3 - 2\eta_3 \quad (5.110)$$

and from (5.107a), we obtain

$$\pi_3 = n_3 - 2\eta_3 \quad (5.111)$$

Substitute (5.111) in (5.110) to obtain

$$\pi_2 = n_2 - 2\eta_2 + n_3 - 2\eta_3 - 2n_3 + 4\eta_3 = n_2 - n_3 - 2\eta_2 + 2\eta_3$$
$$= (n_2 - 2\eta_2) - (n_3 - 2\eta_3)$$
$$(5.112)$$

Similarly, by substituting (5.112) and (5.111) in (5.109), we obtain

$$\pi_1 = (n_1 - 2\eta_1) - (n_2 - 2\eta_2) \qquad (5.113)$$

Note that if there is no second zero-row polynomial, then from (5.108),

$$\pi = (n_1 - 2\eta_1) = 2n - 2\eta \qquad (5.114)$$

which checks with the strict positivity condition requiring that $\pi = 0$—that is, when $\eta = n$.

In summarizing this discussion we can restate the circle positivity condition as follows.

THEOREM 5.7 [12]. *A real rational function $G(z)$ is circle positive real if and only if*

(i) Number $\quad p_k(f) < 0 = m, \quad f(1) > 0, \quad (-1)^m f(-1) > 0 \qquad (5.115)$

(ii) *for $g(z) \not\equiv b_0$, $\pi_\mu(g) = 0$, for odd $\mu = 1, 2, \ldots, \zeta$ and $g(z_1) > 0$ for some $z_1 : |z_1| = 1$; or $g(z) \equiv b_0 \geq 0$* $\qquad (5.116)$

Example 5.4. Let $g(z)$ be given as

$$g(z) = 1 - 2.172z - 1.89z^2 + 1.16z^3 + 6.72z^4 + 1.16z^5$$
$$- 1.89z^6 - 2.172z^7 + z^8 = 0 \qquad (5.117)$$

In the above polynomial, (5.93) is satisfied.

Form the table as in (5.95).

$g(z)$	1	-2.172	-1.89	1.62	6.72	1.16	-1.89	-2.172	1
$g^*(z)$	1	-2.172	-1.89	1.62	6.72	1.16	-1.89	-2.172	1
$g_1(z)$ $\quad \delta_1 \equiv 0$	0	0	0	0	0	0	0	0	

$$(5.118)$$

To proceed, form $g_1(z)$ using (5.97) to obtain

$$g_1(z) = \left[\frac{dg(z)}{dz}\right]^* = 8 - 15.204z - 11.34z^2 + 8.1z^3$$
$$+ 30.88z^4 + 4.86z^5 - 3.78z^6 - 2.172z^7 \qquad (5.119)$$

Form the table again for $g_1(z)$.

	8	-15.204	-11.34	8.1	30.1	4.9	-3.8	-2.18
	-2.18	-3.8	4.9	30.1	8.1	-11.34	-15.2	8
$\delta_1 =$	59.28	-129.84	-80.16	121.87	264.63	14.25	-63.26	
	-63.26	14.25	264.63	121.87	-80.16	-129.84	59.28	
$\delta_2 =$	-488.53	-796.03	11990.83	14935.48	10616.35	-7370.21		
	-7370.21	10616.35	14935.48	11990.83	-796.30	-488.53		
$\delta_3 =$	-5408.14	8213.15	10419.73	8213.15	-5408.14			
$g_2(z) \equiv$	-5408.14	8213.15	10419.73	8213.15	-5408.14			
$\delta_4 =$	0	0	0	0				

$$(5.120)$$

Now we obtain a second zero-row polynomial. Hence we proceed as before to obtain a new polynomial.

$$g_3(z) = \left[\frac{dg_2(z)}{dz}\right]^* = (-21.63 + 24.63z + 20.83z^2 + 8.21z^3) \times 10^3$$

$$(5.121)$$

Proceed with the table form in the same manner.

	-21.63	24.63	20.83	8.21		
	8.21	20.83	24.63	-21.63		
$\delta_1 =$	4.005	-7.04	-6.53		$\Big\} \times 10^2$	(5.122)
	-6.53	-7.04	4.005			
$\delta_2 = $	-26.62	-74.11				
	-74.11	-26.22				
$\delta_3 < 0$						

From (5.108), we have

$$\pi_1 = (n_1 - 2\eta_1) - (n_2 - 2\eta_2) \tag{5.123}$$

$$\pi_2 = (n_2 - 2\eta_2) \tag{5.124}$$

In this case, $n_1 = 8 - 4 = 2n - n_2$, and $n_2 = 4$. Also, $\delta_1 > 0$, $\delta_2 < 0$, $\delta_3 < 0$ for the first sequence. Hence $p_1 > 0$, $p_2 < 0$, $p_3 > 0$, and $\eta_1 = 1$. Similarly, for the second sequence, $\delta_1 > 0$, $\delta_2 < 0$, $\delta_3 < 0$. Hence $p_1 > 0$, $p_2 < 0$, $p_3 > 0$, and thus $\eta_2 = 1$. Therefore from (5.123), we obtain

$$\pi_1 = (4 - 2) - (4 - 2) = 0 \tag{5.125}$$

$$\pi_2 = (4 - 2) = 2 \tag{5.126}$$

The above indicates that there exist no roots on the unit circle with multiplicity 1 (or odd) and there exist two roots on the unit circle with multiplicity 2 (even multiplicity). Therefore the polynomial $g(z)$ is nonnegative. Indeed, $g(z)$ is equal to

$$g(z) \equiv (z - 0.5)^2 (z - 2)^2 \left(z + \frac{1}{\sqrt{2}} - \frac{j}{\sqrt{2}}\right)^2 \left(z + \frac{1}{\sqrt{2}} + \frac{j}{\sqrt{2}}\right)^2 = 0$$

$$(5.127)$$

which verifies the results of the algorithm test. We can also ascertain the root distribution from the test. Since $\eta_1 + \eta_2 = 2$, $g(z) = 0$ has two zeros inside the unit circle, and being self-inversive, it has two reciprocal zeros outside the unit circle.

Remarks
1.   If π_1 in (5.108) is not zero, then the strict positivity test can be discontinued, for the polynomial is not positive.

2.   For the case of strict positivity, we can simplify the table test by noting that if in the table test we first find that $g_1(z)$ has $(n-1)$ zeros inside the unit circle (for n-even) and $g(1) > 0, g(-1) > 0$, then it has n zeros inside the unit circle. Therefore $g(z)$ has n zeros inside the unit circle and n zeros outside the unit circle and no zeros on the unit circle. It follows that strict positivity is ensured. Similarly, as noted before, if $g(1) > 0$ and $g(-1) < 0$, then $g(z)$ has at most $(n-1)$ zeros inside the unit circle and $g(z)$ is not positive. Similar conclusions can be arrived at when n is odd. The above simplifies the table test by not requiring all the δ_k's to be computed. This is analogous to Theorem 5.3 of Section 5.1.

3.   For strict positivity, we can establish another necessary condition by noting that $g(z_1) > 0$ when $z_1 = \pm j$.

MATRIX GENERALIZATION [5,12,25]

In the following, we will present the positive and the strictly positive conditions for rational matrices for both the continuous and discrete cases.

Continuous Case.   Let us consider a real rational $m \times m$ matrix

$$G(s) = \frac{Q(s)}{p(s)} \tag{5.128}$$

where $Q(s)$ is a real polynomial $m \times m$ matrix and $p(s)$ is a real scalar polynomial prime to $Q(s)$. We state [5] Theorem 5.8.

THEOREM 5.8   *A rational $m \times m$ matrix $G(s)$ defined in (5.128) is positive real if and only if:†*

   (i)   *The polynomial $f(s)$, obtained from*

$$\frac{F(s)}{f(s)} = [I + G(s)]^{-1}$$

where $F(s)$ is a real polynomial $m \times m$ matrix and $f(s)$ is a real scalar polynomial prime to $F(s)$, is Hurwitz.

   (ii)   *The Hermitian $m \times m$ polynomial matrix*

$$\Gamma(j\omega) \triangleq p^*(j\omega)Q(j\omega) + p(j\omega)Q^*(j\omega) \tag{5.129}$$

---

† The usual positivity conditions of $G(s)$ are as follows [37]: $G(s)$ is positive real if and only if the corresponding real rational scattering matrix

$$S(s) = [G(s) - I][G(s) + I]^{-1}$$

is bounded real; that is,
   (i) $S(s)$ is analytic in Re $s \geq 0$
   (ii) $I - S^*(j\omega)S(j\omega) \geq 0$, for all real ω

satisfies the inequality

$$\Gamma(j\omega) \geq 0 \qquad \text{for all real } \omega \geq 0 \quad (5.130)\dagger$$

Let us assume that the matrix $\Gamma(j\omega)$ is of the rank r; that is, there is an rth-order principal minor of $\Gamma(j\omega)$, that is not identically zero and all the principal minors of order $r + 1$ vanish identically. We denote by $g^{(r)}(\omega^2)$ this rth-order minor and by $g^{(\theta)}(\omega^2)$ the associated principal minors of order $\theta = 1, 2, \ldots, r - 1$. As is clear from (5.129), the minors are real even polynomials, which can be written

$$g^{(\theta)}(\omega^2) = \sum_{k=0}^{n\theta} g_{2k}^{(\theta)} \omega^{2k} \qquad (5.131)$$

Following the discussions of Section 5.1, we can state the corresponding algorithm for positivity test.

THEOREM 5.9 *A real rational $m \times m$ matrix $G(s)$ is positive real if and only if:*

(i) *All the elements in the first column of the Routh array equation (5.56) produced by the polynomial $f(s)$ are different from zero and positive.*

(ii) *All the numbers $\pi_\nu^{(\theta)}$ from (5.62) produced by the Routh array (5.54) and each principal minor $g^{(\theta)}(\omega^2) \not\equiv g_0, \theta = 1, 2, \ldots, r$, of the corresponding matrix $\Gamma(j\omega)$ of (5.143), are equal to zero for odd $\nu = 1, 3, \ldots$ \lfloor odd $(t_{\nu+1})\rfloor$, where r is the rank of $\Gamma(j\omega)$ and t_θ is the total number of the zero rows in the array (5.54) generated by $g^{(\theta)}(\omega^2)$. In addition, $g^{(\theta)}(\omega_\theta^2) > 0$, for some $\omega_\theta \geq 0$ and all $\theta = 1, 2, \ldots, r$. If $g^{(\theta)}(\omega_\theta^2) \equiv g_0^{(\theta)}$, then, for $r > 1$, $g_0^{(\theta)} > 0$, and, for $r = 1$, $g_0^{(1)} \geq 0$.*

Discrete Case. Let us consider a real rational $m \times m$ matrix

$$G(z) = \frac{Q(z)}{p(z)} \qquad (5.132)$$

where $Q(z)$ is a real polynomial $m \times m$ matrix and $p(z)$ is a real scalar polynomial prime to $Q(z)$. We state Definition 5.3.

DEFINITION 5.3 [12]. *A real rational $m \times m$ matrix $G(z)$ is circle positive real if and only if*

(i) *$G(z)$ is analytic for all $z : |z| > 1$*   (5.133)

(ii) *$G^*(z) + G(z) \geq 0$ for all $z : |z| \geq 1$*   (5.134)

If (ii) is replaced by

$$G^*(z) + G(z) > 0 \quad \text{for all } z : |z| \geq 1 \quad (5.135)$$

the matrix $G(z)$ is said to be strictly circle positive real.

† For the case of strictly positive real, this equation is replaced by $\Gamma(jw) > 0$.

We start with the expression

$$\frac{F(z)}{f(z)} = [G(z) + I]^{-1} \qquad (5.136)$$

where $F(z)$ is a real polynomial $m \times m$ matrix and $f(z)$ is a real scalar polynomial relatively prime to $F(z)$, and prove Theorem 5.10.

THEOREM 5.10 [12]. *A real rational $m \times m$ matrix $G(z)$ is circle positive real if and only if*

(i) $f(z) = 0$ *implies* $|z| < 1$ \hspace{3cm} (5.137)

(ii) $G^*(z) + G(z) \geq 0$ *for all $z : |z| = 1$* \hspace{1.5cm} (5.138)

In analogy to the proof of Theorem 5.3 for the rational function, we define, following (5.81), the scattering matrix as

$$S(z) = [G(z) - I][G(z) + I]^{-1} \qquad (5.139)$$

Equation (5.139) can be rewritten as

$$S(z) = I - 2[G(z) + I]^{-1} \qquad (5.140)$$

Following the procedure of proving the necessity condition of Theorem 5.5, the necessary part of Theorem 5.10 is automatic.

Conversely, under the conditions of Theorem 5.10, $S(z)$ is analytic for all $z : |z| \geq 1$, and det $|G(z) + I| \neq 0$, so that by the relation

$$2[G^*(z) + G(z)] = [G(z) + I][I - S^*(z)S(z)][G^*(z) + I] \quad (5.141)$$

we get

$$I - S^*(z)S(z) \geq 0 \quad \text{for all } z : |z| = 1 \quad (5.142)$$

and by invoking the Maximum Modulus theorem [3], we obtain

$$I - S^*(z)S(z) \geq 0 \quad \text{for all } z : |z| \geq 1 \quad (5.143)$$

Therefore (5.139), together with (5.143), implies conditions of the above definition, and the proof of Theorem 5.10 is complete.

For testing purposes, we replace (5.138) of Theorem 5.10 by the polynomial inequality

$$\Gamma \equiv p(z)p(z^{-1})[G^*(z) + G(z)] \geq 0$$
$$\text{for all } z : |z| = 1 \quad (5.144)$$

where Γ is an $m \times m$ Hermitian matrix with a principal minor of order θ denoted by $g^{(\theta)}(z)$. Each minor polynomial $g^{(\theta)}(z)$ can be represented by a self-inversive (reciprocal) polynomial $g^{(\theta)}(z)$ [following (5.91) and (5.92)] of order $2n_\theta$. We assume that the Hermitian matrix $\Gamma(z)$ (of rank $r > 0$) is nonnegative definite for all $z : |z| = 1$; that is, inequality (5.144) is satisfied. Following (5.115) and (5.116), we can state a similar algorithm (as far as a real function) for testing a real rational $m \times m$ matrix $G(z)$ as circle positive as follows:

1. Number $\quad p'_k(f) < 0 = m, \quad (-1)^m f(-1) > 0$ $\hspace{2cm}$ (5.145)

2. all numbers $\pi_\mu^{(\theta)}$ for $g^{(\theta)}(z) \not\equiv b_0^{(\theta)}$, $\theta = 1, 2, \ldots, r$, of $\Gamma(z)$ in (5.144) are equal to zero for odd $\mu = 1, 2, \ldots, \zeta$, and $g^{(\theta)}(z) > 0$ for some $z : |z_\theta| = 1$ and all $\theta = 1, 2, \ldots, r$.

$\hspace{10cm}$ (5.145a)

If $g^{(\theta)}(z) \equiv b_0^{(\theta)}$, then, for $r > 1$, $b_0^{(\theta)} > 0$ and, for $r = 1$, $b_0^{(1)} > 0$. The numbers $\pi_\mu^{(\theta)}$ and ζ are as defined in (5.108) for each self-inversive polynomial $g^{(\theta)}(z)$. The scalar polynomial f is defined in (5.136).

CRITICAL CONSTRAINT CASES [12,26]

In the case of strict positive reality conditions for both continuous and discrete cases, we can simplify these conditions by assuming that the system corresponding to positivity conditions is initially positive real. That is, for certain parameters in $\Gamma(j\omega)$, or $\Gamma(z)$, it is known that $\Gamma(j\omega_1) > 0$ for $\omega_1 \geq 0$ and $\Gamma(z_1) > 0$, for a certain z on the unit circle.

Under the above assumptions, referred to as the critical cases, the test for strict positivity reduces to

$$\tilde{g}(\omega^2) = |\Gamma(j\omega)| > 0$$
$$\text{for all } \omega \text{ and } \Gamma(0) > 0 \quad (5.146)$$

for the continuous case, or

$$\tilde{g}(z) = |\Gamma(z)| > 0$$
$$\text{for all } z : |z| = 1 \text{ and } \Gamma(1) > 0 \quad (5.147)$$

for the discrete case.

To show the above, we note first that $\Gamma(z)$ or $\Gamma(j\omega)$ is a Hermitian matrix and that there exists unitary matrices $U(z)$ and $U(j\omega)$ such that

$$\Lambda(z) = U^*(z)\Gamma(z)U(z) \hspace{3cm} (5.148)$$

$$\Lambda(j\omega) = U^*(j\omega)\Gamma(j\omega)U(j\omega) \hspace{2cm} (5.149)$$

where $\hspace{2cm} \Lambda(z) = \text{diag } \{\lambda_1(z), \lambda_2(z), \ldots, \lambda_m(z)\} \hspace{1.5cm}$ (5.150)

$$\Lambda(j\omega) = \text{diag } \{\lambda_1(j\omega), \lambda_2(j\omega), \ldots, \lambda_m(j\omega)\} \hspace{1cm} (5.151)$$

It is evident from the above that $\Gamma(z)$ or $\Gamma(j\omega)$ is positive definite if and only if $\Lambda(z)$ and $\Lambda(j\omega)$ are also positive definite. Therefore if the condition for strict positivity is violated, it first shows when $|\Gamma(z)| = 0$ or $|\Gamma(j\omega)| = 0$.

Hence the conditions $|\Gamma(z)| = 0$ and $|\Gamma(j\omega)| = 0$ are referred to as critical constraints.

5.3 APPLICATIONS [7]

In this section we will indicate some of the applications of the positivity and nonnegativity conditions discussed earlier. In particular, we will mention applications to absolute stability of nonlinear continuous and discrete systems, to optimality conditions, and to positivity problems that arise in network synthesis. In the next chapter we will also discuss the application of positivity to the stability of polynomials with two or more variables.

a. Absolute Stability [27]†

CONTINUOUS CASE

Let us consider the nonlinear system shown in Figure 5.1 and described by

$$\frac{dx}{dt} = Ax + b\phi(\sigma) \qquad\qquad \sigma = c'x \quad (5.152)$$

where $x = x(t)$ is a real n vector, the state of the system, b, c are real constant

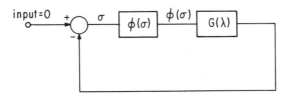

Figure 5.1. System description of (5.152).

n vectors, A is a real $n \times n$ constant matrix, and $\phi(\sigma)$ is a real-valued continuous function of a real scalar σ that belongs to class

$$\Gamma_k : \phi(0) = 0 \quad 0 \le \sigma\phi(\sigma) \le K\sigma^2, \quad K < +\infty$$

Luré [27] introduced the following definition. "The system in (5.152) is said to be absolutely stable if the equilibrium $x = 0$ of (5.152) is globally asymptotically stable for any $\phi(\sigma) \in \Gamma_k$."

† A comprehensive study of this problem has appeared in the following: K. S. Narendra and J. H. Taylor, *Frequency Domain Criteria for Absolute Stability*, New York: Academic Press, 1973.

The Luré problem consists in finding conditions on A, b, c, and K under which (5.152) is absolutely stable. A solution to this problem was given by Popov [29] in terms of the frequency characteristic of the linear plant:

$$G(\lambda) = c'(A - \lambda I)^{-1}b \qquad (5.153)$$

which is the transfer function of the linear part of the system in (5.152) from the input ϕ to the output $-\sigma$ and $\lambda = \delta + j\omega$ is the complex variable. By assuming that A is a Hurwitz matrix, that is, all zeros of the polynomial

$$\Delta(\lambda) = |(A - \lambda I)| \qquad (5.154)$$

lie in the open left half of the λ-plane, and the rational function $G(\lambda)$ has no common cancelable factors [or, equivalently, the linear part in (5.152) is completely controllable and observable [28], Popov proved Theorem 5.11 [29].

THEOREM 5.11   *The system in (5.152) is absolutely stable if*

$$K^{-1} + \mathrm{Re}\,[1 + j\omega q]G(j\omega) > 0$$
$$\textit{for all real } \omega \geq 0 \quad (5.155)$$

holds for some real number q.

The function $G(j\omega)$ can be written as

$$G = \frac{C_1 + jC_2}{B_1 + jB_2} \qquad (5.156)$$

where
$$B_1 = \sum_{k=0}^{n} b_k X_k \qquad B_2 = \sum_{k=0}^{n} b_k Y_k \qquad (5.157)$$

$$C_1 = \sum_{k=0}^{n} c_k X_k \qquad C_2 = \sum_{k=0}^{n} c_k Y_k \qquad (5.158)$$

Substituting relationships (5.156) and (5.158) in (5.155) (and noting that the denominator is always positive), we obtain the following equivalent inequality.

$$\Pi(\omega^2) = S(\omega)\frac{1}{K} + Q(\omega)q + T(\omega) > 0$$
$$\text{for all } \omega \geq 0 \quad (5.158a)$$

where $S(\omega)$, $Q(\omega)$, and $T(\omega)$ are polynomials in ω^2.

$$S = B_1^2 + B_2^2, \qquad Q = \omega(B_2 C_1 - B_1 C_2), \qquad T = B_1 C_1 + B_2 C_2$$
$$(5.159)$$

If we assume that q and K are given, then (5.159) can be written

$$\Pi(\omega^2) = \sum_{k=0}^{n} a_{2k}\omega^{2k} > 0$$

for all real $\omega \geq 0$ (5.160)

This inequality reduces to the positivity condition discussed in Section 5.1. Hence (5.160) offers an algebraic alternative to the useful graphical solution of Popov criterion.

DISCRETE CASE [30]

A free nonlinear discrete system can be described by the difference equation

$$x_{n+1} = Ax_n + b\phi(\sigma_n) \qquad \sigma_n = c'x_n \quad (5.161)$$

where x, b, and c are real vectors, A is a real $n \times n$ matrix, and $\phi(\sigma)$ is a continuous real scalar function of the real variable σ that belongs to the class $\Phi_k : \phi(0) = 0, 0 \leq \sigma\phi(\sigma) \leq K\sigma^2, K < +\infty$.

Denote the transfer function of the linear part of (5.161) by $G(z) = c'(A - zI)^{-1}b$, where z is the complex variable. Let A be a stable matrix; that is, all zeros of $\Delta(z) = |(A - zI)|$ are inside the unit circle $|z| = 1$. We have the following theorem due to Tsypkin [31].

THEOREM 5.12 *The system of (5.161) is absolutely stable if*

$$K^{-1} + \text{Re}\,[G(z)] > 0$$

for all $z : |z| = 1$ (5.162)

If we let

$$G(z) = \frac{p(z)}{q(z)} \tag{5.163}$$

then we can easily verify that

$$\text{Re}\,[G(z)] = \frac{1}{2}\left[\frac{p(z)q(z^{-1}) + p(z^{-1})q(z)}{q(z)q(z^{-1})}\right] \tag{5.164}$$

Assuming K in (5.162) is given, the latter inequality [noting that $q(z)q(z^{-1})$ is always positive for $|z| = 1$] can be written

$$\tilde{g}(z, z^{-1}) = \sum_{k=0}^{n} b_k(z^k + z^{-k}) > 0$$

for all $z : |z| = 1$ (5.165)

where n is the degree in z of $q(z)$.

The polynomial $\tilde{g}(z, z^{-1})$ in (5.165) can also be written

$$\tilde{g}(z, z^{-1}) = z^{-n}g(z) \tag{5.166}$$

where the self-inversive (reciprocal) polynomial $g(z)$ is given by

$$g(z) = \sum_{k=0}^{n} b_k(z^{n+k} + z^{n-k}) \tag{5.167}$$

Hence the stability theorem is equivalent to the following positivity condition.

$$g(z) = \sum_{k=0}^{n} b_k(z^{n+k} + z^{n-k}) > 0$$
$$\textit{for all } z : |z| = 1 \quad (5.168)$$

The test of this inequality is discussed in the preceding section.

If for the nonlinear discrete system expressed in (5.161) we impose the following additional condition on the nonlinearity:

$$-\infty \leq \frac{d\phi(\sigma)}{d\sigma} \leq K_1 \tag{5.169}$$

we have the following.

THEOREM 5.13 [32].   *A nonlinear discrete system of (5.161) is absolutely stable if a nonnegative q exists such that the inequality*

$$\text{Re } G(z)[1 + q(z - 1)] + \frac{1}{K} - \frac{K_1|q|}{2} |(z - 1)G(z)|^2 > 0 \quad (5.170)$$

is satisfied for all $z : |z| = 1$.

If q, K, and K_1 are given, the inequality of (5.170) can be similarly reduced to testing the positivity of a reciprocal polynomial as given in (5.168) for all $z : |z| = 1$.

It is of interest to note that the preceding theorem reduces to Popov's theorem when the sampling period $T \to 0$ in the limit. Hence it represents the solution of the Luré problem for the discrete case.

Before concluding the discussion of absolute stability, we mention that the analytical tests can also be extended to exponential absolute stability [33], to hyperstability [34], and to absolute stability of multinonlinear continuous and discrete systems [26,35]. In the latter application, the positivity test for the matrix inequalities discussed earlier can be effectively used. Furthermore, the critical constraint plays an important role in the design of such systems [26].

b. Optimality [8]

Let us consider again the system in (5.152) in the form

$$\frac{dx}{dt} = Ax + qv(t) \qquad\qquad v(t) = -r'x \quad (5.171)$$

where the nonlinearity $\phi(\sigma)$ in (5.152) is replaced by a continuous real-valued function of time $v(t)$—the control function. The performance index of the system is

$$J = \int_0^\infty (x'Qx + v^2)\, dt \qquad (5.172)$$

where $Q = R'R \geq 0$ is a nonnegative definite matrix. Under appropriate controllability and observability conditions [the rational function $R(\lambda I - A)^{-1}q$ has no common cancelable factors], Kalman proved Theorem 5.14.

THEOREM 5.14 [8]. *The control function $v(t)$ of the system in (5.171) is optimal with respect to the performance index J, defined in (5.172), if and only if*

(i) *$v(t)$ is a stable control; that is, $(A - qr')$ is Hurwitz*

(ii) *$|1 + r'(j\omega I - A)^{-1}q|^2 \geq 1$* \qquad\qquad *for all $\omega \geq 0$* \quad (5.173)

It is of interest to note that in (5.173) one recognizes the frequency characteristic $1 + r'(j\omega I - A)^{-1}q$ as the well-known Bode [36] return difference, which may serve as a measure of system sensitivity to parameter variations. It is shown by Kalman that

$$|1 + r'\phi(j\omega)q|^2 = 1 + \|H\phi(j\omega)q\|^2 \qquad (5.174)$$

where H is a pre-Hamiltonian function and $\phi(j\omega) = (j\omega I - A)^{-1}$. Also, the double lines in the right-hand side of (5.174) indicate the usual Euclidean norm. Moreover, it is shown by Kalman that

$$|\psi_k(j\omega)|^2 = |j\omega I - A + qr'|^2 = |\psi(j\omega)|^2 |1 + r'\phi(j\omega)q|^2 \qquad (5.175)$$

where

$$\psi_k(j\omega) = |(j\omega I - A + qr')| \qquad (5.176)$$

$$\psi(j\omega) = |(j\omega I - A)| \qquad (5.177)$$

Substituting (5.174) in (5.175), we obtain

$$|\psi_k(j\omega)|^2 = |\psi(j\omega)|^2(1 + \|H\phi(j\omega)q\|^2) \qquad (5.178)$$

or

$$\frac{|\psi_k(j\omega)|^2 - |\psi(j\omega)|^2}{|\psi(j\omega)|^2} = \|H\phi(j\omega)q\|^2 \qquad (5.179)$$

Substituting (5.179) in (5.174), we obtain

$$|1 + r'(j\omega I - A)^{-1}q|^2 - 1 = \|H\phi(j\omega)q\|^2 = \frac{|\psi_k(j\omega)|^2 - |\psi(j\omega)|^2}{|\psi(j\omega)|^2}$$

$$(5.180)$$

Hence we obtain the identical inequalities

$$|1 + r'(j\omega I - A)^{-1}q|^2 - 1 \geq 0 \Leftrightarrow \frac{|\psi_k(j\omega)|^2 - |\psi(j\omega)|^2}{|\psi_k(j\omega)|^2} \geq 0$$

$$\text{for all real } \omega \geq 0 \quad (5.181)$$

It is shown by Lemma 3 of Kalman's work that

$$|\psi_k(j\omega)|^2 - |\psi(j\omega)|^2 = \Pi(\omega^2) = \sum_{k=0}^{n-1} a_{2k}\omega^{2k} \qquad (5.182)$$

Substituting (5.182) in (5.179), and noting that $|\psi(j\omega)|^2 > 0$ for all real $\omega \geq 0$ we finally obtain

$$|1 + r'(j\omega I - A)^{-1}q|^2 - 1 \geq 0 \Leftrightarrow \Pi(\omega^2) \geq 0$$

$$\text{for all real } \omega \geq 0 \quad (5.183)$$

Without loss of generality, we can readily apply the nonnegativity test of Section 5.1. Note that (5.183) is the same as (5.173).

A similar condition for optimality was also obtained by Popov [9] for a stationary stochastic input process. In both cases, the conditions for optimality reduce to that of testing even polynomials for the nonnegativity condition.

c. Passivity [37]

Let us consider again essentially the same system as (5.171), described by the linear equation

$$\frac{dx}{dt} = Ax + qv(t) \qquad\qquad \sigma(t) = r'x \quad (5.184)$$

where $v(t)$ and $\sigma(t)$ are the scalar input and output functions. From linear network synthesis it is known that, by an appropriate choice of the input-output pair (v, σ), the passivity of the system in (5.184) corresponds to the transfer function

$$G(s) = r'(sI - A)^{-1}q = \frac{\Lambda(s)}{\Delta(s)} \qquad (5.185)$$

that is positive real. Hence, for continuity of discussion, we introduce Definition 5.4 (discussed in Section 5.1).

DEFINITION 5.4 *A rational function* $G(s) = \Lambda(s)/\Delta(s)$ *with real and relatively prime polynomials* $\Lambda(s)$ *and* $\Delta(s)$ *is positive real if and only if*

(a)   *the polynomial* $\Gamma(s) = \Lambda(s) + \Delta(s)$ *is Hurwitz*   (5.186)

(b)   Re $G(j\omega) \geq 0$ *for all real* $\omega \geq 0$ (5.187)

We immediately notice that (5.187) can be rewritten as a polynomial inequality

$$\Pi(\omega^2) \equiv \Lambda_1(\omega)\Delta_1(\omega) + \Lambda_2(\omega)\Delta_2(\omega) \geq 0$$
$$\text{for all real } \omega \geq 0 \quad (5.188)$$

where $\Lambda(j\omega) = \Lambda_1(\omega) + j\Lambda_2(\omega)$
and $\Delta(j\omega) = \Delta_1(\omega) + j\Delta_2(\omega)$ (5.189)

Obviously the test of nonnegativity discussed in Section 5.1 is also applicable to (5.188) and hence to the passivity condition.

Needless to say, the passivity condition plays an important role in the physical realizability condition for passive networks. It is thoroughly discussed in the network synthesis literature [2,3] and need not be elaborated further in this text. However, it is of interest to note that several useful graphical procedures [38] have been developed by network theorists to test the nonnegativity condition of (5.189).

5.4   STABILITY OF NONUNIFORMLY DISCRETE AND DISTRIBUTED PARAMETER SYSTEMS [14,15]

In this section we will present a stability criterion for distributed parameter and nonuniformly discrete systems. Specifically, a finite algorithm will be presented which tests whether all the zeros of a function of the form

$$F(s) = \sum_{n=0}^{N} c_n e^{su_n}$$
$$c_n, u_n \text{ are real,} \quad u_0 = 0, \quad u_{n+1} > u_n \quad (5.190)$$

lie within the interior of the left half of the s-plane.

The algorithm also tests the stability of those systems whose system functions are ratios of finite sums of exponentials. Included in such systems are all distributed systems whose components are uniform, lossless transmission lines and all sampled systems with a periodically varying sampling rate.

STABILITY CRITERION

In (5.190), $F(s)$ has all its zeros in the interior of the left half s-plane if and only if

$$\Psi_0(s) = \frac{F(s) - F(-s)e^{su_N}}{F(s) + F(-s)e^{su_N}} = \frac{\sum_{n=0}^{m} a_n e^{sx_n}}{\sum_{n=0}^{m} b_n e^{sx_n}}$$

$$x_0 = 0, \quad x_{n+1} > x_n \quad (5.191)$$

reduces to $0/K$ (K is a constant), under the repeated application of the following algorithm:

$$\Psi_{k-1}(s) = \frac{\sum_n a_n e^{sx_n}}{\sum_n b_n e^{sx_n}}$$

$$x_0 = 0, \quad x_{n+1} > x_n \quad (5.192)$$

1. Terminate the algorithm unless $0 < R_k < \infty$, where

$$R_k = -\frac{a_0}{b_0} \qquad (5.193)$$

If R_k is negative, then $F(s)$ of (5.190) has zeros in the right half s-plane (i.e., it is unstable).

2. Decompose each of the coefficients in the numerator and denominator into two parts, a_{nu}, a_{nv} and b_{nu} and b_{nv} defined as follows:

$$a_{nu} = R_k b_{nv} = a_n + R_k b_n \qquad (5.194)$$

$$a_{nv} = -R_k b_{nv} = a_n - R_k b_n \qquad (5.195)$$

The decomposed function becomes

$$\Psi_{k-1}(s) = \frac{\sum_n a_{nu} e^{sx_n} + \sum_n a_{nv} e^{sx_n}}{\sum_n b_{nu} e^{sx_n} + \sum_n b_{nv} e^{sx_n}} \qquad (5.196)$$

3. Identify the lowest x_n with nonzero q_{nu} as τ_k.

4. Obtain

$$\Psi_k(s) = \frac{\sum_n a_{nu} e^{s(x_n - \tau_k)} + \sum_n a_{nv} e^{sx_n}}{\sum_n b_{nu} e^{s(s_n - \tau_k)} + \sum_n b_{nv} e^{sx_n}} \qquad (5.197)$$

The algorithm is repeated for $k = 1, 2, \ldots, M$ until it terminates. The number of steps M is always finite, provided the exponents u_n are rationally related.

Before presenting the proof procedure, we first illustrate the application of the algorithm to a specific example.

Example 5.5. Determine whether

$$F(s) = 10e^{14.3s} - 2e^{8.2s} - e^{6.1s} + 5 \qquad (5.198)$$

has all its zeros in the open left half s-plane. From (5.191), determine

$$\Psi_0(s) = \frac{5e^{14.3s} - e^{8.2s} + e^{6.1s} - 5}{15e^{14.3s} - 3e^{8.2s} - 3e^{6.1s} + 15} \qquad (5.199)$$

We observe that no decomposition is necessary, since all the ratios of corresponding coefficients in $\Psi_0(s)$ equal $\pm R_1 = \frac{1}{3}$; also, $\tau_1 = 8.2$. We then have from (5.197),

$$\Psi_1(s) = \frac{5e^{(14.3-8.2)s} - e^{(8.2-8.2)s} + e^{6.1s} - 5}{15e^{(14.3-8.2)s} - 3e^{(8.2-8.2)s} - 3e^{6.1s} + 15} \qquad (5.200)$$

$$= \frac{6e^{6.1s} - 6}{12e^{6.1s} + 12}$$

Again no decomposition is required, and $R_2 = \frac{1}{2}$, $\tau_2 = 6.1$. Thus

$$\Psi_2(s) = \frac{6e^{(6.1-6.1)s} - 6}{12e^{(6.1-6.1)s} + 12} = \frac{0}{24} \qquad (5.201)$$

We therefore conclude that $F(s)$ is the characteristic function of a stable system.

If $\Psi_0(s)$ is looked upon as a driving-point impedance, the network realized by the algorithm is shown in Figure 5.2, where the elements are uniform transmission lines with characteristic impedance equal to R_k and one-way delays equal to $\frac{\tau_k}{2}$.

Proof of the Stability Criterion: The proof of the criterion is based primarily on the following: "$F(s)$ is stable (i.e., all zeros are in the open left half s-plane) if and only if $\Psi_0(s)$ in (5.191) is a positive real function of s." To show this, we make use of the following theorem, due to Hadamard [39].

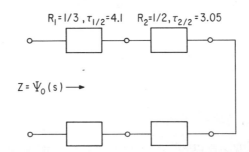

Figure 5.2. Network of cascade transmission lines.

Since $F(s)$ in (5.190) has no singularities except at infinity, it can be expanded by the Hadamard factorization theorem [39] as

$$F(s) = e^{Q(s)} \prod_{n=1}^{\infty} \left[\left(\frac{1-s}{s_n} \right) e^{s+s^2/2+\cdots+s^p/p} \right] \quad (5.202)$$

where the s_n are zeros of $F(s)$ and $Q(s)$ is a polynomial in s. Hence, $e^{su_n}[F(-s)/F(s)]$ is analytic in the right half s-plane if and only if $F(s)$ has no zeros in the right half-plane. In addition $e^{su_n}[F(-s)/F(s)]$ is bounded by one on the $j\omega$-axis for any $F(s)$ as above. By the use of the Maximum Modulus theorem [15] $e^{su_n}[F(-s)/F(s)]$ is bounded by one in the right half-plane if and only if $F(s)$ has no zeros in the right half-plane. Noting that $\Psi_0(s)$ is real for real s and that Re $\Psi_0(s) \geq 0$, Re $s \geq 0$ if and only if the above is satisfied, we readily ascertain that $\Psi_0(s)$ of (5.191) is positive real (pr) if and only if $F(s)$ has no zeros in the right half of the s-plane.

Alternatively, for real problems (i.e., u_n rational), a mapping $\lambda = \tanh s\tau$ can be used to reduce $\Psi_0(s)$ to a rational function which then becomes positive real in λ and we then conclude that $F(s) = $ numerator + denominator of $\Psi_0(s)$ is a Hurwitz polynomial.

The above discussion proves that the stability of $F(s)$ reduces to that of testing $\Psi_0(s)$ for positive realness. For this we make use of the following results due to Kinariwala [15].

1. An odd function of the form (5.191) is the impedance function of a SCULL (short-circuited cascade of uniform, lossless transmission lines).

2. A function of the form (5.191) is the impedance function of a SCULL if and only if it reduces to (zero/constant) under repeated application of the algorithm presented in the stability criterion. This is actually a synthesis algorithm for the short-circuited cascade structure as shown in the example.

This concludes the proof of the stability criterion and the algorithm for its test.

Remarks [40]

1. For uniformly (sampled or distributed parameter) systems, we let $u_N = n$ in (5.190) to obtain

$$F(s) = a_0 + a_1 e^s + a_2 e^{2s} + \cdots + a_n e^{ns} \quad a_n > 0 \quad (5.203)$$

Using the stability test mentioned in the preceding algorithm, we notice that it is identical to the table form of stability of linear discrete systems mentioned in (5.95). We can also use the simplified determinantal criterion discussed in Section 2.3 to test the stability of such systems. We only need to substitute z for e^s in (5.203).

2. If we can find that a coefficient of the exponent term in (5.190) is a

factor† of all exponent terms of $F(s)$, then we can also use as remarked in (1), the simplified stability criterion of Section 2.3 or the table form in (5.95) to test stability. For instance in (5.198) a common factor [41] is $e^{0.1s}$. Hence, we can write (5.198) by denoting $e^{0.1s}$ as z, as follows:

$$F_1(z) = 10z^{143} - 2z^{82} - z^{61} + 5 \qquad (5.204)$$

We can readily use the table form of (5.95) to test the root distribution of $F_1(z) = 0$. Although the degree of $F_1(z)$ is very high, the stability test is very simple, since one can readily verify that certain of the δ_k's in the table form are related by the condition [42] $\delta_k = \delta_{k-1}^2$. This fact would enable us essentially to use $F_1(z)$ as a third-degree polynomial with the same coefficients as in (5.204) for the stability test. We can also set s in the exponential in (5.190) equal to $[(1 + \lambda)/(1 - \lambda)]$. Then $F(s)$ is transformed into a polynomial which can be tested for its Hurwitz character [41]. In this transformation, the testing for the Hurwitz character might be exceedingly complicated.

3.   The constraint $F(s)|_{s=0} > 0$ in (5.190) constitutes a necessary condition for the stability of nonuniform systems. It should be noted that before applying this constraint, the coefficient of e^{ks} in (5.190), with k the largest real value, should be positive. The proof is similar to the necessary stability requirement of zeros of $F(z) = 0$ in (2.38) to lie inside the unit circle—namely $F(z)|_{z=1} > 0$.

4.   The relationship between the physical realization of the impedance function SCULL and the table form [(5.95)] is in analogy to the physical realizability condition of a passive LC network and Routh table [(5.56)] for the continuous case.

REFERENCES

[1]   O. BRUNE, "Synthesis of a finite two-terminal network whose driving-point impedance is a prescribed function of frequency," *J. Math. Phys.*, **10**, 191–236 (Oct. 1931).

[2]   G. A. GUILLEMIN, *Synthesis of Passive Networks*. New York: Wiley, 1957, Chap. J.

[3]   L. WEINBERG, *Network Analysis and Synthesis*. New York: McGraw-Hill, 1962, pp. 257–269.

[4]   D. D. ŠILJAK, *Nonlinear Systems*. New York: Wiley, 1969, pp. 352–367.

† For rational exponents or when they are rationally related we can always find such a factor. A discussion of this is presented in E. I. Jury, "Remarks on physical realization of the impedance function of short-circuited cascade of nonuniform lossless transmission line (SCULL)," *IEEE Trans. on Circuit and Systems* (3) (May 1974), pp. 459–461.

[5] D. D. ŠILJAK, "New algebraic criteria for positive realness," *Proceedings of the 4th Annual Princeton Conference on Information Sciences and Systems*, Princeton University, March 1970.

[6] M. V. MEEROV, "Analytic conditions for positivity of a real function" (in Russian), *Dokl. Akad. Nauk USSR*, **165**, 780–782 (1965).

[7] D. D. ŠILJAK, "Algebraic criterion for absolute stability, optimality and passivity of dynamic systems," *Proc. IEEE (London)*, **117** (10), (Oct. 1970).

[8] R. E. KALMAN, "When is linear control system optimal?" *J. Basic Eng. ASME*, **86** (D), 51–60 (March 1964).

[9] V. M. POPOV, "Incompletely controllable positive systems and application to optimization and stability of automatic control systems," *Rev. Roumanian, Sci., Techn., Serie Electr. Energ.* (1967).

[10] E. I. JURY, "The three *p*'s in system theory." Paper presented at the Inst. Symp. in Circuit Theory, Los Angeles, Cal., April 1972. *Symposium Proceedings*, pp. 32–37.

[11] E. J. ROUTH, *A Treatise on the Stability of a Given State of Motion—Adams Prize Essay.* New York: Macmillan, 1877.

[12] D. D. ŠILJAK, "Algebraic criteria for positive realness relative to the unit circle," *J. Franklin Inst.*, **295** (6), 469–476 (April 1973).

[13] E. I. JURY, *Theory and Application of the z-Transform Method.* New York: Wiley, 1964, Chap. 3.

[14] B. K. KINARIWALA and A. GERSHO, "A stability criterion for nonuniformly sampled and distributed parameter systems," *Bell. Syst. Tech. J.*, 1153 (Sept. 1966).

[15] B. K. KINARIWALA, "Theory of cascaded structures: lossless transmission lines," *Bell. Syst. Tech. J.*, **45**, 631–649 (April 1966).

[16] D. D. ŠILJAK, "Algebraic test for passivity," *IEEE Trans. on Circuit Theory*, **CT-18** (2), 285–286 (March 1971).

[17] M. E. VAN VALKENBURG, *Introduction to Modern Network Synthesis.* New York: Wiley, 1964, p. 78.

[18] F. R. GANTMACHER, *The Theory of Matrices.* New York: Chelsea, 1959, Vol. II, Chap. XV.

[19] B. D. O. ANDERSON, "On the computation of the Cauchy index," *Quart. J. Appl. Math.*, **29** (4), 577–582 (Jan. 1972).

[20] E. I. JURY, "A note on the analytical absolute stability test," *Proc. IEEE*, **58** (5), 823–824 (May 1970).

[21] A. T. FULLER and R. M. MACMILLIAN (Discussion), *Brit. J. Appl. Phys.*, **6**, 450–451, eq. (1) (Dec. 1955).

[22] Z. BURINGTON, *Handbook of Mathematical Tables.* 4th ed. New York: McGraw-Hill Book Co., 1965.

[23] W. D. FRYER, "Applications of Routh's algorithm to network-theory problems," *IRE Trans. on Circuit Theory*, **CT-6**, 144–149 (1959).

[24] E. I. JURY, "A note on the reciprocal zeros of a real polynomial with respect to unit circle," *IEEE Trans. on Circuit Theory*, **CT-11** (2), 292–294 (June 1964).

[25] B. D. O. ANDERSON and S. VONGPNITLERD, *Network Analysis and Synthesis: A Modern Systems Approach*. Englewood Cliffs, N.J.: Prentice-Hall, 1973.

[26] E. I. JURY and B. W. LEE, "A stability theory for multinonlinear control systems." Paper No. 289, Third IFAC Congress, London (June 1966).

[27] A. I. LURÉ, *Some Nonlinear Problems in the Theory of Automatic Control*. London: Her Majesty's Stationery Office, 1954.

[28] R. E. KALMAN, "Mathematical description of linear dynamical systems," *SIAM J. Control* (1963).

[29] V. M. POPOV, "Absolute stability of nonlinear systems of automatic control," *Autom. i. Talenekh.*, **22, 961**–974 (1961).

[30] C. K. SUN and D. D. ŠILJAK, "Absolute stability test for discrete systems," *Electronics Letters*, **5** (11), 236 (May 1969).

[31] Y. Z. TSYPKIN, "On the stability in the large of nonlinear sampled-data systems," *Dokl. Akad. Nauk*, **145**, 52–55 (July 1962).

[32] E. I. JURY and B. W. LEE, "On the stability of a certain class of nonlinear sampled-data systems," *IEEE Trans. on Automatic Control*, **AC-9** (1), 51–61 (Jan. 1964).

[33] D. D. ŠILJAK and C. K. SUN, "Exponential absolute stability of discrete systems," *ZAMM*, **51** (4), 271–275 (July 1971).

[34] V. M. POPOV, "The solution of a new stability problem for controlled systems," *Automation and Remote Control*, **24**, 1–23 (Jan. 1963).

[35] B. D. O. ANDERSON, "Stability of control systems with multiple non-linearities, *J. Franklin Inst.*, **282**, 155–160 (1966).

[36] F. W. BODE, *Network Analysis and Feedback Amplifier Design*. New York: Van Nostrand, 1945.

[37] R. W. NEWCOMB, *Linear Multiport Synthesis*. New York: McGraw-Hill, 1966.

[38] E. S. KUH and D. O. PEDERSON, *Principles of Circuit Synthesis*. New York: McGraw-Hill, 1959.

[39] E. C. TITCHMARK, *The Theory of Functions*. London: Oxford University Press, 1939, Chap. VIII, p. 250.

[40] E. I. JURY, "Physical realization of the impedance function of SCULL," *IEEE Trans. on Circuit Theory*, **CT-13** (1), 100–103 (Jan. 1972).

[41] L. WEINBERG, "Approximation and stability-test procedures suitable for digital filters." Paper presented at the Workshop of Digital Filtering, Harriman, N.Y. (Jan. 11–14, 1970).

[42] E. I. JURY, "Further remarks on a paper, 'Über die Wurzelverteilung von

linearen Abtastsystemes'," by M. Thoma. *Regelungstechnik* **2** (12), 75–79
(see footnote 6) (1964).

[43] R. N. GADENZ and C. C. LI, "On positive definiteness of quartic forms of two
variables," *IEEE Trans. on Automatic Control* (correspondence) **AC-9** (April
1964), pp. 187–188.

[44] W. H. KU, "Explicit Criterion for the Positive Definiteness of a General
Quartic Form," *IEEE Trans. on Aut. Contr.*, **AC-10,** (3) (July, 1965), pp. 372–
373.

Stability of Two-Dimensional and Multidimensional Filters

Stability of two-dimensional continuous filters arises in providing a test for a driving-point impedance realizability condition using commensurate-delay transmission lines and lumped reactances [1–3]. On the other hand, stability of two-dimensional digital filters occurs in the useful design of two-dimensional digital filters. If the filter is unstable, any noise (including roundoff errors in computation) will propagate through the output and be amplified. Two-dimensional digital filters find widespread application in many fields, such as image processing and geophysics for processing of seismic, gravity, and magnetic data [4–6].

In this chapter the stability conditions, as well as the tests for checking them, will be thoroughly discussed. In particular, in the first section, stability of two-dimensional digital filters will be discussed, while the continuous counterpart will be covered in the second section.

In the last section we will extend the two-dimensional stability conditions, first, to multidimensional digital filters, and by using the bilinear transformation, we will obtain the conditions for the stability of multidimensional continuous filters. Furthermore, we will discuss the computational procedures for checking the stability conditions. It is of interest to mention at this point that the positivity conditions discussed in the preceding chapter will play an important role in the stability tests discussed here.

6.1 STABILITY OF TWO-DIMENSIONAL DIGITAL FILTERS

In this section we will derive the necessary and sufficient condition for stability of two-dimensional digital filters, as well as the procedures for its test.

THEOREM 6.1. *A causal recursive filter* [6] *(a filter that recurses in the $+m$, $+n$ direction) with the z-transform $H(z_1, z_2) = A(z_1, z_2)/B(z_1, z_2)$, where A and B are noncancelable polynomials in z_1 and z_2, is stable if and only if there*

*are no values of z_1 and z_2 such that $B(z_1, z_2) = 0$ and $|z_1| \leq 1$ and $|z_2| \leq 1$
[or, equivalently, $B(z_1, z_2) \neq 0$ for $|z_1|$ and $|z_2|$ simultaneously less than or
equal to unity].*

Proof: Let

$$H(z_1, z_2) = \frac{A(z_1, z_2)}{B(z_1, z_2)} = \sum_{m=0}^{\infty} \sum_{n=0}^{\infty} h_{mn} z_1^m z_2^n \tag{6.1}†$$

where the coefficients h_{mn} represent the impulse response of the filter. To
prove Theorem 6.1, we want to show the necessary and sufficient condition
that

$$\sum_{m=0}^{\infty} \sum_{n=0}^{\infty} |h_{mn}| < \infty \tag{6.2}$$

is that $H(z_1, z_2)$ is analytic in the region $D = \{(z_1, z_2); |z_1| \leq 1 \cap |z_2| \leq 1\}$.‡

The Sufficiency Part.   If $H(z_1, z_2)$ is analytic in D, we can find $\varepsilon > 0$ such
that $H(z_1, z_2)$ is analytic in $D_1 = \{(z_1, z_2); |z_1| < 1 + \varepsilon \cap |z_2| < 1 + \varepsilon\}$,
which implies that

$$\sum_m \sum_n h_{mn} z_1^m z_2^n \tag{6.3}$$

is absolutely convergent in D_1. Therefore

$$\sum_m \sum_n |h_{mn}| < \infty \tag{6.4}$$

The Necessary Part.   Let

$$\sum_m \sum_n |h_{mn}| < \infty \tag{6.5}$$

Since

$$|h_{mn} z_1^m z_2^n| \leq |h_{mn}| \qquad\qquad z_1, z_2 \in D, \tag{6.5a}$$

by the Weierstrass M test, (6.5) shows that

$$\sum_m \sum_n h_{mn} z_1^m z_2^n \tag{6.6}$$

is absolutely convergent uniformly in D, which implies, in turn, that $H(z_1, z_2)$
is analytic in D.

One way to test the preceding stability condition is to map the unit disk
$d_1 \equiv z_1$ ($|z_1| \leq 1$) in the z_1-plane into the z_2-plane by the implicit mapping

† It may be noted that in the usual definition of the two-dimensional z-transform the
powers of z_1 and z_2 are taken to be negative integers [7,8]. To follow the derivation of
the stability test proposed by many authors, we present the above form [5,9].
‡ Using the conventional definition of two-dimensional z-transform [7], this condition
becomes $D = \{(z_1, z_2); |z_i| \geq 1 \cap |z_2| \geq 1\}$. The symbol $\cap$ denotes intersection.

relation $B(z_1, z_2) = 0$. The filter is stable if and only if the image of d_1 in the z_2-plane does not overlap the unit disk $d_2 \equiv (z_2, |z_2| \leq 1)$. The test of this procedure is difficult and not very practical. Hence we prove the following simplified test.

THEOREM 6.2 [6]. *A causal filter with a z-transform $H(z_1, z_2)$, defined in (6.1), is stable if and only if*

(i) *the map of $d_1 = (z_1; |z_1| = 1)$ in the z_2-plane, according to $B(z_1, z_2) = 0$, lies outside $d_2 \equiv (z_2; |z_2| \leq 1)$*

(ii) *no point in $d_1 = (z_1; |z_1| \leq 1)$ maps into the point $z_2 = 0$ by the relation $B(z_1, z_2) = 0$.*

Proof: We want to show that the necessary and sufficient condition of stability discussed earlier in Theorem 6.1 and those of Theorem 6.2 are equivalent. It is evident that the stability conditions of Theorem 6.1 imply the two conditions in Theorem 6.2. Hence we proceed to prove the converse.

We can write $B(z_1, z_2)$ as follows:

$$B(z_1, z_2) = T_0(z_1)z_2^n + T_1(z_1)z_2^{n-1} + \cdots + T_n(z_1) = 0 \qquad (6.7)$$

The two-variable polynomial defines an algebraic function $z_2 = f(z_1)$. We first modify the unit-circle contour in the z_1-plane to exclude any of the finite number of singular points of f inside the contour, thereby resulting in a modified contour ∂d_1^1, as shown in Figure 6.1. We use d_1^1 to denote the closed region enclosed by ∂d_1^1. A point $z_1 = z_1^0$ is called a singular point† of $z_2 = f(z_1)$,

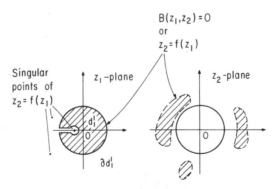

Figure 6.1. Mapping of $z_2 = f(z_1)$.

† We may note that the number of singular points is finite. Hence they do not form a connected graph.

1.   if $B(z_1^0, z_2) = 0$, considered as an equation in z_2, has multiple (finite or infinite) roots, or
2.   if $f(z_1^0) = 0$.

According to the theory of algebraic functions [10], in d_1^1 the function $z_2 = f(z_1)$ has a number of branches, each of which is holomorphic. Therefore, from the Maximum Modulus theorem [10], the maximum of $|f(z_1)|$ over d_1^1 occurs on ∂d_1^1, and the minimum of $|f(z_1)|$ over d_1^1 can occur in the interior only if the minimum is zero. However, condition (ii) of Theorem 6.1 says that $f(z_1)$ is never zero in d_1^1. Hence the minimum of $f(z_1)$ occurs on ∂d_1^1. That is,

$$|f(d_1^1)| \leq \min |f(\partial d_1^1)| \qquad (6.7a)$$

which implies that if $|f(\partial d_1^1)| > 1$, then $|f(d_1^1)| > 0$. In other words, to ensure that $f(d_1^1)$ lies outside the unit disk $d_2 = (z_2, |z_2| \leq 1)$, it is sufficient to ensure that $f(\partial d_1^1)$ lies outside d_2. To complete the proof, we need to show that if $|f(\partial d_1)| > 1$, then $|f(d_1)| > 1$, where $d_1 \equiv (z_1, |z_1| \leq 1)$ and $\partial d_1 = (z_1, |z_1| = 1)$. Since the detour in ∂d_1^1 can be any path leading from ∂d_1 to the singular point, what is left to show is simply that $|f(s)| > 1$, where s is the singular point. For the first type of singularity, each branch of $z_2 = f(z_1)$ is continuous at $z_1 = s$, and since $|f(s + \varepsilon e^{j\theta})| > 1$ for arbitrary small ε and any θ, we have $|f(s)| > 1$. The same conclusion is obtained if the second type of singular point exists, since in this case $f(z_1)$ has at most a pole of finite order at s [10].

In order to test stability, using Theorem 6.2, we must map $\partial d_1 = (z_1, |z_1| = 1)$ into the $\bar{z}_2$-plane according to $B(z_1, z_2) = 0$ and observe whether the image lies outside $d_2 = \{z_2, |z_2| \leq 1\}$. Also, we need to solve $B(z_1, 0) = 0$ and see whether there is any root with magnitude less than unity. These two conditions can be represented as follows:

$$B(z_1, 0) \neq 0 \qquad\qquad |z_1| \leq 1 \quad (6.8)$$

$$B(z_1, z_2) \neq 0 \qquad |z_1| = 1, \quad |z_2| \leq 1 \quad (6.9)$$

OUTLINE OF THE TEST PROCEDURE [11]

First, the checking of (6.8) is straightforward because $B(z_1, 0)$ is a single-variable polynomial, and as discussed in the preceding chapters, there are a number of tests for determining whether or not its zeros all lie outside the unit circle. These tests include the inner determinants, as in (2.142), or the table form, as in (5.95). In addition, they include the Schur–Cohn symmetric matrix, as in (3.16). The latter matrix is negative definite (i.e., odd-order principal minors are negative and even-order principal minors are positive) if and only if $B(z_1, 0)$ has all its zeros in $|z_1| > 1$ as required by condition (6.8).

Finally, if we replace the one-variable polynomial $B(z_1, 0) = F(z_1)$ by its reciprocal—that is, $F_1(z) = z_1^n F(1/z_1)$—the condition in (6.8) reduces to the necessary and sufficient condition for all the roots (zeros) of $F_1(z) = 0$ to be inside the unit circle—namely, the stability condition of a one-variable polynomial for a one-dimensional digital filter. Therefore any of the various stability tests for the discrete system can be applied.

Example 6.1. Consider

$$B(z_1, z_2) = 12 + 10z_1 + 6z_2 + 5z_1 z_2 + 2z_1^2 + z_1^2 z_2 \qquad (6.10)$$

Then
$$B(z_1, 0) = 12 + 10z_1 + 2z_1^2 \qquad (6.11)$$

From (3.16), the Schur–Cohn symmetric matrix is

$$C = \begin{bmatrix} a_2^2 - a_0^2 & a_1 a_2 - a_1 a_0 \\ a_1 a_2 - a_1 a_0 & a_2^2 - a_0^2 \end{bmatrix} = \begin{bmatrix} -140 & -100 \\ -100 & -140 \end{bmatrix} \qquad (6.12)$$

which is easily seen to have a negative 1×1 leading principal minor and a positive 2×2 leading principal minor. Thus condition (6.8) is satisfied for $B(z_1, z_2)$. To apply the stability condition, we replace $B(z_1, 0)$ by its reciprocal polynomial

$$B_1(z_1, 0) = z_1^2 B(z_1^{-1}, 0) = 12z_1^2 + 10z_1 + 2 \qquad (6.13)$$

The necessary and sufficient condition for the zeros of $B_1(z_1, 0) = 0$ to lie inside the unit circle [which is equivalent to all the roots of $B(z_1, 0)$ lying outside the unit circle] is

$$B_1(1, 0) = 12 + 10 + 2 > 0 \qquad (6.14)$$

$$B_1(-1, 0) = 12 - 10 + 2 > 0 \qquad (6.15)$$

and
$$2 < 12 \qquad (6.16)$$

Checking of (6.9): We can apply the Schur–Cohn test for checking (6.9). The idea is to think of z_1 as a parameter, so that $B(z_1, z_2)$ is a polynomial in z_2, whose coefficients are functions of a parameter z_1. For stability [or for satisfying (6.9)], the polynomial cannot be zero inside $|z_2| \leq 1$; this means that the associated Schur–Cohn matrix for complex coefficients must be negative definite. In this case, the entries of the Schur–Cohn matrix are polynomials in z_1 and/or $\bar{z}_1$ (z_1 conjugate).

Checking the negative definiteness of the Schur–Cohn matrix requires determination of the principal minors, which must have certain signs.†

---

† A simpler test, which requires only that the Schur–Cohn determinant be negative and a certain auxiliary condition, has been indicated by Šiljak (see References 22 and 24). Also, this simplification is mentioned in (5.147).

These minors are again polynomial in z_1 and $\bar{z}_1$ and are *real* because the Schur–Cohn symmetric matrix is *Hermitian*. This fact is also true for multi-dimensional stability conditions, as will be shown in a later section.

Accordingly, condition (6.9) is equivalent to requiring the sign definiteness on $|z_1| = 1$ of a set of functions that take real values and are polynomial in z_1 and $\bar{z}_1$ Bearing in mind that on $|z_1| = |e^{j\theta}| = 1$ we have $\bar{z}_1 = e^{-j\theta} = z_1^{-1}$, it follows that the functions are of the form

$$f_i(z_1, z_1^{-1}) = \sum_{j=1}^{N_i} c_j z_1^j + 2c_0 + \sum_{j=1}^{N_i'} d_j z_1^{-j} = \sum_{j=0}^{N_i} c_j(z_1^j + z_1^{-j}) \qquad (6.17)\dagger$$

on $|z_1| = 1$. To complete the checking of (6.9) we still require a test for the polynomial in (6.17) to be positive,$\ddagger$ as required for all z_1 on $|z_1| = 1$, and this condition can be readily handled by noting the discussion of Section 5.2. It is of interest to note the similarity between (6.17) and (5.89). Because $f_i(z_1)$ in (6.17) is a self-inversive (or reciprocal) polynomial, there are as many zeros of $z_1^{N_i} f_i(z_1)$ inside $|z_1| < 1$ as there are outside. This point has also been discussed in (5.91) and (5.92). Therefore $f_i(z_1)$ is positive on $|z_1| = 1$ if and only if $f_i(1) > 0$ [or $f_i(z_1)$ is positive at any one point of $|z_1| = 1$] and $z_1^{N_i} f_i(z_1)$ has N_i zeros inside $|z_1| < 1$ [for then $z_1^{N_i} f_i(z_1)$ has N_i zeros outside the unit circle and thus no zeros on $|z_1| = 1$]. Hence $f_i(z_1)$ is positive (or has constant sign on $|z_1| = 1$), and the stability condition is satisfied.

The test for positivity can be identically carried out as shown in Lemma 5.2 and by the use of the table in (5.95). Furthermore, in order to simplify the checking for negative definiteness of the Schur–Cohn matrix, we can replace the polynomial $B(z_1, z_2)$, considered as a polynomial in z_2, by its reciprocal in a way similar to that shown for checking (6.8). Therefore the condition for negative definiteness becomes one for positive definiteness (or the stability condition for a complex polynomial).§ The Hermitian matrix $C = (\gamma_{ij})$ with

$$\gamma_{ij} = \sum_{p=1}^{i} (a_{n-i+p}\bar{a}_{n-j+p} - \bar{a}_{i-p}a_{j-p}) \qquad i \leq j \quad (6.18)$$

is the Schur–Cohn symmetric matrix [16] [see (3.123)] associated with the polynomiâl

$$F(z) = \sum_{i=0}^{n} a_i z^i \qquad (6.19)$$

$\dagger$ The second equality follows because $f_i(z_1)$ is real on $|z_1| = 1$.
$\ddagger$ If it is required to be negative, we can always multiply by minus one to be positive.
§ See also Reference 23.

To check (6.9), we replace $F(z)$ by $B(z_1, z_2)$ written as a polynomial in z_2:

$$B(z_1, z_2) = \sum_{j=0}^{q} \left(\sum_{i=0}^{p} b_{ij} z_1^i \right) z_2^j \tag{6.20}$$

The coefficient a_j is therefore $\sum_{i=0}^{p} b_{ij} z_1^i$.

We can also use the innerwise matrix of (2.39) instead of the Schur–Cohn symmetric matrix.

Example 6.1 (continued). In this case, we rewrite (6.10) as

$$B(z_1, z_2) = (12 + 10z_1 + 2z_1^2) + (6 + 5z_1 + z_1^2)z_2 \tag{6.21}$$

Noting (6.18) and (6.20), we obtain the 1×1 Schur–Cohn matrix:

$$\begin{aligned} C &= [(6 + 5z_1 + z_1^2)(6 + 5z_1^{-1} + z_1^{-2}) \\ &\quad - (12 + 10z_1 + 2z_1^2)(12 + 10z_1^{-1} + 2z_1^{-2})] \\ &= [-18z_1^2 - 105z_1 - 186 - 105z_1^{-1} - 18z_1^{-2}] \end{aligned}$$
$$\hspace{6cm} |z_1| = 1 \tag{6.22}$$

and C is negative definite if

$$f(z_1, z_1^{-1}) = 18z_1^2 + 105z_1 + 186 + 105z_1^{-1} + 18z_2^{-2} > 0$$
$$\text{for } |z_1| = 1 \tag{6.23}$$

Note if we replace (6.21) as a function of z_2 by its reciprocal, we obtain condition (6.23) directly because, in this case, C must be positive definite. Also, $f(z_1, z_1^{-1})$ can be obtained from the innerwise matrix Δ_2 in (2.39).

To test the positivity of (6.23), we obtain

$$g(z_1) = z_1^2 f(z_1, z_1^{-1}) = 18z_1^4 + 105z_1^3 + 186z_1^2 + 105z_1 + 18 > 0$$
$$\text{for } |z_1| = 1 \tag{6.24}$$

From (5.97), we have

$$g_{\text{new}}(z_1) = \left[\frac{dg(z_1)}{dz_1} \right]^* = 105z_1^3 + 372z_1^2 + 315z_1 + 72 \tag{6.25}$$

For positivity [since $g(1) > 0$], we require that g_{new} have two zeros inside the unit circle. Although we can use the table in (5.95) to ascertain whether it does, positivity is ensured here because $72 < 105$ (which represents at least one root inside the unit circle) and $g_{\text{new}}(1) > 0$, $g_{\text{new}}(-1) > 0$ (which ensures an even number and hence precisely two roots inside the unit circle by noting the preceding condition).

Therefore Example 6.1 represents a stable two-dimensional digital filter.

In concluding the discussion of this section, we may note the following:

1. The conditions in (6.8) and (6.9) can be replaced by conditions involving interchange of z_1 and z_2,

$$B(0, z_2) \neq 0 \qquad\qquad\qquad |z_2| \leq 1 \quad (6.26)$$

$$B(z_1, z_2) \neq 0 \qquad |z_1| \leq 1, \quad |z_2| = 1 \quad (6.27)$$

The preceding interchange could be advantageous for computational purposes.

2. The conditions in (6.8) and (6.9) can be readily computed using one algorithm based on the inners double-triangularization method. This algorithm will be discussed in detail in Chapter 7. Also, this algorithm has the advantage of checking the stability conditions for both the discrete and the continuous cases in a unified form.

6.2 STABILITY OF TWO-DIMENSIONAL CONTINUOUS FILTERS

As pointed out earlier in this chapter, polynomials in two complex variables $H(s_1, s_2)$ occur in continuous systems containing both lumped and distributed elements. The test of $H(s_1, s_2)$ for stability or Hurwitz character is required to ascertain whether a real function of two variables has the two-variable reactance property. This property is of importance in the realizability condition of networks containing lumped and distributed elements.

The conditions for stability of two-dimensional continuous filters can be readily derived in the same way as the stability conditions for digital filters discussed earlier. However, in order not to duplicate the proof of Theorem 6.2, we will obtain the continuous stability conditions from the digital filter by using a bilinear transformation. This transformation defines an isomorphism [12] mapping that matches the stability condition of continuous filters with those of digital filters. The bilinear transformations from s_1, s_2 to z_1, z_2 are [6]

$$\frac{s_1}{\alpha_1} = \frac{1 - z_1}{1 + z_1} \qquad\qquad (6.28)$$

$$\frac{s_2}{\alpha_2} = \frac{1 - z_2}{1 + z_2} \qquad\qquad (6.29)$$

for any desired positive α_1 and α_2.

Substituting (6.28) and (6.29) in (6.1), we obtain

$$H_c(s_1, s_2) = \frac{A_c(s_1, s_2)}{B_c(s_1, s_2)} \qquad\qquad (6.30)$$

The stability conditions for $H_c(s_1, s_2)$ are readily obtainable from (6.8) and (6.9) by noting (6.28) and (6.29):

$$B_c(s_1, 1) \neq 0 \qquad \qquad \text{Re } s_1 \geq 0 \quad (6.31)$$

$$B_c(s_1, s_2) \neq 0 \quad s_1 = j\omega, \quad \text{Re } s_2 \geq 0 \quad (6.32)$$

These two conditions constitute the necessary and sufficient conditions for the two-dimensional continuous filter to be stable.

OUTLINE OF THE TEST PROCEDURE

The checking of (6.31) is straightforward, since $B_c(s_1, 1)$ is a polynomial of one variable, and hence testing its Hurwitz character is straightforward. Any of the methods discussed in Section 2.1 can be used. In particular, the inner form of the Liénard–Chipart criterion is the simplest.

The checking of (6.32) involves two stages similar to the digital filter. The first stage requires that the complex polynomial $B_c(j\omega, s_2)$ be Hurwitz. This condition requires that the Hermite matrix associated with $B_c(j\omega, s_2)$ be positive definite or, alternatively, that the innerwise matrix Δ_{2n} of (2.4) be positive innerwise. When the coefficient of the highest power of s_2 in B_c is unity, we can use the Hermite matrix as given in (3.108). When it is not, we can use the following matrix associated [13] with $F(s) = \sum_{i=0}^{n} b_i s^i$.

$$\frac{1}{2h_{pq}} = (-1)^{(p+q)/2} \sum_{k=1}^{p} (-1)^k \text{ Re } (b_{n-k+1}\bar{b}_{n-p-q+k})$$

$$p + q = \text{even}, \quad p \leq q \quad (6.33)$$

$$\frac{1}{2h_{pq}} = (-1)^{(p+q-1)/2} \sum_{k=1}^{p} (-1)^k \text{ Im } (b_{n-k+1}\bar{b}_{n-p-q+k})$$

$$p + q = \text{odd}, \quad p \leq q \quad (6.34)$$

The second stage requires that all the leading principal minors† of the Hermite matrix (or the determinants of the innerwise matrix) be positive for all ω. Since the polynomials of the principal minors are even in ω, for positivity, we can readily apply Theorem 5.3 and the table of (5.54). Example 6.2 illustrates the procedure.

Example 6.2. We test whether or not the following polynomial is Hurwitz.

$$B_c(s_1, s_2) = (4s_1 + 4)s_2^2 + (s_1^2 + 5s_1 + 1)s_2 + s_1^2 + s_1 + 4 \quad (6.35)$$

From (6.31), we have

$$B_c(s_1, 1) = 4s_1 + 4 + s_1^2 + 5s_1 + 1 + s_1^2 + s_1 + 4 = 2s_1^2 + 10s_1 + 9 \quad (6.36)$$

---

† See (5.146) for further simplification of this condition.

This polynomial is obviously Hurwitz, since all coefficients of s_1 are positive. Hence condition (6.31) is satisfied.

To test (6.32), we let $s_1 = j\omega$ in (6.35) and obtain either the Hermite matrix from (6.33) and (6.34) or the innerwise matrix associated with $B_c(j\omega, s_2)$ From (2.3) and (2.4), the innerwise matrix is

$$\Delta_4 = \begin{bmatrix} -4\omega & 1 - \omega^2 & \omega & 0 \\ 0 & -4\omega & 1 - \omega^2 & \omega \\ 0 & -4 & -5\omega & 4 - \omega^2 \\ -4 & -5\omega & 4 - \omega^2 & 0 \end{bmatrix} \tag{6.37}$$

The conditions for $B_c(s_1, s_2)$ to be Hurwitz for all ω are

$$|\Delta_2| = 16\omega^2 + 4 > 0 \qquad\qquad \text{for all } \omega \quad (6.38)$$

$$|\Delta_4| = 16 - 80\omega^2 + 100\omega^4 = (10\omega^2 - 4)^2 > 0$$
$$\text{for all } \omega \quad (6.39)\dagger$$

Obviously (6.38) is satisfied for all ω, whereas (6.39) is zero when $\omega = 2/\sqrt{10}$. Hence $|\Delta_4|$ is nonnegative or not positive. Therefore the polynomial $B_c(s_1, s_2)$ is not Hurwitz, and the system associated with this polynomial—that is, $H_c(s_1, s_2)$—is not stable.

Using (6.38) and (6.39), we obtain the Hermitian matrix associated with $B_c(j\omega, s_2)$ as follows:

$$H = \begin{bmatrix} \frac{1}{2}[16\omega^2 + 4] & \frac{1}{2}[4\omega^2 - 12\omega] \\ \frac{1}{2}[4\omega^3 - 12\omega] & \frac{1}{2}[4 + \omega^2] \end{bmatrix} \tag{6.40}$$

Within a positive constant, the leading principal minors of (6.40) are equivalent to the inners determinants of (6.37). Hence we also ascertain that condition (6.32) is not satified and that $B_c(s_1, s_2)$ is not Hurwitz.

6.3 EXTENSION TO MULTIDIMENSIONAL FILTERS [14]

In this section we will generalize the stability conditions to multivariable polynomials. These conditions are developed in detail for the digital filter, and by using a bilinear transformation, we present the stability conditions

† By using (5.146), the checking of $|\Delta_2| > 0$ is redundant. We replace it by $|\Delta_2(0)| > 0$ and $|\Delta_4(0)| > 0$.

for continuous filters. It seems feasible, in technology development, that three- and four-dimensional digital filters will have many applications in the future.

STABILITY CONDITIONS FOR MULTIDIMENSIONAL DIGITAL FILTERS

Many n-dimensional digital filters can be described by a multivariable transfer function of the form

$$H(z_1, z_2, \ldots, z_n) = \frac{A(z_1, z_2, \ldots, z_n)}{B(z_1, z_2, \ldots, z_n)} \qquad (6.41)$$

We can associate with (6.41) a causal, multivariable impulse response that is a n-indexed sequence $\{h_{i_1, i_2, \ldots, i_n}\}$, where $i_j = 0, 1, \ldots$ for $j = 1, 2, \ldots, n$. If $A(\cdot, \cdot, \ldots, \cdot)$ and $B(\cdot, \cdot, \ldots, \cdot)$ have no common zeros, we can obtain the stability condition by a direct generalization of Theorem 6.2 in Section 6.1. In this case, the stability condition is given by

$$B(z_1, z_2, \ldots, z_n) \neq 0 \qquad \bigcap_{i=1}^{n} |z_i| \leq 1 \quad (6.42)$$

Another form of the stability condition that yields more general conditions than (6.42) is also available [14]. This form involves the replacement of some of the regions $|z_i| \leq 1$ by either $|z_i| = 1$ or $|z_i| \geq 1$. However, the stability test to be developed for (6.42) can be also applied to these more general conditions [15].

We present the following two theorems, which provide simplification of condition (6.42) in a manner similar to what was done for the stability of two-dimensional digital filters.

THEOREM 6.3. *Let $B(z_1, z_2, \ldots, z_n)$ be a polynomial in n variables. Then condition (6.42) is equivalent to*

$$B(z_1, z_2, \ldots, z_{n-1}, 0) \neq 0 \qquad \bigcap_{i=1}^{n-1} |z_i| \leq 1 \quad (6.43)$$

and $$B(z_1, z_2, \ldots, z_n) \neq 0$$

$$\left\{ \bigcap_{i=1}^{n-1} |z_i| = 1 \right\} \cap \{|z_n| \leq 1\} \quad (6.44)$$

Proof: Obviously (6.43) and (6.44) are implied by (6.42), since conditions (6.43) and (6.44) require $B(\cdot, \cdot, \ldots, \cdot)$ to be nonzero over a region contained in that defined by (6.42). It remains to prove the converse, which we shall do by induction.

The result is valid for $n = 2$, as noted in Section 6.1. Assume that it is true for $n = 3, 4, \ldots, m - 1$. We shall prove it for $n = m$. Fix $z_1 = z_1^0$, with $|z_1^0| \le 1$. Then by the inductive argument applied to the $(m - 1)$-variable polynomial $B(z_1^0, z_2, z_3, \ldots, z_m)$, it follows that

$$B(z_1^0, z_2, \ldots, z_{m-1}, 0) \ne 0 \qquad \bigcap_{i=2}^{m-1} |z_i| \le 1 \quad (6.45)$$

$$B(z_1^0, z_2, \ldots, z_{m-1}, z_m) \ne 0$$

$$\left\{ \bigcap_{i=2}^{m-1} |z_i| = 1 \right\} \cap \{|z_m| \le 1\} \quad (6.46)$$

imply
$$B(z_1^0, z_2, \ldots, z_m) \ne 0 \qquad \bigcap_{i=1}^{m} |z_i| \le 1 \quad (6.47)$$

Consequently, noting the arbitrary nature of z_1^0, we see that

$$B(z_1, z_2, \ldots, z_{m-1}, 0) = 0 \qquad \bigcap_{i=1}^{m-1} |z_i| \le 1 \quad (6.48)$$

and
$$B(z_1, z_2, \ldots, z_{m-1}, z_m) \ne 0$$

$$\{|z_1| \le 1\} \cap \left\{ \bigcap_{i=2}^{m-1} |z_i| = 1 \right\} \cap \{|z_m| \le 1\} \quad (6.49)$$

imply
$$B(z_1, z_2, \ldots, z_m) \ne 0 \qquad \bigcap_{i=1}^{m} |z_i| \le 1 \quad (6.50)$$

Now temporarily fix $z_2 = z_2^0, \ldots, z_{m-1} = z_{m-1}^0$ in the region $\bigcap_{i=2}^{m-1} |z_i| = 1$ so that the two-variable theorem can be applied in (6.49), thinking of z_1 and z_m as the variables. Then (6.49) is implied by

$$B(z_1, z_2^0, \ldots, z_{m-1}^0, 0) \ne 0 \qquad\qquad |z_1| \le 1 \quad (6.51)$$

and
$$B(z_1, z_2^0, \ldots, z_{m-1}^0, z_m) \ne 0$$

$$|z_1| = 1 \cap |z_m| \le 1 \quad (6.52)$$

or using the arbitrary nature of $z_2^0, \ldots, z_{m-1}^0$, (6.49) is implied by

$$B(z_1, z_2, \ldots, z_{m-1}, 0) \ne 0$$

$$\{|z_1| \le 1\} \cap \left\{ \bigcap_{i=2}^{m-1} |z_i| = 1 \right\} \quad (6.53)$$

and
$$B(z_1, z_2, \ldots, z_{m-1}, z_m) \ne 0$$

$$\{|z_1| = 1\} \cap \left\{ \bigcap_{i=2}^{m-1} |z_i| = 1 \right\} \cap \{|z_m| \le 1\} \quad (6.54)$$

Hence conditions (6.48), (6.53), and (6.54) imply (6.50). However, (6.48) implies (6.53), so that (6.48) and (6.32) alone imply (6.50). This verifies the inductive hypothesis for the case $n = m$ and therefore proves the theorem.

The second theorem rests on the observation that $B(z_1, z_2, \ldots, z_{n-1}, 0)$ is an $(n-1)$-variable polynomial, so that condition (6.43) may therefore be replaced by

$$B(z_1, z_2, \ldots, z_{n-2}, 0, 0) \neq 0 \qquad \bigcap_{i=1}^{n-2} |z_i| \leq 1 \qquad (6.55)$$

and
$$B(z_1, z_2, \ldots, z_{n-2}, z_{n-1}, 0) \neq 0$$

$$\left\{ \bigcap_{i=1}^{n-2} |z_i| = 1 \right\} \cap \{|z_{n-1}| \leq 1\} \qquad (6.56)$$

by reapplication of Theorem 6.3. The same idea can then be applied to the $(n-2)$-variable polynomial appearing in (6.55), and so on. In this way, one proves Theorem 6.4.

THEOREM 6.4. *Let $B(z_1, z_2, \ldots, z_n)$ be a polynomial in n variables. Then condition (6.42) is equivalent to*

$$B(z_1, z_2, \ldots, z_n) \neq 0$$

$$\left\{ \bigcap_{i=1}^{n-1} |z_i| = 1 \right\} \cap \{|z_n| \leq 1\} \qquad (6.57)$$

$$B(z_1, z_2, \ldots, z_{n-1}, 0) \neq 0$$

$$\left\{ \bigcap_{i=1}^{n-2} |z_i| = 1 \right\} \cap \{|z_{n-1}| \leq 1\} \qquad (6.58)$$

$$B(z_1, z_2, \ldots, z_{n-2}, 0, 0) \neq 0$$

$$\left\{ \bigcap_{i=1}^{n-3} |z_i| = 1 \right\} \cap \{|z_{n-2}| \leq 1\} \qquad (6.59)$$

$$\vdots$$

$$B(z_1, z_2, 0, \ldots, 0) \neq 0$$
$$\{|z_1| = 1\} \cap \{|z_2| \leq 1\} \qquad (6.59a)$$

$$B(z_1, 0, 0, \ldots, 0) \neq 0 \qquad |z_1| \leq 1 \qquad (6.60)$$

Of course, the ordering of the variables z_i is arbitrary, so that there are many alternatives to the indexing in the conditions appearing in Theorem 6.3. Some alternatives may involve conditions that are more easily checked than others.

FURTHER REFORMULATION OF THE STABILITY CONDITIONS

According to Theorem 6.3, the problem of checking the stability condition (6.42) for an n-variable polynomial is equivalent to the problem of checking conditions

$$B_m(z_1, z_2, \ldots, z_m) \neq 0 \qquad \left\{ \bigcap_{i=1}^{m-1} |z_i| = 1 \right\} \cap \{|z_m| \leq 1\} \qquad (6.61)$$

for $m = 1, 2, \ldots, n$, where $B_m(\cdot, \cdot, \ldots, \cdot)$ is $B(\cdot, \cdot, \ldots, \cdot)$ with the last $n - m$ variables set to zero. In this section we show that for each m, (6.61) is equivalent to a set of inequalities of the form

$$C(z_1, \bar{z}_1, z_2, \bar{z}_2, \ldots, z_{m-1}, \bar{z}_{m-1}) > 0$$
$$\bigcap_{i=1}^{m-1} |z_i| = 1 \quad (6.62)$$

where C is a polynomial in $2(m - 1)$ variables taking real values on $\bigcap_{i=1}^{m-1} |z_i| = 1$, and the number of such inequalities associated with each $B_m(\cdot, \cdot, \ldots, \cdot)$ depends on the degree of B_m regarded as a polynomial in z_m.

To establish this, we recall the Schur–Cohn criterion (see Section 3.2). Suppose that $f(z)$ is a pth-degree polynomial

$$f(z) = \sum_{i=0}^{p} a_i z^i \qquad a_p \neq 0 \quad (6.63)$$

and associate with $f(z)$ the $p \times p$ Hermitian matrix $\Gamma = (\gamma_{ij})$, defined similar to (6.18) by

$$\gamma_{ij} = \sum_{k=1}^{i} (a_{p-i+k}\bar{a}_{p-j+k} - \bar{a}_{i-k}a_{j-k}) \qquad i \leq j \quad (6.64)$$

Then $f(z) \neq 0$ for $|z| \leq 1$ if and only if Γ is negative definite.[†]

Because the negative definiteness of a matrix can be checked by examining the signs of the leading principal minors,[‡] and because each entry of Γ is an integral function of the coefficients of $f(\cdot)$, it follows that $f(z) \neq 0$ for $|z| \leq 1$ if and only if a set of sign-definite inequalities that are integral in the a_j is satisfied.

Now apply these ideas to the m-variable polynomial $B_m(z_1, z_2, \ldots, z_m)$ in (6.61). Thinking of $z_1, z_2, \ldots, z_{m-1}$ as parameters taking values somewhere on the unit circle, $B_m(z_1, z_2, \ldots, z_m)$ can be regarded as a polynomial in z_m. Thus

$$B_m(z_1, z_2, \ldots, z_m) = \sum_{i=0}^{p} a_i(z_1, z_2, \ldots, z_{m-1})z_m^i \quad (6.65)$$

And it is nonzero inside $|z_m| \leq 1$ if and only if the Schur–Cohn matrix is negative definite—that is, if and only if a number of inequalities that are integral in the a_i are satisfied. Since the a_i are polynomials in $z_1, z_2, \ldots, z_{m-1}$ and since the complex conjugates of the a_i appear in the Schur–Cohn matrix, it follows that these inequalities, p in number, have the form of (6.62). Realness of the function C on $\bigcap_{i=1}^{m-1} |z_i| = 1$ follows from the Hermitian nature of Γ, which implies realness of the principal minors.

[†] By working with the reciprocal of $f(z)$, as discussed in Section 6.1, the condition on Γ becomes positive definite. This offers minor simplifications.
[‡] By noting (5.147), we can simplify the negative-definiteness condition.

Notice that the arguments $\bar{z}_1, \bar{z}_2, \ldots, \bar{z}_{m-1}$ in (6.62) can be replaced if desired by $z_1^{-1}, z_2^{-1}, \ldots, z_{m-1}^{-1}$, since $\bar{z}_1 = z_1^{-1}$ for $|z_1| = 1$. Notice also that if the original polynomial $B(z_1, z_2, \ldots, z_n)$ is real under test, each polynomial (6.62), regarded as a polynomial in $2m - 2$ variables, will have real coefficients; for the realness of $B(\cdot, \cdot, \ldots, \cdot)$ implies realness of $B_m(\cdot, \cdot, \ldots, \cdot)$, which, in turn, implies that each entry of Γ, and therefore each principal minor, is a real multivariable polynomial in $z_1, \bar{z}_1, \ldots, \bar{z}_{m-1}$.

STABILITY CONDITIONS FOR THREE- AND FOUR-VARIABLE POLYNOMIALS

First note that stability testing of a two-variable polynomial $B(z_1, z_2)$ leads, via Theorem 6.2, to a two-variable polynomial—actually, $B(z_1, z_2)$—whose properties are of interest on $|z_1| = 1$, $|z_2| \leq 1$ and a one-variable polynomial —actually, $B(z_1, 0)$. By the preceding discussion, the two-variable polynomial generates a number of polynomials in z_1 and $\bar{z}_1$ and a number of constants, for all of which sign definiteness must be checked. Checking the constants is immediate, and checking the polynomial in z_1 and $\bar{z}_1$ is not much more difficult, as shown in Section 6.1.

Generally checking stability of an n-variable polynomial requires checking positivity of polynomials in up to $2n - 2$ variables, $z_1, \bar{z}_1, z_2, \ldots, \bar{z}_{n-1}$. Specifically, a three-variable polynomial produces constants (where positivity is easily checked), polynomials in z_1 and $\bar{z}_1$ (where positivity is checked via the Sturm test), and polynomials in $z_1, \bar{z}_1, z_2$, and $\bar{z}_2$. The checking for positivity of this last set of polynomials can reasonably proceed by plotting values of the polynomials over a two-dimensional grid in a $(\theta_1 - \theta_2)$-plane, taking $z_1 = e^{j\theta_1}$ and $z_2 = e^{j\theta_2}$.

This idea could conceivably be extended to a four-variable polynomial $B(z_1, z_2, z_3, z_4)$, which would demand the checking for positivity of polynomials across a $(\theta_1, \theta_2, \theta_3)$-space.

For many variable polynomials, the use of the computer may become necessary.† An algorithm for checking the positivity condition for one variable will be presented in the next chapter.

Example 6.3. In order to obtain an example that requires but little computation, an exceptionally simple polynomial is needed. We shall choose

$$B(z_1, z_2, z_3) = z_1^2 z_2 + z_1 - z_2 + z_3 + 4 \qquad (6.66)$$

---

† N. K. Bose and E. I. Jury, "Positivity and stability tests for multidimensional filters (discrete continuous)," *IEEE Trans. on Acoustics, Speech and Signal Processing,* **ASSP-22** (3) (June 1974), pp. 174–180.

To demonstrate stability, we need to check that

$$B(z_1, z_2, z_3) \neq 0$$
$$|z_1| = |z_2| = 1, \quad |z_3| \leq 1 \quad (6.67)$$

$$B(z_1, z_2, 0) = z_1^2 z_2 + z_1 - z_2 + 4 \neq 0$$
$$|z_1| = 1, \quad |z_2| \leq 1 \quad (6.68)$$

$$B(z_1, 0, 0) = z_1 + 4 \neq 0 \qquad\qquad |z_1| \leq 1 \quad (6.69)$$

Equation (6.69) is immediate, and (6.68) almost immediate, by direct verification. [In considering (6.68), observe that $|z_1^2 z_2 + z_1 - z_2| \leq |z_1|^2 |z_2| + |z_1| + |z_2| \leq 3$]. For (6.67), applying the Schur–Cohn procedure to $B(z_1, z_2, z_3)$, regarded as a polynomial in z_3, or proceeding immediately using the linear nature of $B(z_1, z_2, z_3)$ in z_3, we obtain the fact that (6.67) is equivalent to

$$|z_1^2 z_2 + z_1 - z_2 + 4| > 1$$
$$|z_1| = 1, \quad |z_2| = 1 \quad (6.70)$$

We could plot out values of the left side of (6.70) for various z_1 and z_2. However, in this case, (6.70) may be directly verified by ad hoc means. Since $|z_1^2 z_2 + z_1 - z_2| \leq 3$, as noted earlier, it follows that (6.70) fails if and only if for some z_1 and z_2 on $|z_1| = 1$, $|z_2| = 1$, we have

$$z_1^2 z_2 + z_1 - z_2 = -3 \qquad\qquad (6.71)$$

To see that this result is impossible, notice that this relation implies

$$z_2 = \frac{3 + z_1}{1 - z_1^2} \qquad\qquad (6.72)$$

Now $|z_2| = 1$ implies $|1 - z_1^2| = |3 + z_1|$; for z_1 on $|z_1| = 1$, one has $0 \leq |1 - z_1^2| \leq 2$ and $2 \leq |3 + z_1| \leq 4$. So we require $2 = |3 + z_1| = |1 - z_1^2|$, which cannot be satisfied by z_1 with $|z_1| = 1$, as is easily seen. Therefore the polynomial $B(z_1, z_2, z_3)$ is nonzero inside $\bigcap_{i=1}^{3} |z_i| \leq 1$.

STABILITY CONDITIONS FOR MULTIDIMENSIONAL CONTINUOUS FILTERS [14]

This discussion extends the case $n = 2$, mentioned in the preceding section, to any n. This step can be readily achieved by applying the bilinear transformation $s_i/\alpha_i = (1 - z_i/(1 - z_i)$, similar to (6.28) and (6.29), to (6.41) to obtain

$$H_c(s_1, s_2, \ldots, s_n) = \frac{A_c(s_1, s_2, \ldots, s_n)}{B_c(s_1, s_2, \ldots, s_n)} \qquad\qquad (6.73)$$

Following (6.72), the stability condition for the multidimensional continuous filter becomes

$$B_c(s_1, s_2, \ldots, s_n) \neq 0 \qquad \bigcap_{i=1}^{n} \text{Re } s_i \geq 0 \qquad (6.74)$$

The s-plane conditions correspond to the simplified stability conditions derived in (6.44), (6.58), (6.59), and (6.60). Thus (6.74) is equivalent to

$$B_c(s_1, s_2, \ldots, s_n) \neq 0$$
$$\left\{ \bigcap_{i=1}^{n-1} \text{Re } s_i = 0 \right\} \cap \{\text{Re } s_n \geq 0\} \quad (6.75)$$

$$B_c(s_1, s_2, \ldots, s_{n-1}, 1) \neq 0$$
$$\left\{ \bigcap_{i=1}^{n-2} \text{Re } s_i = 0 \right\} \cap \{\text{Re } s_{n-1} \geq 0\} \quad (6.76)$$

$$B_c(s_1, s_2, \ldots, s_{n-2}, 1, 1) \neq 0$$
$$\vdots \left\{ \bigcap_{i=1}^{n-3} \text{Re } s_i = 0 \right\} \cap \{\text{Re } s_{n-2} \geq 0\} \quad (6.77)$$

$$B_c(s_1, 1, \ldots, 1, 1) \neq 0 \qquad \text{Re } s_1 \geq 0 \quad (6.78)$$

Each of these conditions can be reformulated as a set of positivity conditions over smaller regions. Thus (6.76) can be replaced by a set of positivity conditions of the form

$$C(s_1, \bar{s}_1, \ldots, s_{n-2}, \bar{s}_{n-2}) > 0$$
$$\bigcap_{i=1}^{n-2} \text{Re } s_i = 0 \quad (6.79)$$

where C takes on only real values in the region of interest. The polynomials C follow by applying the Hermite criterion [16] to the polynomial $B_c(s_1, s_2, \ldots, s_{n-1}, 1)$ viewed as a polynomial in s_{n-1} with s_1 through s_{n-2} as parameters. (The Hermite criterion plays the same role here as played by the Schur–Cohn criterion for the discrete case.)

By analogy to the discussion in Section 6.2, it may well be that the discussion of this section has relevance in obtaining realizability properties of impedances of networks consisting of inductors, capacitors, transformers, and transmission lines, where the transmission lines may be of incommensurate lengths.†

Recent work [17] has shown that one can obtain a finite algorithm for testing the stability of n-dimensional recursive and continuous filters. The

---

† See the following reference: N. K. Bose, K. Zaki, and R. W. Newcomb, "A multivariable bounded reality criterion." *Franklin Inst.* (1974), Vol. 297, No. 6, pp. 479–484.

development of such an algorithm is based on applying decision algebra [18–20] to problems of system theory. Present research on the decision algebra approach gives promise of many applications of this concept to many engineering problems [21].

REFERENCES

[1] H. OZAKI and T. KASAMI, "Positive real functions of several variables and their application to variable networks," *IRE Trans. on Circuit Theory*, **CT-7,** 251–260 (Sept. 1960).

[2] H. G. ANSELL, "On certain two-variable generalizations of circuit theory with applications to networks of transmission lines and lumped reactances," *IEEE Trans. on Circuit Theory*, **CT-11** (2), 214–223 (June 1964).

[3] L. WEINBERG, "Approximation and stability-test procedures suitable for digital filters." Paper presented at the Arden House Workshop of Digital Filtering, Harriman, N.Y., Jan. 1970.

[4] J. L. SHANKS, "Two-dimensional recursive filters," *1969 SWIEECO Rec.*, 19E1–19E8.

[5] J. L. SHANKS, S. TREITEL, and J. H. JUSTICE, "Stability and synthesis of two-dimensional recursive filters," *IEEE Trans. on Audio and Electroacoustics*, **AU-20** (2), 115–128 (June 1972).

[6] T. S. HUANG, "Stability of two-dimensional recursive filters," *IEEE Trans. on Audio and Electroacoustics*, **AU-20** (2), 158–163 (June 1972).

[7] E. I. JURY, *Theory and Application of the z-Transform Method*. New York: Wiley, 1964, Chap. 2.

[8] J. V. HU and L. R. RABINER, "Design techniques for two-dimensional digital filters," *IEEE Trans. on Audio and Electroacoustics*, **AU-20** (4), 249–257 (1972).

[9] C. FARMER and J. B. BEDNAR, "Stability of spatial digital filters," *Mathematical Biosciences*, **14,** 113–119 (1972).

[10] G. A. BLISS, *Algebraic Functions*. New York: American Math. Society, 1933.

[11] B. D. O. ANDERSON and E. I. JURY, "Stability test for two-dimensional recursive filters," *IEEE Trans. on Audio and Electroacoustics*, **AU-21,** 366–372 (Aug. 1973).

[12] K. STEIGLITZ, "The equivalence of digital and analog signal processing," *Information and Control*, **8** (5), 455–467 (Oct. 1965).

[13] S. H. LEHNIGH, *Stability Theorems for Linear Motions with an Introduction to Lyapunov's Direct Method*. Englewood Cliffs, N.J.: Prentice-Hall, 1966.

[14] B. D. O. ANDERSON and E. I. JURY, "Stability of multidimensional digital filters," *IEEE Trans. on Circuits and Systems*, **CAS-21** (2), 300–304 (March 1974).

[15] J. H. JUSTICE and J. L. SHANKS, "Stability criterion for n-dimensional digital filters," *IEEE Trans. on Automatic Control*, **AC-18** (3), 284–286 (June 1973).

[16] S. BARNETT, *Matrices in Control Theory*. London: Van Nostrand Reinhold, 1971.

[17] N. K. BOSE and P. S. KAMAT, "Algorithm for stability test of multidimensional filters," *IEEE Trans. on Acoustics, Speech and Signal Processing* (1974).

[18] A. TARSKI, *A Decision Method for Elementary Algebra*. Berkeley, Ca.: University of California Press, 1951.

[19] A. SEIDENBERG, "A new decision method for elementary algebra," *Ann. Math.*, **60**, 365–374 (1954).

[20] B. E. MESERVE, "Inequalities of higher degree in one unknown," *Am. J. Math.*, **49**, 357–370 (1947).

[21] B. D. O. ANDERSON, N. K. BOSE, and E. I. JURY, "Output feedback stabilization and related problems," *IEEE Trans. on Automatic Control* (1975).

[22] D. D. ŠILJAK, "Algebraic criteria for positive realness relative to the unicircle," *J. Franklin Inst.*, **295** (6), 469–476 (April 1973).

[23] G. A. MARIA and M. M. FAHMY, "On the stability of two-dimensional digital filters," *IEEE Trans. on Audio and Electroacoustics*, 470–472 (Oct. 1973).

[24] D. D. ŠILJAK, "Stability Criteria For Two-Variable Polynomials," Report No. NGR 05–07–010–7401, School of Engineering, University of Santa Clara, Santa Clara, Calif. (Jan. 1974).

CHAPTER 7 | Computational Algorithm for Inners Determinants

In the preceding chapters we have discussed in detail the root-clustering and root-distribution problems as related to the stability of dynamic systems. We further discussed the total integral square and sums of a signal that occur in the quadratic optimization problem. In all these discussions we emphasized the inners approach. In particular, we formulated most of the discussed problems in terms of the sign of an innerwise matrix. A common feature of the innerwise matrices is the left triangle of zeros. Based on this unifying form, we will develop here a computational algorithm to compute the sign, as well as the magnitude if needed, of the inners determinants in a recursive fashion. This algorithm, which is a variant of the Gaussian elimination algorithm, is programmed on a digital computer and can be used for the solution of any of the problems discussed in this book.

In the first section we will develop the computational algorithm, which is referred to as double-triangularization algorithm when none of the inners determinants is zero. This algorithm is mainly useful for stability study, as well as in the evaluation of the integrals discussed in Chapter 4. In the second section we will relax this condition, and we will discuss in detail the critical cases—that is, when some of the inners determinants are zero. This extension will be useful mainly in general root-distribution problems and, in particular, in the checking of the positivity and nonnegativity conditions discussed in detail in chapters 5 and 6. Again, a computer flowchart will be presented, as mentioned in the last example of Section 7.4, for this modified double-triangularization algorithm. In the third section we will present combinatorial rules for certain transformations. In particular, a simplified rule will be presented for the bilinear transformation from the left half to the inside unit disk of the complex plane, and vice versa. Such a transformation has been used in several earlier discussions. Finally, several computer examples will be given in the last section.

205

7.1 GAUSSIAN ELIMINATION ALGORITHM [1–3]

In this section we will present the algorithm in two forms. In the first form we assume that the innerwise matrix A has no zero elements, and we will compute the inner determinants. In the second form we assume that the matrix Δ has a left triangle of zeros, as commonly obtained in the preceding discussions. In both forms we will assume that the inners determinants are nonzero.

FIRST FORM [3]

Consider the following algorithm of a square matrix A of even dimension:

Step 1: Set $i = \frac{1}{2}n$, $k = \frac{1}{2}n$, $\ell = \frac{1}{2}n$, $m = 1$.

Step 2: If all elements of row i to the left of $a_{i\ell}$ and to the right of a_{ik} are zero, go to step 5. If some elements are not zero, go to step 3.

Step 3: If $a_{ii} = 0$, select j such that either $j - 1$ and $a_{ij} \neq 0$ or $j > k$ and $a_{ij} \neq 0$. Interchange columns i and j; go to step 4. (To calculate the determinants, we can do it before we interchange the columns.)

Step 4: With $a_{ii} \neq 0$, multiply column i by a_{ij}/a_{ii} and subtract from column j; do so for all j such that $j < 1$ or $j > k$; go to step 5.

Step 5: If $i \leq \frac{1}{2}n$, replace i by $i + 2m - 1$, k by $k + 1$. If $m = \frac{1}{2}n$, stop; otherwise go to step 2. If $i > \frac{1}{2}n$, replace i by $i - 2m$, ℓ by $\ell - 1$, m by $m + 1$; go to step 2.

Call the resulting matrix $\tilde{A}$; then $\tilde{A}$ has the property

$$\tilde{a}_{ij} = 0 \qquad \text{for} \begin{cases} j = 1, 2, \ldots, \frac{1}{2}n - 1 \\ i = j + 1, j + 2, \ldots, n - j \end{cases}$$

$$\text{and for} \begin{cases} j = \frac{1}{2}n + 1, \frac{1}{2}n + 2, \ldots, n \\ i = n - j + 1, \ldots, j - 1 \end{cases}$$

When A is of odd dimension, the same algorithm can be applied after all elements of row $(n + 1)/2$ are made zero except $a_{\frac{n+1}{2}, \frac{n+1}{2}}$.

Observing the resulting matrix of the preceding linear transformation, we are led to define the matrix of the form of $\tilde{A}$ as a double-triangular matrix. The determinants of the inners of A, as well as of the matrix itself, can be obtained by this algorithm.

The use of this algorithm,† besides computing the inners determinants of any general matrix, lies in determining the stability conditions of a general-system matrix A. This point is illustrated by Example 7.1.

---

† Another use of this algorithm lies in the nonsingular test of doubly infinite matrix: "Some new results on a class of periodically varying systems." by S. N. Basuthakur, J. M. Milne, and J. K. Sen, *Int. J. Electronics*, **35**, 257–266 (1973).

Example 7.1. We wish to present the computation of the necessary and sufficient condition of the following 3×3 A matrix to have all its eigenvalues in the open left half of the complex plane. Let $A \in \mathbb{R}^{3 \times 3}$ be presented as

$$A = \begin{bmatrix} a_{11} & a_{12} & a_{13} \\ a_{21} & a_{22} & a_{23} \\ a_{31} & a_{32} & a_{33} \end{bmatrix} \tag{7.1}$$

From (3.248) to (3.250), the necessary and sufficient conditions (without the diagonally dominant assumption) can be equivalently represented as†

$$\sum_{i=1}^{3} a_{ii} < 0 \tag{7.2}$$

$$\prod_{i=1}^{3} \left(a_{ii} - \sum_{\substack{i=1 \\ i \neq j}}^{3} a_{ij} \right) = |A| < 0 \tag{7.3}$$

$$\prod_{i=1}^{3} \left(\tilde{a}_{ii} - \sum_{\substack{j=1 \\ i \neq j}}^{3} \tilde{a}_{ij} \right) = |\tilde{A}| < 0 \tag{7.4}$$

The matrix $\tilde{A}$ is given by (3.251). We will show that by applying the double-triangularization algorithm to the matrix A in (7.1), we obtain the stability conditions for the A matrix without computing the determinant of $\tilde{A}$—that is, $|\tilde{A}|$. Condition (7.2) is readily computed from A, for it represents the trace of A. That is,

$$\text{trace } A = \sum_{i=1}^{3} a_{ii} = a_{11} + a_{22} + a_{33} = x_1 < 0 \tag{7.5}$$

Perform double triangularization on the matrix A.

Step 1:‡ Make a_{21} and a_{23} zero by pivoting on a_{22}. We obtain

$$\begin{bmatrix} a_{11} - \dfrac{a_{12}a_{21}}{a_{22}} & a_{12} & a_{13} - \dfrac{a_{12}a_{23}}{a_{22}} \\ 0 & a_{22} & 0 \\ a_{31} - \dfrac{a_{32}a_{23}}{a_{22}} & a_{32} & a_{33} - \dfrac{a_{32}a_{23}}{a_{22}} \end{bmatrix} \tag{7.6}$$

Store the entries of the matrix (7.6) as follows:

$$A1(i) = p_k \left[A(1, 1) - \frac{a_{12}a_{21}}{a_{22}} \right]$$

$$\text{when } i = 1, \quad A(1, 1) = a_{11}$$

† See Appendix C for proofs of Theorem 3.15 and the definition of $\tilde{A}$.
‡ The steps indicated in this example follow from the algorithm but are not identical to its step numbering.

$$A1(i) = p_k\left[A(3, 3) - \frac{a_{23}a_{32}}{a_{22}}\right] \tag{7.7}$$

$$\text{when } i = 2, \quad A(3, 3) = a_{33}$$

The factor p_k is the pivoted entry of the A matrix. For instance, if $i = n$, $p_k = 1$. During the first iteration, the pivoted entry is a_{22}. Hence $p_k = a_{22}$ in (7.7). Also,

$$a'_{13} = \left[A(1, 3) - \frac{a_{12}a_{23}}{a_{22}}\right] \qquad A(1, 3) = a_{13}$$

$$a'_{31} = \left[A(3, 1) - \frac{a_{21}a_{32}}{a_{22}}\right] \qquad A(3, 1) = a_{31}$$

Entries that are stored are written in capital letters (7.6) as follows:

$$\begin{bmatrix} A(1, 1) - \dfrac{a_{12}a_{21}}{a_{22}} & a_{12} & A(1, 3) - \dfrac{a_{12}a_{23}}{a_{22}} \\ 0 & a_{22} & 0 \\ A(3, 1) - \dfrac{a_{21}a_{32}}{a_{22}} & a_{32} & A(3, 3) - \dfrac{a_{23}a_{32}}{a_{22}} \end{bmatrix} \tag{7.8}$$

Step 2: Make a'_{13} zero by pivoting at a'_{11}, to obtain from (7.8)

$$\begin{bmatrix} A1(1) & a_{12} & 0 \\ 0 & a_{22} & 0 \\ a'_{31} & a'_{32} & a'_{33} \end{bmatrix} \tag{7.9}$$

where

$$a'_{33} = A(3, 3) - \left\{\frac{a_{23}a_{32}}{a_{22}} - \frac{[A(1, 3) - (a_{12}a_{23}/a_{22})][A(3, 1) - (a_{21}a_{32}/a_{22})]}{A(1, 1) - (a_{21}a_{12}/a_{22})}\right\}$$

$$= \frac{\begin{aligned}&[A(1, 1)A(3, 3) - A(1, 3)A(3,1)] \\ &- [A(3, 3)a_{12}a_{21} - A(1, 1)a_{23}a_{32} + A(1, 3)a_{21}a_{32} + A(3, 1)a_{12}a_{23}]/a_{22}\end{aligned}}{[A(1, 1) - (a_{12}a_{21}/a_{22})]} \tag{7.10}$$

Store $A1(3)$, where

$$A1(3) = p_k[A(1, 1)A(3, 3) - A(1, 3)A(3, 1)] = [a_{11}a_{33} - a_{31}a_{13}] \tag{7.11}$$

In this equation we substituted for $p_k = 1$ because $i = 3 = n$.

Step 3: Multiply diagonal entries of (7.9) to obtain condition (7.3) as follows:

$$|A| = A1(1)a_{22}a'_{33} = x_2 < 0 \qquad (7.12)$$

Step 4: By utilizing the stored elements, it can be readily verified, by actual expansion of terms, that the stability condition in (7.4) is equivalent to

$$|\tilde{A}| = x_1 x_3 - x_2 < 0 \qquad (7.13)\dagger$$

where $A1(1) + A1(2) + A1(3) = x_3$ and x_1, x_2 are given by (7.5) and (7.12).

It should be noted that the same procedure can be followed for the general dimension of the A matrix and for other regions of eigenvalue clustering, such as the unit disk or the negative real axis.

SECOND FORM [3]

In this form the innerwise matrix has a special pattern of left triangle of zeros. As discussed before, this form has the unifying feature for the root-clustering and root-distribution problem. This special pattern makes the double-triangularization algorithm discussed in the first form very effective, since zeros already exist. The matrix Δ has the following pattern.

$$\Delta = \begin{bmatrix} a_{11} & a_{12} & a_{13} & a_{14} & \cdots & a_{1n} \\ 0 & a_{22} & a_{23} & a_{24} & \cdots & a_{2n} \\ 0 & 0 & a_{33} & a_{34} & \cdots & a_{3n} \\ 0 & 0 & & a_{44} & \cdots & a_{4n} \\ \vdots & \vdots & \vdots & \vdots & \cdots & \vdots \\ 0 & 0 & & a_{n-3,4} & \cdots & a_{n-3,n} \\ 0 & 0 & a_{n-2,3} & a_{n-2,4} & \cdots & a_{n-2,n} \\ 0 & a_{n-1,2} & a_{n-1,3} & a_{n-1,4} & \cdots & a_{n-1,n} \\ a_{n1} & a_{n2} & a_{n3} & a_{n4} & \cdots & a_{nn} \end{bmatrix} \qquad (7.14)$$

† It may be noted that the characteristic polynomial of A, i.e., $F(s)$ can be written

$$F(s) = |sI - A| = s^3 - \sum_{i=1}^{3} a_{ii}s^2 + \left\{ \begin{vmatrix} a_{11} & a_{12} \\ a_{21} & a_{22} \end{vmatrix} + \begin{vmatrix} a_{13} & a_{13} \\ a_{31} & a_{33} \end{vmatrix} + \begin{vmatrix} a_{22} & a_{23} \\ a_{32} & a_{33} \end{vmatrix} \right\} s - |A|$$

Hence (7.13) follows from the Hurwitz determinant.

Assume that n, the dimension of Δ, is even and that $|\Delta_{2k}| \neq 0$, the determinants of inners for $k = 1, 2, \ldots, \frac{1}{2}n$. The following algorithm can be used to compute the $|\Delta_{2k}|$'s:

Step 1: Set $i = \frac{1}{2}n$, $k = \frac{1}{2}n$, $m = 1$, $|\Delta_0| = 1$.

Step 2: If all elements of row i to the right of a_{ik} are zero, go to step 4; otherwise go to step 3.

Step 3: Multiply column i by a_{ij}/a_{ii} and subtract from column j; do so for all j such that $j > k$; go to step 4.

Step 4: If $i \leq \frac{1}{2}n$, calculate $C_m = a_{ii}a_{i+2m-1,i+2m-1}$. Replace i by $i + 2m - 1$, k by $k + 1$. Calculate $|\Delta_{2m}| = C_m |\Delta_{2m-2}|$. If $m = \frac{1}{2}n$, stop; otherwise go to step 2. If $i > \frac{1}{2}n$, replace i by $i - 2m$, m by $m + 1$; go to step 2.

When n is odd, the second form can be also used as for the even case.

The algorithm presented in the foregoing two forms can be explained as follows. By choosing a permutation matrix P, form $\bar{\Delta} = P\Delta P'$ such that the corresponding minors of $\bar{\Delta}$ are the same as those determinants of the inners of Δ. This procedure has been demonstrated in Section 1.4. Perform the Gaussian elimination on $\bar{\Delta}$ and obtain $\tilde{\Delta}$ (where $\tilde{\Delta}$ is obtainable from Δ by the double-triangularization algorithm).

Example 7.2. To explain the steps involved in the algorithm for the second form, double triangularize the following (5×5) matrix.

$$\Delta_5 = \begin{bmatrix} a_{11} & a_{12} & a_{13} & a_{14} & a_{15} \\ 0 & a_{22} & a_{23} & a_{24} & a_{25} \\ 0 & 0 & a_{33} & a_{34} & a_{35} \\ 0 & a_{42} & a_{43} & a_{44} & a_{45} \\ a_{51} & a_{52} & a_{53} & a_{54} & a_{55} \end{bmatrix} \tag{7.15}$$

Step 1: Make a_{34} and a_{35} by pivoting on a_{33} to obtain

$$\begin{bmatrix} a_{11} & a_{12} & a_{13} & a_{14} - \dfrac{a_{13}}{a_{33}} a_{34} & a_{15} - \dfrac{a_{13}}{a_{33}} a_{35} \\[2ex] 0 & a_{22} & a_{23} & a_{24} - \dfrac{a_{23}}{a_{33}} a_{34} & a_{25} - \dfrac{a_{23}}{a_{33}} a_{35} \\[2ex] 0 & 0 & a_{33} & a_{34} - \dfrac{a_{33}}{a_{33}} a_{34} & a_{35} - \dfrac{a_{33}}{a_{33}} a_{35} \\[2ex] 0 & a_{42} & a_{43} & a_{44} - \dfrac{a_{43}}{a_{33}} a_{34} & a_{45} - \dfrac{a_{43}}{a_{33}} a_{35} \\[2ex] a_{51} & a_{52} & a_{53} & a_{54} - \dfrac{a_{53}}{a_{33}} a_{34} & a_{55} - \dfrac{a_{53}}{a_{33}} a_{35} \end{bmatrix} \tag{7.16}$$

The matrix in (7.16) can be rewritten as

$$
\begin{bmatrix}
a_{11} & a_{12} & a_{13} & \tilde{a}_{14} & \tilde{a}_{15} \\
0 & a_{22} & a_{23} & \tilde{a}_{24} & \tilde{a}_{25} \\
0 & 0 & a_{33} & 0 & 0 \\
0 & a_{42} & a_{43} & \tilde{a}_{44} & \tilde{a}_{45} \\
a_{51} & a_{52} & a_{53} & \tilde{a}_{54} & \tilde{a}_{55}
\end{bmatrix}
\tag{7.17}
$$

where the $\tilde{a}_{ij}$ are the proper entries in (7.16).

Step 2: Make $\tilde{a}_{24}$ and $\tilde{a}_{25}$ zeros by pivoting on a_{22} to obtain, from (7.17),

$$
\begin{bmatrix}
a_{11} & a_{12} & a_{13} & \tilde{a}_{14} - \dfrac{a_{12}}{a_{22}}\tilde{a}_{24} & \tilde{a}_{15} - \dfrac{a_{12}}{a_{22}}\tilde{a}_{25} \\[2mm]
0 & a_{22} & a_{23} & \tilde{a}_{24} - \dfrac{a_{22}}{a_{22}}\tilde{a}_{24} & \tilde{a}_{25} - \dfrac{a_{22}}{a_{22}}\tilde{a}_{25} \\[2mm]
0 & 0 & a_{33} & 0 & 0 \\[2mm]
0 & a_{42} & a_{43} & \tilde{a}_{44} - \dfrac{a_{42}}{a_{22}}\tilde{a}_{24} & \tilde{a}_{45} - \dfrac{a_{42}}{a_{22}}\tilde{a}_{25} \\[2mm]
a_{51} & a_{52} & a_{53} & \tilde{a}_{54} - \dfrac{a_{52}}{a_{22}}\tilde{a}_{24} & \tilde{a}_{55} - \dfrac{a_{52}}{a_{22}}\tilde{a}_{25}
\end{bmatrix}
\tag{7.18}
$$

The above matrix can be rewritten as

$$
\begin{bmatrix}
a_{11} & a_{12} & a_{13} & \tilde{\tilde{a}}_{14} & \tilde{\tilde{a}}_{15} \\
0 & a_{22} & a_{23} & 0 & 0 \\
0 & 0 & a_{33} & 0 & 0 \\
0 & a_{42} & a_{43} & \tilde{\tilde{a}}_{44} & \tilde{\tilde{a}}_{45} \\
a_{51} & a_{52} & a_{53} & \tilde{\tilde{a}}_{54} & \tilde{\tilde{a}}_{55}
\end{bmatrix}
\tag{7.19}
$$

where $\tilde{\tilde{a}}_{ij}$ are the proper entries in (7.18).

Steps 3 and 4: Make $\tilde{\tilde{a}}_{45}$ zero by pivoting on $\tilde{\tilde{a}}_{44}$, and $\tilde{\tilde{a}}_{15}$ zero by pivoting on a_{11}. We finally obtain

$$
\tilde{\Delta}_5 =
\begin{bmatrix}
a_{11} & a_{12} & a_{13} & \tilde{\tilde{a}}_{14} & 0 \\
0 & a_{22} & a_{23} & 0 & 0 \\
0 & 0 & a_{33} & 0 & 0 \\
0 & a_{42} & a_{43} & \tilde{\tilde{a}}_{44} & 0 \\
a_{51} & a_{52} & a_{53} & \tilde{a}_{54} & \dot{a}_{55}
\end{bmatrix}
\tag{7.20}
$$

The determinants of the innerwise matrix Δ_5 of (7.15) are

$$|\Delta_1| = a_{33}$$
$$|\Delta_3| = a_{22}a_{33}\tilde{\tilde{a}}_{44} \qquad\qquad (7.21)$$
$$|\Delta_5| = a_{11}a_{22}a_{33}\tilde{\tilde{a}}_{44}\dot{a}_5$$

7.2 COMPUTATIONAL ALGORITHM FOR TESTING POSITIVITY AND NONNEGATIVITY CONDITIONS

In this section we will extend the computational algorithm discussed in Section 7.1 to cover the critical cases where some of the inners determinants are zero. In particular, we will modify the algorithm of the second form to yield information on the root distribution of the system-characteristic equation, and as a result of this modification, we obtain the conditions of positivity and nonnegativity discussed in Chapters 5 and 6. We first discuss the root-distribution problems for the discrete case—that is, root distribution with respect to the unit circle—and later we will discuss the continuous case—that is, root distribution with respect to the imaginary axis in the complex plane.

ROOT DISTRIBUTION WITH RESPECT TO THE UNIT CIRCLE [4]

It is well known that the table form discussed in (5.95) can be effectively utilized to obtain the root distribution of the real polynomial in (5.94). In the table form there are two critical cases which have been dealt with. In the first critical case all the entries of a certain row in the table are zeros. This case indicates that the polynomial in (5.94) has either roots on the unit circle or reciprocal roots with respect to the unit circle, or both. The second critical case arises when the row of the first zero δ_k in the table of (5.95) is not identically zero. This case represents some roots outside the unit circle. In order to deal with the first critical case, we generate a new polynomial $g_{new}(z)$ by obtaining the reciprocal of the derivative of a polynomial generated from the table entries of the second row preceding the identically zero row. This is given by (5.97) and illustrated by the example of (5.98). To deal with the second critical case, we replace the first δ_k, which is zero in the table, by [4] a small positive or negative value ϵ and proceed with the computation as needed to obtain all the information on the root distribution.† The above

---

† It may be noted that the perturbation method based on ϵ is only valid if the polynomial has no zeros on the unit circle. For the more general case one may use the procedure indicated on p. 131 of Reference 4.

procedures deal with all the critical cases that arise either individually or simultaneously.

We can similarly modify the computational algorithm of the second form to deal with the above two critical cases. This can be done based on the following remarks:

1.   In the table form the δ_k's obtained are related to the inner determinants $|\Delta_{2k}|$'s of the Schur–Cohn matrix. Indeed, the table form can be slightly modified [4] to yield the δ_k's that are identical to the $|\Delta_{2k}|$'s. In particular, if δ_k is zero, it follows that $|\Delta_{2k}|$ is also zero and vice versa. This indicates that when the critical cases arise in the calculation of δ_k's, they also arise in the computation of $|\Delta_{2k}|$'s, using the algorithm.

2.   If the table form can be utilized to deal with both the critical cases, it is evident that the computational algorithm can also be utilized to deal with those critical cases. This is a foregone conclusion, since both the table form and the double-triangularization algorithm represent equivalent methods for computing the $|\Delta_{2k}|$'s (i.e., the inners determinants).

Based on the above remarks, we will indicate how to distinguish between the two critical cases by using the algorithm, as well as how to proceed to obtain the root distribution of the polynomial. To illustrate the procedure, we will deal first with a low-order polynomial, and the results for the general case will follow.

FIRST CRITICAL CASE

This case is best illustrated by Example 7.3.

Example 7.3.   In this example we will deal with the first critical case using the table form. Later we indicate the relationships with the inner form by using the double-triangularization algorithm. Let $f(z)$ be given as

$$f(z) = z^4 + a_2 z^3 + a_1 z^2 + a_2 z + 1 \qquad (7.22)$$

The above polynomial represents a reciprocal (or self-inversive) one. Hence, following (5.97), we obtain

$$f_{\text{new}}(z) = \left[\frac{df(z)}{dz} \right]^*$$

$$= [a_2 + 2a_1 z + 3a_2 z^2 + 4z^3]^* = 4 + 3a_2 z + 2a_1 z^2 + a_2 z^3 \qquad (7.23)$$

and [ ]* represents a polynomial whose roots are reciprocal with respect to the unit circle of [ ]. We generate the table form from (5.95) as follows:

Row	z^0	z^1	z^2	z^3
1	4	$3a_2$	$2a_1$	a_2
2	a_2	$2a_1$	$3a_2$	a_4
3	$\delta_1 = (16 - a_2^2)$	$12a_2 - 2a_1a_2$	$8a_1 - 3a_2^2$	
4	$8a_1 - 3a_2^2$	$12a_2 - 2a_1a_2$	$16 - a_2^2$	
5	$\delta_2 = 0$	0		(7.24)

We have assumed that the fifth row has all zeros, which indicates the first type of critical case. To obtain these zeros, we must have

$$(16 - a_2^2)^2 = (8a_1 - 3a_2^2)^2 \tag{7.25}$$

and
$$(12a_2 - 2a_1a_2)[(16 - a_2^2) - (8a_1 - 3a_2^2)] = 0 \tag{7.26}$$

To satisfy (7.26), we have two possibilities:†

$$12a_2 - 2a_1a_2 = 0 \quad \text{and} \quad 16 - a_2^2 - 8a_1 + 3a_2^2 \neq 0 \tag{7.27}$$

or
$$12a_2 - 2a_1a_2 \neq 0 \quad \text{and} \quad 16 - a_2^2 - 8a_1 + 3a_2^2 = 0 \tag{7.28}$$

To proceed with the table form, we generate a new polynomial based on the entries of row 4 according to (5.97):

$$f_{1new}(z) = \left[\frac{df_1(z)}{dz}\right]^* \tag{7.29}$$

In this case,

$$f_1(z) = (16 - a_2^2)z^2 + (12a_2 - 2a_1a_2)z + (8a_1 - 3a_2^2) \tag{7.30}$$

so that

$$f_{1new}(z) = 2(16 - a_2^2) + (12a_2 - 2a_1a_2)z \tag{7.31}$$

We can proceed with the table form using (7.31). If, in the process of the table reduction, we encounter a similar critical case, we again follow the same procedure. Thus we can calculate the root distribution when the critical case arises.

INNER FORM

We can generate the innerwise matrix corresponding to (7.23), by applying (2.39), as follows:

---

† This is in addition to both equations being simultaneously equal to zero. In this case the polynomial $f(z)$ in (7.22) has two roots on the unit circle at $\pm j$.

$$\Delta_6 = \begin{bmatrix} a_2 & 2a_1 & 3a_2 & 0 & 0 & 4 \\ 0 & a_2 & 2a_1 & 0 & 4 & 3a_2 \\ 0 & 0 & a_2 & 4 & 3a_2 & 2a_1 \\ 0 & 0 & 4 & a_2 & 2a_1 & 3a_2 \\ 0 & 4 & 3a_2 & 0 & a_2 & 2a_1 \\ 4 & 3a_2 & 2a_1 & 0 & 0 & a_2 \end{bmatrix} \tag{7.32}$$

We apply the algorithm of (7.14) to double triangularize (7.32). First we pivot on a_2 to obtain

$$\tilde{\Delta}_6 = \begin{bmatrix} a_2 & 2a_1 & 3a_2 & -12 & -9a_2 & 4-6a_1 \\ 0 & a_2 & 2a_1 & -\dfrac{8a_1}{a_2} & 4-6a_1 & 3a_2 - \dfrac{4a_1^2}{a_2} \\ 0 & 0 & a_2 & 0 & 0 & 0 \\ 0 & 0 & 4 & a_2 - \dfrac{16}{a_2} & 2a_1 - 12 & 3a_2 - \dfrac{8a_1}{a_2} \\ 0 & 4 & 3a_2 & -12 & -8a_2 & -4a_1 \\ 4 & 3a_2 & 2a_1 & -\dfrac{8a_1}{a_2} & -6a_1 & a_2 - \dfrac{4a_1^2}{a_2} \end{bmatrix} \longleftarrow \tag{7.33}$$

From (7.33), we obtain

$$|\Delta_2| = a_2^2 - 16 = -\delta_1 \tag{7.33a}$$

To proceed with the algorithm of (7.14), we pivot on the entry $a_2 - 16/a_2$ in the fourth row of (7.33). By so doing we obtain the fifth and sixth column for the new matrix $\tilde{\tilde{\Delta}}_6$ as follows:

Fifth Column	**Sixth Column**
$\dfrac{-9a_2^3 + 24a_2a_1}{a_2^2 - 16}$	$\dfrac{40a_2^2 - 6a_2^2a_1 - 64}{a_2^2 - 16}$
$\dfrac{4a_2^2 + 16a_1^2 - 6a_1a_2^2 - 64}{a_2^2 - 16}$	$\dfrac{a_2(3a_2^2 - 48 - 4a_1^2 + 24a_1)}{a_2^2 - 16}$
0	0
0	0
$\dfrac{-8a_2^3 - 16a_2 + 24a_2a_1}{a_2^2 - 16}$	$\dfrac{36a_2^2 - 4a_1a_2^2 - 32a_1}{a_2^2 - 16}$
$\dfrac{16a_1^2 - 6a_1a_2^2}{a_2^2 - 16}$	$\dfrac{a_2^3 - 4a_1^2a_2 + 24a_2a_1 - 16a_2}{a_2^2 - 16}$

$$\tag{7.34}$$

Let us form the following ratios of elements in (7.33) and (7.34):

$$\frac{\text{Second row and second column of (7.33)}}{\text{Fifth row and second column of (7.33)}}$$

$$= \frac{\text{Second row and fifth column of (7.34)}}{\text{Fifth row and fifth column of (7.34)}} \quad (7.35)$$

which gives

$$\frac{a_2}{4} = \frac{4a_2^2 + 16a_1^2 - 6a_1a_2^2 - 64}{a_2(-8a_2^2 - 16 + 24a_1)} \quad (7.36)$$

or
$$(16 - a_2^2)^2 = (8a_1 - 3a_2^2)^2 \quad (7.37)$$

It is of interest to note that (7.37) is exactly (7.25). Furthermore, the next step in double triangulation will produce a fifth column of the following form, if and only if (7.35), and therefore (7.37), is satisfied:

Column Five

$$
\begin{matrix}
x \\
0 \\
0 \\
0 \\
0 \\
0 \\
x
\end{matrix}
\Big\} \text{zero values} \quad (7.38)
$$

Hence we have shown that a zero occurs in the first column, row five of table (7.24) if and only if $|\Delta_4| = 0$.

We can easily establish which critical case has occurred. We return to the step when Δ_2 is double triangularized in (7.33) and generate a new polynomial from the entries indicated by the arrow in the right side of (7.33), to give

$$f_1(z) = \left(a_2 - \frac{16}{a_2}\right)z^2 + (2a_1 - 12)z + \left(3a_2 - \frac{8a_1}{a_2}\right) = 0 \quad (7.39)$$

$$f_1(z) = I_1 z^2 + I_2 z + I_3 = 0 \quad (7.40)$$

Next we differentiate (7.39) and obtain the conjugate, and proceed in the same fashion as in (7.29). Note that (7.39) is equivalent to (7.30).

Thus to distinguish the first critical case, we must have from (7.40),

$$\frac{I_1}{I_2} = \frac{I_3}{I_2} \quad (7.41)$$

or equivalently, the polynomial of (7.39) is self-inversive (reciprocal).

This procedure is easily generalized. For the general case when $|\Delta_{2m}|$ is zero, check all the entries in the $(n + m - 1)$th row beginning from the

$(n + m - 1)$th column in the matrix where $\Delta_{2(m-1)}$ was triangularized. If the entries are those representing a self-inversive polynomial, then the first critical case is established.

SECOND CRITICAL CASE [4]

In this case, δ_k in the table of (5.95) is zero, but not all the entries in that row are zero. We can distinguish this case in the inner form by simply noticing that the extracted polynomial of (7.39) is not self-inversive. To proceed with this critical case, apply the known procedure indicated in other texts [4,5]† to this generated polynomial.

The procedures for dealing with both critical cases are included in the modified computer algorithm for the double triangularization. Many examples will be illustrated in Section 7.4, where a flow chart is given.

ROOT DISTRIBUTION WITH RESPECT TO
THE IMAGINARY AXIS [2,6]

It is well known that we can obtain the root distribution of any polynomial (real or complex) with respect to the imaginary axis with the use of the Routh table. Similar to the discrete case (i.e., root distribution with respect to the unit circle), we have to distinguish between two critical cases. The first critical case for real polynomials arises when complex conjugate roots lie on the $j\omega$ axis in the s-plane or when they are reciprocal of each other with respect to the $j\omega$ axis, or both. In this critical case, the polynomial in s is an even one. The modification in the Routh table needed for this case is explained in (5.53) and (5.54). The second critical case arises when the first entry in a particular row of the Routh table is zero and not all other entries of this row are zeros. It is well known that to account for this critical case, we multiply the generated (or original) polynomial by $s + c$ (c = positive constant) and go on with the table procedure [6].

In order to show how to use the algorithms of (7.14) to obtain all the information on the root distribution with respect to the imaginary axis, we have to generate the innerwise matrix for the tested polynomial. In this case, we must start with the matrix Δ_{2n} of (2.4) by placing zeros in the proper array of the matrix for the imaginary coefficients of $F(s)$. This matrix dimension is necessitated by the fact that we need n conditions for the n roots of $F(s) = 0$ from the Δ_{2n} innerwise matrix. This doubling of the size of the Hurwitz matrix (for real polynomials) does not increase significantly the computation burden, because of the zero entries in Δ_{2n}.

It is of interest to note that for stability, the dimension of the innerwise

---

† In this case the extracted polynomial is multiplied by $z + c$, where c is a constant which can be determined. See p. 131 of Reference 4.

matrix (2.10) is $(n - 1)$. This matrix gives about $n/2$ conditions, and the other $n/2$ conditions are the sign of the coefficients of $F(s)$ in (2.9). This is because we utilized the simpler form of the Liénard–Chipart stability criterion.

To establish the connection between the Routh table and the double-sized innerwise matrix, we discuss Example 7.4 in detail, and the procedure for the general case immediately follows.

FIRST CRITICAL CASE

This case is best illustrated by Example 7.4.

Example 7.4. In this example we will show the relationships between the Routh table and the inner form when we have the first type of critical case (i.e., reciprocal roots with respect to imaginary axis). Let

$$F(s) = s^4 + a_2 s^2 + a_4 \tag{7.42}$$

We proceed with the table form of (5.56) by inserting in the second rows the coefficients of the derivative of $F(s)$ as follows:†

Table Form

Row 1	1	a_2	a_4
Row 2	4	$2a_2$	
Row 3	$\dfrac{a_2}{2}$	a_4	
Row 4	0		

$$\tag{7.43}$$

For the zero to exist in the fourth row, we must have

$$4a_4 = a_2^2 \tag{7.44}$$

We generate a new polynomial from the entry of the third row to obtain

$$F_1(s) = \frac{a_2}{2} s^2 + a_4 \tag{7.45}$$

Differentiate the above polynomial and proceed with the table form to get

Row 3	$\dfrac{a_2}{2}$	a_4
Row 4	a_2	
Row 5	a_4	

$$\tag{7.46}$$

From the entries of the first column of the above tables, we obtain all the information on the roots distribution when the first critical case arises. The determination of the first critical case is ascertained (i.e., when the only entry

† See (5.54).

in row 4 is zero)† from rows 2 and 3 in (7.43). Then two rows are proportional to each other by noting (7.44).

INNER FORM

To obtain the inner matrix corresponding to (7.42), we obtain the first derivative of $F(s)$ of (7.42) as follows:

$$F'(s) = 4s^3 + 2a_2s \qquad (7.47)$$

The inner matrix is obtained for the total polynomial $F_0(s)$ as follows:

$$F_0(s) = F(s) + F'(s) = s^4 + 4s^3 + 2a_2s^2 + 2a_2s + a_4 \qquad (7.48)$$

To generate the innerwise matrix of $F_0(s)$ by utilizing (2.4), we must first obtain $F_0(js)$:

$$F_0(js) = s^4 - j4s^3 - a_2s^2 + 2ja_2s + a_4 \qquad (7.49)$$

Using (2.4) and reversing the bottom and the upper rows, we obtain‡

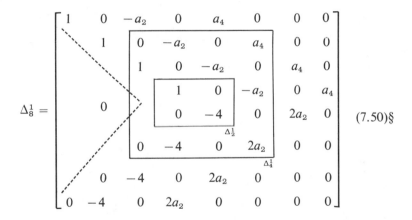

We note that by changing the rows in (2.4), the value of the inner determinant $|\Delta_{2m}|$ becomes $(-1)^m |\Delta_{2m}^1|$. The reason for changing the rows is to enable us to double triangularize the matrix (7.50) by using the algorithm applied to (7.14); otherwise we have to pivot with zero entry, which is to be avoided, as indicated in the algorithm.

† In a general case the whole row is zero.
‡ This reversal is needed to apply the algorithm only for the polynomial whose highest degree is even. It is not needed for the odd degree.
§ We can also use (2.3) to obtain the root distribution. In this case, the innerwise matrix is $(2n - 1) \times (2n - 1)$.

We apply the first step in triangularizing the matrix in (7.50) to obtain

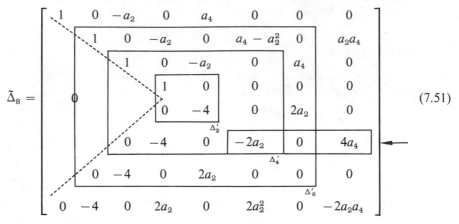

$$(7.51)$$

We continue the double triangularization of (7.51) by pivoting on another entry to obtain

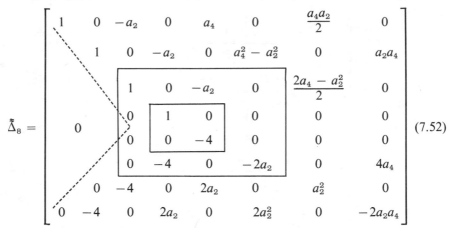

$$(7.52)$$

We continue the process until we observe the following:

$$
\begin{array}{c}
\widehat{x} \\
\begin{bmatrix}
0 \\
0 \\
0 \\
0 \\
0 \\
0
\end{bmatrix} \\
0_{\Delta'_6}
\end{array}
$$

$$(7.53)$$

In order that $|\Delta'_6|$ be zero, we have, from (7.52),

$$\frac{(2a_4 - a_2^2)/2}{a_2^2} = \frac{-1}{4} \tag{7.54}$$

or
$$a_2^2 + \frac{4(2a_4 - a_2^2)}{2} = 0 \tag{7.55}$$

The above equation gives

$$4a_4 = a_2^2 \tag{7.56}$$

It is noticed that (7.56) is the same as (7.44), which produced the first critical case. To generate the new polynomial, we go to the step when Δ'_4 is double triangularized. We check the entries in the sixth row from the sixth column (i.e., entries corresponding to a_{66}, a_{67}, and a_{68}). From (7.51), this polynomial is obtained from the right side indicated by the arrow as follows:

$$F_1(js) = -2a_2s^2 + 4a_4 \tag{7.57}$$
or
$$F_1(s) = 2a_2s^2 + 4a_4 \tag{7.58}$$

The polynomial in (7.58) is the same as that in (7.45) obtained from the table of (7.43). Hence, similar to the table form, we are able to detect the first critical case and to generate the new polynomial. To proceed with the inner test, we repeat the above procedure to the polynomial $F_1(s)$ and continue in the same fashion.

A similar situation arises in any general even polynomial, and hence we can tackle the root distribution accordingly. It is of interest to note that the polynomial generated from the right side of (7.51) has alternate zeros. This fact distinguishes the first critical case from the second one, as will be explained in a later example. Finally, we may mention that the table form and the double triangularization for the discrete case are identical within a constant. However, it is not so for the continuous case. This is due to the fact that the Routh table corresponds to a Hurwitz matrix which is half the size of the innerwise matrix of (7.50). The number of computations is almost the same, since the matrix in (7.50) has many zeros (sparse matrix). Also, it may be noted that the Routh scheme is similar to the lower-triangularization method applied to a Hurwitz matrix. In our scheme we use the double-triangularization method to unify the computation of the root distribution for both the continuous and the discrete cases in one algorithm. This represents a decisive advantage over the use of two tables.†

Finally, we mention that $(-1)^m$ multiplied by the inner determinants of

---

† In the following reference S. Barnett has shown the duality property of the two tables. "Interchangeability of the Routh and Jury Tabular Algorithms for Linear System Zero Location." Papers presented at IEEE Decision and Control Conf., San Diego, Ca., Dec. 1973.

(7.50) corresponds to the Hurwitz determinants. The latter can be obtained from the Routh table, using the well-known relationships.

SECOND CRITICAL CASE [6]

This critical case arises when the first entry in any Routh row is zero but not all the entries of this row are zero. This indicates that there exists at least one root in the right half-plane (unstable case). To distinguish this critical case and to indicate the procedure for continuing the test for the root distribution, we illustrate the method using Example 7.5.

Example 7.5. Let $F(s)$ in this case be given as

$$F(s) = s^4 + a_3 s^3 + a_2 s^2 + a_1 s + a_0 \tag{7.59}$$

We generate the Routh table:

Row 1	1	a_2	a_0
Row 2	a_3	a_1	
Row 3	0	$a_3 a_0$	

$$(7.60)$$

We observe from the first entry of row 3 that

$$a_3 a_2 - a_1 = 0 \tag{7.61}$$

and

$$a_3 a_0 \neq 0 \tag{7.62}$$

Conditions (7.61) and (7.62) constitute the second critical case. To proceed with the test, we first generate a new polynomial from rows 2 and 3 of (7.60):

$$f_1(s) = a_3 s^3 + 0 s^2 + a_1 s + a_3 a_0 \tag{7.63}$$

and then multiply $f_1(s)$ by $(s + c)$ where c is a positive constant which in certain cases can be unity. Therefore the new polynomial becomes

$$F_1(s) = (s + 1)f_1(s) = a_3 s^4 + a_3 s^3 + a_1 s^2 + (a_1 - a_3 a_0)s + a_3 a_0 \tag{7.64}$$

Now we generate the Routh table for the polynomial in (7.64) and continue the test.

INNER FORM

To obtain the equivalent procedure shown in the table method for the inner form, we generate the innerwise matrix following (2.4) as applied to (7.59) to obtain, after reversal of rows, the following:

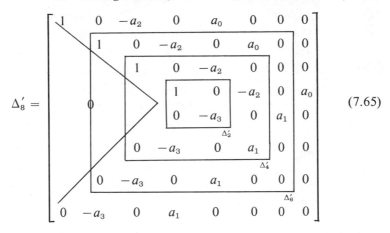

$$(7.65)$$

Apply the double-triangularization procedure of (7.14) to (7.65) to obtain

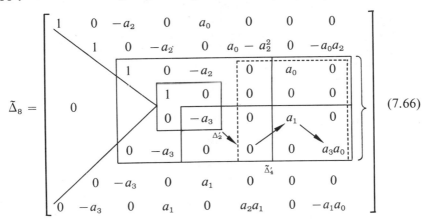

$$(7.66)$$

We observe from (7.65) (in order that $|\Delta_4'| = 0$), that we must have $a_1 = a_3 a_2$ and that the dotted block in (7.66) cannot be reduced to zero. Therefore, this indicates the second type of critical case. Note in this case that $|\Delta_4'|$ is zero but the other higher-order inner determinants are not zero. This distinguishes the second critical case from the first.

To generate the new polynomial, we follow the arrows in (7.66) to obtain

$$f_1(js) = -ja_3 s^3 + 0s^2 + ja_1 s + a_3 a_0 \tag{7.67}$$

and

$$f_1(s) = a_3 s^3 + a_1 s + a_3 a_0 \tag{7.68}$$

We notice that (7.68) is identical to (7.63). To continue the test, we multiply $f_1(s)$ by $(s + 1)$, as indicated before, and obtain $F_1(s)$:

$$F_1(s) = a_3 s^4 + a_3 s^3 + a_1 s^2 + (a_1 + a_3 a_0)s + a_3 a_0 \tag{7.69}$$

Now we generate $F_1(js)$ and use the matrix in (2.4) to obtain the new innerwise matrix, and thus we continue the double-triangularization procedure.

In concluding this discussion, it is pertinent to indicate the following remarks:

1. The application of the double-triangularization method can also be utilized for obtaining the root distribution with respect to more general regions than with respect to the imaginary axis and to the unit circle. Such regions are discussed in Section 3.4.

2. The application of the double-triangularization method for testing positivity conditions for even polynomials for the continuous and reciprocal (or self-inversive) polynomials for the discrete case is straightforward and simple. Furthermore, Šiljak's theorem, discussed, in (5.62) and (5.63) for nonnegativity test for the continuous case, and the corresponding one in (5.108) for the discrete case can be readily incorporated in the computational algorithm. This represents a unified method for testing the positivity and nonnegativity conditions discussed in Chapters 5 and 6 for both continuous and discrete systems.† This offers computational simplifications and finally solves this classical problem in a unified straightforward procedure. Similar results can be achieved for complex polynomials.‡

7.3 COMBINATORIAL RULES FOR CERTAIN TRANSFORMATIONS

In many applications we encounter the problem of converting the polynomial equation (usually the characteristic equation of a given system)

$$F(z) = a_n z^n + a_{n-1} z^{n-1} + \cdots + a_1 z + a_0 = 0 \qquad (7.70)$$

in the z-plane onto the corresponding polynomial equation

$$G(s) = b_n s^n + b_{n-1} s^{n-1} + \cdots + b_1 s + b_0 = 0 \qquad (7.71)$$

in the s-plane, through a given transformation. One such transformation in common use is the bilinear transformation [7].

$$z = \frac{s+1}{s-1} \quad \text{and} \quad \frac{z+1}{z-1} = s \qquad (7.72)$$

The transformation in (7.72) (known as Möbius mapping [8]) constitutes a one-to-one mapping of the open disk $|z| < 1$ into the open left half-plane

---

† If the roots of the polynomial lie very near to the unit circle or the $j\omega$ axis, we can make use of methods given in Reference 27.
‡ For numerical comparison of root-location algorithms see References 28 and 29.

Re $[s] < 0$. The imaginary axis, Re $[s] = 0$, maps one-to-one on the circle $|z| = 1$ with point $z = 1$ deleted.

This bilinear transformation in (7.72) is encountered in many applications, such as in the proof of the simplified Schur–Cohn criterion [9], in the evaluation of total-square integrals in minimization problems of discrete data systems [10], and in digital filter design [11,12]. In this section we will discuss in detail the bilinear transformation of (7.72) and other related transformations used in the preceding chapters, to develop combinatorial rules for generating the mapped polynomials.

a. Bilinear Transformation $z = (s + 1)/(s - 1)$

This transformation, mentioned earlier in (7.72), will generate the polynomial $G(s)$ of (7.71) from the polynomial $F(z)$ in (7.70) as follows:

$$G(s) = (s - 1)^n F\left(\frac{s + 1}{s - 1}\right) = 0 \tag{7.73}$$

which is a polynomial of degree n in s.

The coefficients b_k's of (7.71) are obtained from the coefficients of (7.70) by using the following known relationship [4]:

$$b_{n-k} = \sum_{j=0}^{n} a_{n-j} \left[\binom{n-j}{k} - \binom{j}{1}\binom{n-j}{k-1} + \binom{j}{2}\binom{n-j}{k-2} + \cdots \right.$$
$$\left. + (-1)^{k-1}\binom{j}{k-1}\binom{n-j}{1} + (-1)^k \binom{j}{k} \right] \tag{7.74}$$

where

$$\binom{j}{k} = \frac{j!}{k!\,(j-k)!} \tag{7.75}$$

The relationship between the b_i's and the a_i's can be written in matrix form, as the following example illustrates. Let

$$F(z) = a_4 z^4 + a_3 z^3 + a_2 z^2 + a_1 z + a_0 \tag{7.76}$$

From (7.74) and (7.75), we have

$$\begin{aligned}
b_4 &= a_4 + a_3 + a_2 + a_1 + a_0 \\
b_3 &= 4(a_4 - a_0) + 2(a_3 - a_1) \\
b_2 &= 6(a_4 + a_0) - 2a_2 \\
b_1 &= 4(a_4 - a_0) + 2(a_1 - a_3) \\
b_0 &= a_4 - a_3 + a_2 - a_1 + a_0
\end{aligned} \tag{7.77}$$

Equation (7.77) can be written in matrix form as follows:

$$
\begin{bmatrix} b_4 \\ b_3 \\ b_2 \\ b_1 \\ b_0 \end{bmatrix} = \begin{bmatrix} 1 & 1 & 1 & 1 & 1 \\ 4 & 2 & 0 & -2 & -4 \\ 6 & 0 & -2 & 0 & 6 \\ 4 & -2 & 0 & 2 & -4 \\ 1 & -1 & 1 & -1 & 1 \end{bmatrix} \begin{bmatrix} a_4 \\ a_3 \\ a_2 \\ a_1 \\ a_0 \end{bmatrix} \tag{7.78}
$$

In general, the relationship between the b_i's and the a_i's can be written as [13]

$$b] = Qa] \tag{7.79}$$

where $\qquad b] = [b_n b_{n-1} \cdots b_0]', a = [a_n a_{n-1} \cdots a_0]'$

and Q is an $(n + 1) \times (n + 1)$ matrix.

In recent years the matrix Q has been the subject of investigation by many researchers, such as Power [14], Duffin [8], Fielder [12,15], Halijak [16], Greaves et al. [10], and this writer [17,18]. It has been established that the entries of Q, [i.e., (q_{ij})] can be easily generated by a combinatorial rule given by

$$q_{ij} = q_{i,j-1} - q_{i-1,j-1} - q_{i-1,j} \tag{7.80}\dagger$$

Furthermore, the present writer further simplified the computation of the entries of Q by showing the symmetric properties of the rows and columns of Q. As a result of this simplification, only about one-quarter of the entries of Q need be calculated.

In the following pages we will present proofs of (7.80) and of the symmetry properties of the matrix Q.

Proof of (7.80) [18]: From (7.74), we deduce that the entries of Q in (7.79) can be expressed as

$$q_{ij} = \sum_{k=0}^{j-1} (-1)^k \binom{j-1}{k} \binom{n+1-j}{i-k} \tag{7.81}$$

---

† A generating rule for a general bilinear transformation has been obtained by S. Barnett: "Some applications of matrices to location of zeros of polynomials," *Int. J. Control*, **17** (4), 823–831 (1973).

We note the following properties.

$$\binom{j}{i} = \binom{j-1}{i} + \binom{j-1}{i-1}$$

$$\binom{j}{i} = \binom{j-1}{j-1}$$

$$\binom{j}{0} = \binom{j-1}{0}$$

$$\binom{j}{i} = 0$$

for $i < 0$ or $j > 1$ (7.82)

The properties of (7.82) will enable us to prove (7.80). Using (7.81), the right-hand side of (7.80) becomes

$$I = q_{i,j-1} - q_{i-1,j-1} - q_{i-1,j}.$$

$$= \sum_{k=0}^{j-2}(-1)^k\binom{j-2}{k}\binom{n+2-j}{i-k} - \sum_{k=0}^{j-2}(-1)^k\binom{j-2}{k}\binom{n+2-j}{i-1-k}$$

$$- \sum_{k=0}^{j-1}(-1)^k\binom{j-1}{k}\binom{n+1-j}{i-1-k}$$

$$= \sum_{k=0}^{j-2}(-1)^k\binom{j-2}{k}\binom{n+1-j}{i-k-1} - \sum_{k=0}^{j-2}(-1)^k\binom{j-2}{k}\binom{n+1-j}{i-2-k}$$

$$- \sum_{k=0}^{j-1}(-1)^k\binom{j-1}{k}\binom{n+1-j}{i-1-k} \qquad (7.83)$$

Keeping the summation up to the term $\binom{n+1-j}{i-j+2}$,

$$I = \sum_{k=0}^{j-2}(-1)^k\binom{j-2}{k}\binom{n+1-j}{i-k} - \sum_{k=0}^{j-4}(-1)^k\binom{j-2}{k}\binom{n+1-j}{i-2-k}$$

$$- \sum_{k=0}^{j-3}(-1)^k\binom{j-1}{k}\binom{n+1-j}{i-1-k} - (-1)^{j-3}\binom{j-2}{j-3}\binom{n+1-j}{i-j+1}$$

$$- (-1)^{j-2}\binom{j-1}{j-2}\binom{n+1-j}{i-j+1} \qquad (7.84)$$

Reordering the dummy variable k,

$$I = \sum_{k=0}^{j-2}(-1)^k \binom{j-2}{k}\binom{n+1-j}{i-k} - \sum_{k=2}^{i-2}(-1)^{k-2}\binom{j-2}{k-2}\binom{n+1-j}{i-1}$$
$$- \sum_{k=1}^{j-2}(-1)^{k-1}\binom{j-1}{k-1}\binom{n+1-j}{i-k}$$
$$- (-1)^{j-2}\binom{j-1}{j-2} - \binom{j-2}{j-3}\binom{n+1-j}{i-j+1}$$

$$(7.85)$$

Using (7.82),

$$I = \sum_{k=0}^{j-2}(-1)^k\left[\binom{j-2}{k} - \binom{j-2}{k-2} + \binom{j-1}{k-1}\right]\binom{n+1-j}{i-k}$$
$$- (-1)^{j-2}\binom{j-2}{j-2}\binom{n+1-j}{i-j+1}$$

$$= \sum_{k=0}^{j-2}(-1)^k\left[\binom{j-2}{k} + \binom{j-2}{k-1}\right]\binom{n+1-j}{i-k}$$
$$- (-1)^{j-2}\binom{j-1}{j-1}\binom{n+1-j}{i-j+1}$$

$$= \sum_{k=0}^{j-1}(-1)^k\binom{j-1}{k}\binom{n+1-j}{i-k}$$

$$= q_{ij} \qquad (7.86)$$

We may note from (7.81) that the first row of the Q matrix has the elements of 1; the first column has the binomial coefficients, which can be easily generated by using the Pascal triangle [14]. From these well-known entries of Q, we can generate the whole matrix, as is illustrated by generating the matrix Q:

$$Q = \begin{bmatrix} 1 & 1 & 1 & 1 & 1 \\ 4 & 2 & 0 & -2 & -4 \\ 6 & 0 & -2 & 0 & 6 \\ 4 & -2 & 0 & 2 & -4 \\ 1 & -1 & 1 & -1 & 1 \end{bmatrix} \qquad (7.87)$$

The following rule represents (7.80):

$$\begin{bmatrix} c & d \\ b & a \end{bmatrix} \qquad \text{where } a = b - c - d \quad (7.88)$$

Proof of the Symmetry and the Sign Rule for q_{ij} [17]: In this proof we assume first the case $n = $ odd. The matrix Q in (7.79) can be presented as

$$
Q = \left[
\begin{array}{cccc:cccc}
q_{nn} & q_{n(n-1)} & \cdots & q_{n[(n+1)/2]} & q_{n[(n-1)/2]} & \cdots & q_{n1} & q_{n0} \\
q_{(n-1)n} & q_{(n-1)(n-1)} & \cdots & q_{(n-1)[(n+1)/2]} & q_{(n-1)[(n-1)/2]} & \cdots & q_{(n-1)1} & q_{(n-1)0} \\
\vdots & \vdots & \cdots & \vdots & \vdots & \cdots & \vdots & \vdots \\
q_{[(n+1)/2]n} & q_{[(n+1)/2](n-1)} & \cdots & q_{[(n+1)/2][(n+1)/2]} & q_{[(n+1)/2][(n-1)/2]} & \cdots & q_{[(n+1)/2]1} & q_{[(n+1)/2]0} \\
\hdashline
q_{[(n-1)/2]n} & q_{[(n-1)/2](n-1)} & \cdots & q_{[(n-1)/2][(n+1)/2]} & q_{[(n-1)/2][(n-1)/2]} & \cdots & q_{[(n-1)/2]1} & q_{[(n-1)/2]0} \\
\vdots & \vdots & \cdots & \vdots & \vdots & \cdots & \vdots & \vdots \\
q_{1n} & q_{1(n-1)} & \cdots & q_{1[(n+1)/2]} & q_{1[(n-1)/2]} & \cdots & q_{11} & q_{10} \\
q_{0n} & q_{0(n-1)} & \cdots & q_{0[(n+1)/2]} & q_{0[(n-1)/2]} & \cdots & q_{01} & q_{00}
\end{array}
\right]
\tag{7.89}
$$

The matrix $[Q]$ in (7.89) can be partitioned into four submatrices of order $(n + 1)/2 \times (n + 1)/2$, as indicated by the dotted lines. Now we want to prove that for each odd row (column) the elements are symmetric about the dotted line and for each even row (column) the elements are skew-symmetric about the dotted line, or, equivalently,

$$q_{(n-k)(n-j)} = (-1)^k q_{k(n-j)}$$
$$\text{for all } k, j = 0, 1, \ldots, n \quad (7.90)$$

$$q_{(n-k)(n-k)} = (-1)^j q_{k(n-j)}$$
$$\text{for all } k, j = 0, 1, \ldots, n \quad (7.91)$$

Proof: Each element in matrix Q, following (7.74), can be expressed as

$$q_{(n-k)(n-j)} = \sum_{i=0}^{k} (-1)^i \binom{j}{i} \binom{n-j}{k-i}$$
$$k, j = 0, 1, \ldots, n \quad (7.92)$$

Using (7.92), with j replacing $(n - j)$, we have

$$q_{(n-k)j} = \sum_{i=0}^{k} (-1)^i \binom{n-j}{i} \binom{j}{k-i} \quad (7.93)$$

Change the dummy variable i to $\ell = k - i$; then (7.93) becomes

$$q_{(n-k)j} = \sum_{\ell=0}^{k} (-1)^{k-\ell} \binom{n-j}{k-\ell} \binom{j}{\ell}$$

$$= \sum_{i=0}^{k} (-1)^{k-i} \binom{n-j}{k-i} \binom{j}{i}$$

$$= \sum_{i=0}^{k} (-1)^{-k+i} \binom{j}{i} \binom{n-j}{k-i} \quad (7.94)$$

where we have replaced the dummy variable ℓ by i again and used the identity $(-1)^m = (-1)^{-m}$ for any integer m.
   Then

$$(-1)^k q_{(n-k)j} = (-1)^k \sum_{i=0}^{k} (-1)^{-k+i} \binom{j}{i} \binom{n-j}{k-i}$$

$$= \sum_{i=0}^{k} (-1)^i \binom{j}{i} \binom{n-j}{k-i} \quad (7.95)$$

which is the same as (7.92). This completes the proof for (7.90).
   Replacing $(n - k)$ by k in (7.92), we have

$$q_{k(n-j)} = \sum_{i=0}^{n-k} (-1)^i \binom{j}{i} \binom{n-j}{n-k-i} \quad (7.96)$$

Noticing $\binom{j}{i} = 0$ for $j < i$, the summation in (7.96) can be replaced by

$$q_{k(n-j)} = \sum_{i=0}^{j} (-1)^i \binom{j}{i} \binom{n-j}{n-k-i} \tag{7.97}$$

When we use the identity $\binom{m}{\ell} = \binom{m}{m-\ell}$, (7.97) becomes

$$q_{k(n-j)} = \sum_{i=0}^{j} (-1)^i \binom{j}{j-i} \binom{n-j}{n-j-n+k+i}$$

$$= \sum_{i=0}^{j} (-1)^i \binom{j}{j-i} \binom{n-j}{k+i-j} \tag{7.98}$$

and following the same manipulations as in the proof for (7.90), we obtain

$$q_{k(n-j)} = \sum_{\ell=j}^{0} (-1)^{j-\ell} \binom{j}{j-j+\ell} \binom{n-j}{k+j-\ell-j}$$

$$= \sum_{\ell=0}^{j} (-1)^{j-\ell} \binom{j}{\ell} \binom{n-j}{k-\ell}$$

$$= \sum_{i=0}^{j} (-1)^{-j+i} \binom{j}{i} \binom{n-j}{k-i}$$

$$= \sum_{i=0}^{n-k} (-1)^{-j+i} \binom{j}{i} \binom{n-j}{k-i} \tag{7.99}$$

Hence
$$(-1)^j q_{k(n-j)} = (-1)^j \sum_{i=0}^{n-k} (-1)^{-j+i} \binom{j}{i} \binom{n-j}{k-i}$$

$$= \sum_{i=0}^{n-k} (-1)^i \binom{j}{i} \binom{n-j}{k-i} \tag{7.100}$$

Again noticing that $\binom{n-j}{k-i} = 0$ for $n-j < k-i$—that is, $i < k -$

$(n - j) \leq k$ (since $j \leq n$)—the summation in (7.100) can be replaced by

$$(-1)^j q_{k(n-j)} = \sum_{i=0}^{k} (-1)^i \binom{j}{i} \binom{n-j}{k-i} \tag{7.101}$$

which is identical to (7.92). Equation (7.91) is therefore also true.

For the case when $n = $ even, we can partition Q in (7.79) into four submatrices, each of order $n/2 \times n/2$, with the central, $[(n+2)/2]$th, row and column. Since the previous proof does not depend on the nature of n, relations

(7.90) and (7.91) are also true for this case with $k, j = 0, 1, \ldots, n$ except for the $[(n + 2)/2]$th element in each row [for relation (7.90)] and in each column [for relation (7.92)].

In both cases, the first row and the first column are known, as mentioned in the text. Hence, in obtaining the elements of Q, we need only calculate the elements of the submatrix of order $((n - 1)/2) \times ((n - 1)/2)$ for odd n or $n/2 \times n/2$ for even n.

Example 7.6. Let n in (7.70) be equal to 5. The matrix Q in this case is of dimension 6×6. The elements of the first row are unity and those of the first column are binomial coefficients. Using the combinatorial rule of (7.80) and the symmetry of (7.90) and (7.91), we only need to calculate the entries in the dotted lines of the following Q matrix:

$$Q|_{n=5} = \begin{bmatrix} 1 & 1 & 1 & 1 & 1 & 1 \\ 5 & 3 & 1 & -1 & -3 & -5 \\ 10 & 2 & -2 & -2 & 2 & 10 \\ 10 & -2 & -2 & 2 & 2 & -10 \\ 5 & -3 & 1 & 1 & -3 & 5 \\ 1 & -1 & 1 & -1 & 1 & -1 \end{bmatrix} \qquad \begin{bmatrix} c & d \\ b & a \end{bmatrix} \text{ where } a = b - c - d$$

(7.102)

Note that the arrow in (7.102) indicates the direction of calculating the needed entries from the rule $a = b - c - d$ or in the example $3 = 5 - 1 - 1$. Other needed entries can be obtained similarly.

In the discussion of Section 2.3 it is indicated that for the test of the simplified stability criterion for the discrete case, one needs to calculate the coefficient B_i's in (2.44) and (2.48). These coefficients are the same as the b_i's in (7.74). Hence we can use the matrix Q to generate them. However, for the simplified determinantal criterion, we need calculate only about half the B_i's. This offers a simplification in the calculation of the matrix Q such that only $(n - 2)/2 \times n/2$ elements are needed for $n =$ even. For the case $n =$ odd, we need $(n - 1)/2 \times (n - 1)/2$ elements to be calculated. For instance, in the example of (7.87) we need to calculate only the entries indicated in the dotted bracket of (7.102).

The determinant of the matrix Q in (7.79) is given by [18]

$$|Q| = (-2)^{n(n+1)/2} \tag{7.103}$$

The proof is obtained by repeatedly applying the Gaussian reduction to the following $(n + 1) \times (n + 1)$ matrix generated from the combinatorial rule of (7.80):

$$
Q = \begin{bmatrix}
1 & 1 & 1 & 1 & \cdots \\
n & n - 2 & n - 4 & n - 6 & \cdots \\
\binom{n}{2} & \binom{n}{2} - n + 2 & \binom{n}{2} - 4n + 8 & \binom{n}{2} - 6n + 18 & \cdots \\
\binom{n}{3} & \binom{n}{2} - 2\binom{n}{2} + 2n - 2 & \binom{n}{3} - 4\binom{n}{2} + 8n - 12 & \binom{n}{3} - 6\binom{n}{2} + 18n - 38 & \cdots \\
\vdots & & \vdots & \vdots &
\end{bmatrix}
\tag{7.104}
$$

After repeatedly applying the Gaussian reduction, (7.104) becomes

$$
\begin{bmatrix}
1 & 1 & 1 & 1 & \cdots \\
0 & -2 & -4 & -6 & \cdots \\
0 & -2(n - 1) & -4(n - 2) & -6(n - 3) & \cdots \\
0 & -2\binom{n}{2} + 2n - 2 & -4\binom{n}{2} + 8n - 12 & -6\binom{n}{2} + 18n - 38 & \cdots \\
& \vdots & \vdots & j &
\end{bmatrix}
\tag{7.105}
$$

$$
\begin{bmatrix}
1 & 1 & 1 & 1 & \cdots \\
0 & -2 & -4 & -6 & \cdots \\
0 & 0 & 4 & 12 & \cdots \\
0 & 0 & 4n - 8 & 12n - 32 & \cdots \\
\vdots & \vdots & \vdots & \vdots &
\end{bmatrix}
\tag{7.106}
$$

$$
\begin{bmatrix}
1 & 1 & 1 & 1 & \cdots \\
0 & -2 & -4 & -6 & \cdots \\
0 & 0 & 4 & 12 & \cdots \\
0 & 0 & 0 & -8 & \cdots \\
\vdots & \vdots & \vdots & \vdots &
\end{bmatrix}
\tag{7.107}
$$

Continue in this fashion to the last row. Thus Q is reduced to a triangular matrix with diagonal elements $[1 \quad -2 \quad 4 \quad -8 \quad 16 \quad \cdots]$

or $\quad [(-2)^0 \quad (-2)^1 \quad (-2)^2 \quad (-2)^3 \quad (-2)^4 \quad (-2)^5 \quad \cdots \quad (-2)^n]$

$$\tag{7.108}$$

Hence the determinant of Q is $(-2)^{n(n+1)/2}$.

b. Bilinear Transformation $z = s/(s - 1)$

This transformation lies in the mapping of the real segment $[0, 1]$ of the z-plane onto the negative real axis of the s-plane, as discussed in (2.55) in connection with the aperiodicity condition. It has the property that the Q matrix corresponding to (7.79) is an upper triangular matrix with nonzero elements being the Pascal triangle expansion with the proper sign. For instance, Q for $n = 4$ is given as a 5×5 matrix as follows:

$$Q(5) = \begin{bmatrix} 1 & 1 & 1 & 1 & 1 \\ 0 & -1 & -2 & -3 & -4 \\ 0 & 0 & 1 & 3 & 6 \\ 0 & 0 & 0 & -1 & -4 \\ 0 & 0 & 0 & 0 & 1 \end{bmatrix} \quad \begin{pmatrix} c \\ b & a \end{pmatrix}$$

$$\text{where } a = b - c \quad (7.109)$$

The combinatorial rule for the matrix Q is given from (7.109) as

$$q_{ij} = q_{i,j-1} - q_{i-1,j-1} \quad (7.110)$$

The determinants of these Q matrices are just the product of the diagonal elements.

c. The Nonlinear Transformation $z = \left(\dfrac{1}{4}\right)\dfrac{(s + 1)^2}{s}$:

This transformation, discussed in (2.58), maps the real segment $[0, 1]$ in the z-plane onto the periphery of the unit circle $|s|$ in the s-plane. As mentioned in Section 2.3, it arises in the aperiodicity test of linear discrete systems. The coefficient b_i's and b_n are given in (2.61) and (2.62), respectively. The Q matrix in this is upper triangular with nonzero entries:

$$q_{ij} = \frac{1}{2}\left(\frac{1}{4}\right)^{n+i-j}\binom{2(n + 1 - j)}{n + i - j} \qquad i = 1 \quad (7.111)$$

$$q_{ij} = \left(\frac{1}{4}\right)^{n+1-j}\binom{2(n + 1 - j)}{n + i - j} \qquad i \neq 1 \quad (7.111a)$$

Only the first row need be computed by (7.111); the other nonzero elements can be generated from it by using the following combinatorial rule after multiplying the first row by 2.

$$q_{ij} = \frac{1}{4}(q_{i-1,j+1} + 2q_{i,j+1} + q_{i+1,j+1})$$

$$i = 1, 3, \ldots, n + 1 \quad (7.112)$$

The proof of (7.112) can be obtained by using (7.111) in the right-hand side of (7.112).

$$\left(\frac{1}{4}\right)\cdot\left(\frac{1}{4}\right)^{n-j}\left[\binom{2n-2j}{n+i-j-2}+2\binom{n2-2j}{n+i-j-1}+\binom{2n-2j}{n+i-j}\right]$$

$$=\left(\frac{1}{4}\right)^{n+1-j}\left[\binom{2n-2j+1}{n+i-j-1}+\binom{2n-2j+1}{n+i-j}\right]$$

$$=\left(\frac{1}{4}\right)^{n+1-j}\binom{2n-2j+2}{n+i-j}$$

$$=q_{ij} \tag{7.113}$$

d. Combinatorial Rule for (2.69) and (2.73)

From the discussion of Section 2.4 it is evident that if $a_n = 1$ in $F(z)$ of (2.38), we need to compute the entries of an $(n-1) \times (n-1)$ matrix of (2.68) and (2.71). The distinct nonzero elements of this matrix are, for $n =$ even,

$$A] = [A_{0,r}A_{1,r}A_{2,r-1}A_{3,r-1}\cdots A_{n-2,0}A_{n-1,0}]' \tag{7.114}$$

where $r = (n-2)/2$, and for n-odd,

$$A] = [A_{0,r}A_{1,r}A_{2,r}\cdots A_{n-2,0}A_{n-1,0}]' \tag{7.115}$$

where $r = (n-3)/2$,

$$A_{mv} = \sum_{\mu=0}^{m}(-1)^{\mu}\binom{\mu+v}{\mu}a_{m-\mu} \tag{7.116}$$

where $m = 0, 1, 2, \ldots, n-1$
$v = 0, 1, 2, \ldots, r$ (n-even) or $r+1$ (n-odd)

Let $a] = [a_0a_1\cdots a_{n-1}]'$; then in matrix form

$$A] = Qa] \tag{7.117}$$

where Q is of order $n \times n$ and has the following properties:

1. Q is lower triangular with elements alternating in sign.
2. The elements in the last row have magnitude 1.
3. Diagonal elements are all equal to 1.
4. The odd rows for n-even (or even rows for n-odd) can be obtained from the following row by shifting its elements to the left by one position and filling the rightmost element with 0.

5. The elements in the even rows for n-even (or odd rows for n-odd) are given as

$$q_{ij} = (-1)^{i+j} \binom{\frac{n+i}{2} - j}{i - j} \tag{7.118}$$

and they can be generated from the last row by the following rule.

$$q_{ij} = q_{i+2,j+2} - q_{i,j+1} \tag{7.119}$$

The proof of the combinatorial rule of (7.119) is obtained by inserting in (7.118) the right-hand side of (7.119) as follows:

$$(-1)^{i+j} \binom{\frac{n+i}{2} - j - 1}{i - j} - (-1)^{i+j+1} \binom{\frac{n+i}{2} - j - 1}{i - j - 1}$$

$$= (-1)^{i+j} \left[\binom{\frac{n+i}{2} - j - 1}{i - j} + \binom{\frac{n+i}{2} - j - 1}{i - j - 1} \right]$$

$$= (-1)^{i+j} \binom{\frac{n+i}{2} - j}{i - j}$$

$$= q_{ij} \tag{7.120}$$

Example 7.7. Generate the matrix Q of (7.117) for $n = 7$, using (7.118) and (7.119). Following properties (1) to (5) for generating Q, we obtain for this example the matrix

$$Q(7) = \begin{bmatrix} 1 & 0 & 0 & 0 & 0 & 0 & 0 \\ -3 & 1 & 0 & 0 & 0 & 0 & 0 \\ 6 & -3 & 1 & 0 & 0 & 0 & 0 \\ -4 & 3 & -2 & 1 & 0 & 0 & 0 \\ 5 & -4 & 3 & -2 & 1 & 0 & 0 \\ -1 & 1 & -1 & 1 & -1 & 1 & 0 \\ 1 & -1 & 1 & -1 & 1 & -1 & 1 \end{bmatrix} \tag{7.121}$$

7.4 EXAMPLES

In this section we will present 12 examples that are solved by using the computational algorithms presented in Sections 7.1 and 7.2. These examples are based on the material developed in Chapters 1 to 6. They illustrate the fact that most of the material discussed in the preceding six chapters can be solved by using one algorithm based on double triangularization, thus offering a unified computational approach to most problems connected with system theory.

In order to minimize the roundoff error when using the digital computer, we make the original innerwise or A matrix diagonally dominant† before performing the computation. Such a procedure is a standard one in most computations of matrices [19] and has been briefly discussed in Section 3.5, Equations (3.251i–m).

Example 7.8 Stability of Linear Continuous Systems (Liénard–Chipart Test) [3]. The Liénard–Chipart test has been discussed in Section 2.1, and the following example illustrates the test when using the computational algorithm of Section 7.1. The polynomial equation to be tested for stability (or Hurwitz character) is given by

$$F(s) = s^6 + 3s^5 + 4s^4 + 6s^3 + 15s^2 + 21s + 10 = 0 \qquad (7.122)$$

Since all coefficients are positive, it is necessary and sufficient that for $F(s)$ to have all its roots in the open left half-plane, the matrix below, obtained from (2.10), be positive innerwise:

$$\Delta_5 = \begin{bmatrix} 1 & 4 & 15 & 10 & 0 \\ 0 & 1 & 4 & 15 & 10 \\ 0 & 0 & 3 & 6 & 21 \\ 0 & 3 & 6 & 21 & 0 \\ 3 & 6 & 21 & 0 & 0 \end{bmatrix} \qquad (7.123)$$

After applying the algorithm of Section 7.1, the resulting matrix is

$$\tilde{\Delta}_5 = \begin{bmatrix} 1 & 4 & 15 & -48 & 0 \\ 0 & 1 & 4 & 0 & 0 \\ 0 & 0 & 3 & 0 & 0 \\ 0 & 3 & 6 & -12 & 0 \\ 3 & 6 & 21 & -84 & 120 \end{bmatrix} \qquad (7.124)$$

† See reference [30].

Since $|\Delta_3| = 1 \times 3 \times -12 = -36$ from (7.124) is negative, $\tilde{\Delta}$ or Δ is not positive innerwise; hence the polynomial of (7.122) is unstable (or not Hurwitz).

Example 7.9 Discrete-System Stability Test [3]. Let $F(z)$ be given as

$$F(z) = 8z^5 + 20z^4 + 26z^3 + 13z^2 + 6z + 2 = 0 \qquad (7.125)$$

To test for the necessary and sufficient condition for the roots of $F(z) = 0$ to lie inside the unit circle in the z-plane, apply (2.51) to (2.54). In this case the coefficient of (2.52) to (2.54) are all positive; hence we check for the positive innerwise condition of (2.51). In this case, Δ_4^- equals

$$\Delta_4^- = \begin{bmatrix} 8 & 20 & 26 & 11 \\ 0 & 8 & 18 & 20 \\ 0 & -2 & 2 & 7 \\ -2 & -6 & -13 & -18 \end{bmatrix} \qquad (7.125a)$$

Applying the algorithm to (7.125), we get

$$\tilde{\Delta}_4^- = \begin{bmatrix} 8 & 20 & -19 & 0 \\ 0 & 8 & 0 & 0 \\ 0 & -2 & 6.5 & 0 \\ -2 & -6 & 0.5 & -\dfrac{255}{52} \end{bmatrix} \qquad (7.126)$$

Since $|\tilde{\Delta}_4^-| = |\Delta_4^-| = 8 \times 8 \times 6.5 \times 255/52$ is negative, Δ_4^- in (7.125) is not positive innerwise and the discrete system described by $F(z) = 0$ is unstable.

Example 7.10 Aperiodicity Test for Continuous Systems [3]. Let $F(s)$ be given as

$$F(s) = s^3 + 6s^2 + 11s + 6 = 0 \qquad (7.127)$$

From the discussion of Section 2.2, the condition of aperiodicity requires that the roots of (7.127) be distinct and on the negative real axis. The necessary and sufficient conditions are given by (2.33) and (2.34). Since all the coefficients of $F(s)$ in (7.127) are positive, we must test for the positive innerwise of (2.34) as follows:

$$\Delta_5 = \begin{bmatrix} 1 & 6 & 11 & 6 & 0 \\ 0 & 1 & 6 & 11 & 6 \\ 0 & 0 & 3 & 12 & 11 \\ 0 & 3 & 12 & 11 & 0 \\ 3 & 12 & 11 & 0 & 0 \end{bmatrix} \qquad (7.128)$$

Applying the algorithm to (7.128), we obtain

$$\tilde{\Delta}_5 = \begin{bmatrix} 1 & 6 & 11 & 40 & 0 \\ 0 & 1 & 6 & 0 & 0 \\ 0 & 0 & 3 & 0 & 0 \\ 0 & 3 & 12 & 2 & 0 \\ 3 & 12 & 11 & 112 & 2/3 \end{bmatrix} \qquad (7.129)$$

Since all the diagonal entries of (7.129) are positive, $\tilde{\Delta}_5$ as well as Δ_5 are positive innerwise and the system described by (7.127) is aperiodic stable.

Example 7.11 Relative Stability of Continuous Systems [3]. Let $F(s)$ be given as

$$F(s) = s^3 + 4.464s^2 + 7.464s + 4 = 0 \qquad (7.130)$$

We want to test whether all roots of (7.130) have the relative damping coefficient $\zeta > 1/\sqrt{2}$. This example has been discussed in Section 2.2 and in (2.16) to (2.25). To verify (2.23) to (2.25), we apply the double-triangularization algorithm to (2.22) to obtain

$$\tilde{\Delta}_6 = \begin{bmatrix} 0.707 & -4.464 & 5.28 & 33.33 & -92.2 & 0 \\ 0 & 0.707 & -4.464 & -22.91 & 0 & 0 \\ 0 & 0 & 0.707 & 0 & 0 & 0 \\ 0 & 0 & 0.707 & 4.464 & 0 & 0 \\ 0 & 0.707 & 0 & -5.28 & 12.36 & 0 \\ 0.707 & 0 & -5.28 & -29.34 & -29.97 & 6.10 \end{bmatrix} \qquad (7.131)$$

From (7.131), we observe that $|\tilde{\Delta}_2| = |\Delta_2|$, $|\tilde{\Delta}_4| = |\Delta_4|$, $|\tilde{\Delta}_6| = |\Delta_6|$ are all positive. Hence Δ_6 is positive innerwise and the system described by (7.130) is relatively stable.

Example 7.12 Realizability Condition of a Digital Filter [20]. In digital filter design we are interested in the realizability condition in a form of a ladder of the digital transfer function $G(z)$. The realizability condition is formulated in terms of an innerwise matrix formed from the numerator and denominator of $G(z)$ to be "nonnull." As explained in Chapter 1, this condition requires that none of the inners determinants is zero. To explain the procedure, let $G(z)$ be given as

$$G(z) = \frac{1 + 45z + 270z^2 + 320z^3 + 1200z^4}{10z + 129z^2 + 100z^3 + 600z^4} \tag{7.132}$$

Performing the division of numerator by denominator and normalizing, we arrive at the following form:

$$G(z) = 0.1z^{-1} + 3.21 + zZ(z) \tag{7.133}$$

where

$$Z(z) = \frac{-0.256816 - 0.10166z - 1.21z^2}{0.01666 + 0.215z + 0.1666z^2 + z^3} \tag{7.134}$$

We obtain the generalized Routh–Hurwitz matrix following (2.4) from the coefficients of $Z_1(jz)$, where

$$Z_1(z) = z^6 - 1.21z^5 + 0.1666z^4 - 0.10166z^3 + 0.215z^2$$
$$- 0.256816z + 0.0166 \tag{7.135}$$

The condition for realizability is equivalent to the innerwise matrix of (2.4) to be "nonnull." The accompanying computer printout gives both the matrix and its double triangular form. From the latter we determine that the generalized Routh–Hurwitz matrix is nonnull. Hence physical realization of the digital filter exists. We note from the printout that none of the inners determinants is zero.

Example 7.13 Synthesis of RC Network [21,22]. In Chapter 2 we formulated from (2.75) the physical realization of an RC passive network. In this example we will illustrate how we can synthesize the network by using the double-triangularization algorithm.

Consider the continued fraction expansion of $Z(s)$ in (2.75).

$$Z(s) = \frac{1}{(a_n/b_{n-1})s + f_1(s)/g(s)} \tag{7.136}$$

$$\frac{g(s)}{f_1(s)} = r_1 + \frac{g_1(s)}{f_1(s)}$$

$$\frac{f_1(s)}{g_1(s)} = c_2 s + \frac{f_2(s)}{g_1(s)} \tag{7.137}$$

$$\vdots \qquad \vdots \qquad \vdots$$

It is well known that we obtain the elements of the RC network from the above continued fraction expansion—that is, all the r_i's and c_i's. By applying the double-triangularization algorithm to the generalized Routh–Hurwitz matrix generated from the coefficients of $Z(s)$ in (2.75), we also obtain the r_i's and c_i's, and hence we can synthesize the RC network as explained in the following numerical example: Let $Z(s)$ be given as

$$Z(s) = \frac{720s^2 + 180s + 10}{1440s^3 + 600s^2 + 60s + 1} \tag{7.138}$$

Following (7.136) and (7.137), we obtain for the continued fraction expansion

$$f_1(s) = 240s^2 + 40s + 1 \tag{7.139}$$

$$g_1(s) = 60s + 7 \tag{7.140}$$

$$f_2(s) = 12s + 1 \tag{7.141}$$

$$g_2(s) = 2 \tag{7.142}$$

$$f_3(s) = 1 \tag{7.143}$$

From (7.138) to (7.143), and noting (7.137), we obtain the RC network as shown in Figure 7.1. Now we generate the Routh–Hurwitz matrix from

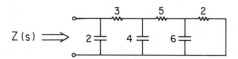

Figure 7.1.   RC network synthesis.

(7.138) by noting that the condition of physical realizability of $Z(s)$ is equivalent to that of the polynomial equation [23]

$$h(s) = a_n s^{2n} + b_{n-1} s^{2n-1} + a_{n-1} s^{2n-2} + b_{n-2} s^{2n-3} + \cdots + b_0 s + a_0 = 0 \tag{7.144}$$

having all its roots (zeros) in the open left half-plane. Applying (2.4), we obtain the generalized Routh–Hurwitz matrix as shown in the accompanying

LET $Z(Z) = 0.2568I6Z - 0.1016Z2 - 1.2IZ3 /0.01666 + 0.166622 + 0.215Z + 0.16662Z + Z3$ BE THE TRANSFER FUNCTION.

THIS IS THE GENERALIZED ROUTH-HURWITZ MATRIX GENERATED FROM COEFFICIENTS OF $Z(Z)$.

$$|\Delta_2| = .12I0E + 0I$$
$$|\Delta_4| = .1209E + 00$$

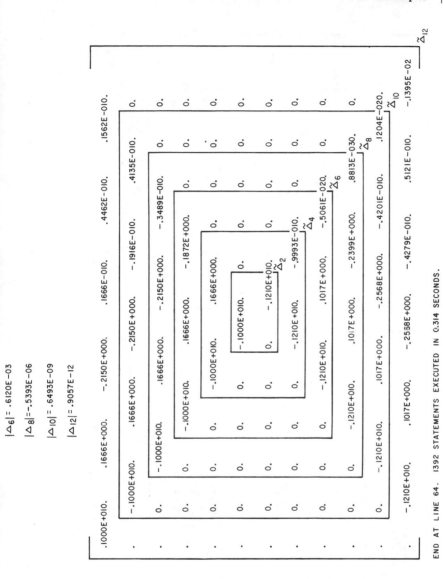

Example 7.12

LET Z(S) = (720S2+180S+10)/(1440S3+600S2+60S+1) BE THE DRIVING POINT IMPEDENCE OF GIVEN NETWORK

THIS IS THE GENERALIZED ROUTH-HURWITZ MATRIX GENERATED FROM THE COEFFICIENTS OF Z(S)

$|\Delta_2| = -.1037E+07$

$|\Delta_4| = .1792E+12$

$|\Delta_6| = -.2580E+16$

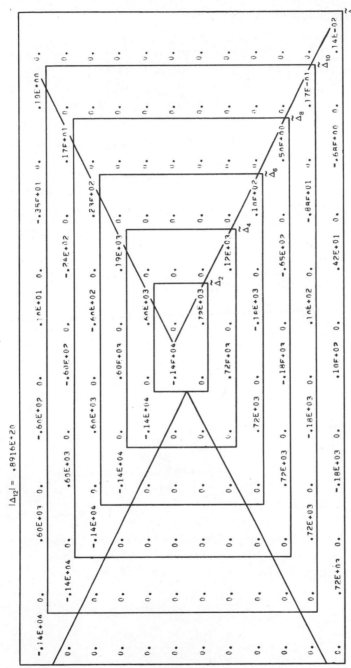

Example 7.13

computer study. By applying the double triangularization, we obtain the inners determinants $|\Delta_2|$ to $|\Delta_{12}|$:

$$
\begin{aligned}
|\tilde{\Delta}_2| &= 0.1037 \times 10^7 = (-1)^1 |\Delta_2| \\
|\tilde{\Delta}_4| &= 0.1792 \times 10^{12} = (-1)^2 |\Delta_4| \\
|\tilde{\Delta}_6| &= 0.2580 \times 10^{16} = (-1)^3 |\Delta_6| \\
|\tilde{\Delta}_8| &= 0.1858 \times 10^{19} = (-1)^4 |\Delta_8| \\
|\tilde{\Delta}_{10}| &= 0.4458 \times 10^{20} = (-1)^5 |\Delta_{10}| \\
|\tilde{\Delta}_{12}| &= 0.8916 \times 10^{20} = (-1)^6 |\Delta_{12}|
\end{aligned}
\tag{7.145}
$$

The change of sign is required because of the reversal of rows in the matrix of generalized Routh–Hurwitz criterion.

If we write the regular Hurwitz matrix for (7.144), using (2.8) for general n, and lower triangularize it, we obtain a relationship between the inners determinants and the diagonal entries of the lower-triangularized Hurwitz matrix.

$$
\frac{|\tilde{\Delta}_{2(n-1)}|}{|\tilde{\Delta}_{2n}|} = |\tilde{\Delta}_{H_n}| \, |\tilde{\Delta}_{H_{n+1}}|
\tag{7.146}
$$

where the Hurwitz determinants are given by

$$
|\Delta_{H_1}| = |\tilde{\Delta}_{H_1}|
\tag{7.147}
$$

$$
|\Delta_{H_2}| = |\tilde{\Delta}_{H_1}| \, |\tilde{\Delta}_{H_2}|
\tag{7.148}
$$

$$
\vdots
$$

$$
|\Delta_{H_n}| = |\tilde{\Delta}_{H_1}| \, |\tilde{\Delta}_{H_2}| \cdots |\tilde{\Delta}_{H_n}|
\tag{7.149}
$$

Now, if we insert the values $|\tilde{\Delta}_{2n}|$ from (7.145) into (7.146), noting that $|\Delta_{H_1}| = |\tilde{\Delta}_{H_1}| = 720$, we obtain

$$
|\tilde{\Delta}_{H_2}| = 240
\tag{7.150}
$$

$$
|\tilde{\Delta}_{H_3}| = 60
\tag{7.151}
$$

$$
|\tilde{\Delta}_{H_4}| = 12
\tag{7.152}
$$

$$
|\tilde{\Delta}_{H_5}| = 2
\tag{7.153}
$$

By using the well-known relationship between the continued fraction expansion and Hurwitz determinants (or, equivalently, Routh first-column entries) we can readily synthesize the RC network. For instance [21],

$$
c_1 = \frac{a_n}{b_{n-1}} = \frac{1440}{720} = 2
\tag{7.154}
$$

$$
r_1 = \frac{|\Delta_{H_2}|}{|\Delta\tilde{H}_2|^2} = \frac{720 \times 240}{240 \times 240} = 3
\tag{7.155}
$$

$$c_2 = \frac{|\Delta \tilde{H}_2|}{|\Delta \tilde{H}_3|} = \frac{240}{60} = 4 \tag{7.155a}$$

$$r_2 = \frac{|\Delta \tilde{H}_3|}{|\Delta \tilde{H}_4|} = \frac{60}{12} = 5 \tag{7.155b}$$

$$c_3 = \frac{|\Delta \tilde{H}_4|}{|\Delta \tilde{H}_5|} = \frac{12}{2} = 6 \tag{7.155c}$$

$$r_3 = |\tilde{\Delta}_{H_5}| = 2 \tag{7.156}$$

The final network is as shown in Figure 7.1.

Example 7.14   Evaluation of Integral Square of Continuous Signal [24].   In Section 4.1 we discussed the evaluation of integral square of signals by the evaluation of complex integrals as given in (4.1). The value of the integral is given in (4.40) to (4.43). To illustrate the use of the double-triangularization method, we present the following example. Let

$$g(s) = s^8 + 3s^6 + 2s^4 + s^2 + 0.2 \tag{7.157}$$

$$h(s) = s^5 + 3s^4 + 4s^3 + 3s^2 + 1.2s + 0.3 \tag{7.158}$$

From (4.35) to (4.39), and using the double-triangularization procedure in the accompanying computer printout, we obtain

$$I_5 = 0.3470 \tag{7.159}$$

We also observe that in the process of double triangularization of the denominator matrix in (4.40), we find that the inners determinants are all positive. Since all the coefficients of $h(s)$ in (7.158) are positive, the system described by $h(s)$ is stable, and thus the value of I_n in the printout is meaningful.

Example 7.15   Evaluation of the Infinite Sum of the Square of Discrete Signal [24].   Following the discussion of Section 4.2, we can evaluate the infinite sum of the square of a discrete signal by using (4.97). Suppose that $E(z)$ is given by

$$E(z) = \frac{z^3 - 0.74z^2 - 0.14z}{z^3 - 0.32z^2 + 0.35z + 0.05} \tag{7.160}$$

The value of

$$I_3 = \sum_{n=0}^{\infty} e^2(nT) \tag{7.161}$$

is obtained from (4.126), (4.131), and (4.132). The value is computed using the double triangularization of both the numerator and the denominator

LET G(S) = S8+3.0S6+2.0S4+S2+0.2

LET H(S) = S5+3.0S4+4.0S3+3.0S2+1.2S+0.3

THIS IS THE MATRIX W.

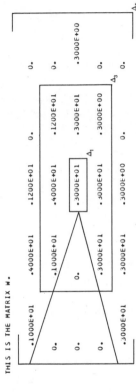

THE DETERMINANTS OF INNERS OF MATRIX W ARE

$|\Delta_1| = .3000E+01$

$|\Delta_3| = .1710E+02$

$|\Delta_5| = .3231E+01$

THE DOUBLE TRIANGULARIZED MATRIX W IS

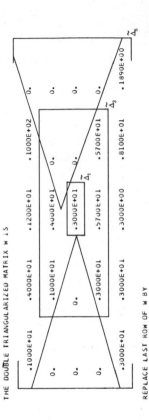

REPLACE LAST ROW OF W BY

(1.0, 3.0, 2.0, 1.0, 0.2)

THIS IS THE W1 MATRIX.

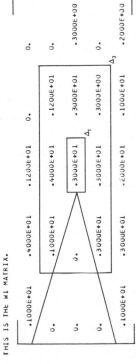

THE DETERMINANTS OF INNERS OF MATRIX W1 ARE

$|\Delta_1| = .3000E+01$

$|\Delta_3| = .1710E+02$

$|\Delta_5| = .2232E+01$

THE DOUBLE TRIANGULARIZED MATRIX W1 IS

I5 = .3473E+00

END AT LINE 108. 144 STATEMENTS EXECUTED IN 0.138 SECONDS.

Example 7.14

matrices of (4.132). The value of the infinite sum

$$I_3 = 1.338 \tag{7.162}$$

is printed by the computer as shown in the accompanying printout. It is noted that the denominator polynomial of (7.160) is stable. This fact is ascertained from the simplified determinantal criterion of Section 2.3 and (2.43).† We may note that by satisfying (2.43) and the positiveness of the inner determinant of $|X_4 + Y_4|$, then Δ_2^- is positive innerwise or equivalently $|\Delta_2^-| > 0$ [see footnote of (4.132)] and hence the system is stable. Therefore the preceding computational method is used simultaneously for checking stability and the value of the infinite sum of the square of the error signal.

Example 7.16   Testing for Negative Real Part of Eigenvalues of the A Matrix.   Similar to the discussion of the illustrative example of the first form of Section 7.1, we can test the stability of the A matrix by using the double-triangularization procedure and storing the relevant entries. This is done by the following testing of the A matrix of dimension 4×4 for eigenvalues with negative real part. From the accompanying computer printout we ascertain that the A matrix is stable; that is, all its eigenvalues have negative real part. Note that in this case the A matrix has no left triangle of zeros, and therefore the first form of the double-triangularization algorithm is used.

To illustrate the computer algorithm for testing the stability of any A matrix of dimension 4×4, we present the following procedure. Let the A matrix of dimension 4×4 be given by

$$A = \begin{bmatrix} -4 & 5 & 1 & -3 \\ -8 & -4 & 3 & -5 \\ -12 & 3 & -4 & 2 \\ 4 & 6 & -5 & 3 \end{bmatrix} \tag{7.162a}$$

From (3.251c) we obtain $|\Delta_1|$ for the above matrix as follows:

$$|\Delta_1| = \text{Sum } A(I, I) = -4 - 4 - 4 + 3 = -9 = -x_1$$

Since $|\Delta_1|$ is negative, we proceed to obtain the other stability conditions by performing the double-triangularization procedure for (7.162a) as discussed in Section 7.1. The following steps illustrate the pivoting required.

† If we denote the denominator of (7.160) as $F(z)$, then from (2.43) we have

$$B_0 = -F(-1) > 0$$
$$B_3 = F(1) > 0$$
$$B_1 = 3(a_3 - a_0) + a_2 - a_1 = 3(1 - 0.05) - 0.32 - 0.35 > 0$$

Hence (2.43) is satisfied.

THE ERROR IS GIVEN BY

$$E(Z) = (Z3 - 0.74 Z2 - 0.14 Z)/(Z3 - 0.32 Z2 + 0.35 Z + 0.05)$$

THIS IS THE MATRIX W.

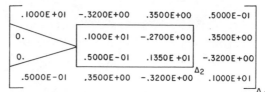

THE DETERMINANTS OF INNERS OF MATRIX W ARE

$|\Delta_2| = .1363E+01$

$|\Delta_4| = .1105E+01$

THE DOUBLE TRIANGULARIZED MATRIX W IS

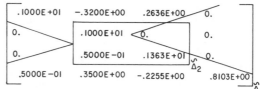

REPLACE LAST ROW OF W BY

$(0.0, -0.28, -0.8, 1.2)$

THIS IS THE WI MATRIX.

THE DETERMINANTS OF INNERS OF MATRIX WI ARE

$|\Delta_2| = .1363E+01$

$|\Delta_4| = .1474E+01$

THE DOUBLE TRIANGULARIZED MATRIX WI IS

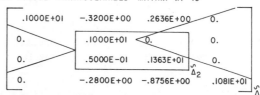

THE ERROR I3 = 1.338

END AT LINE 69. 252 STATEMENTS EXECUTED IN 0.118 SECONDS.

Example 7.15

```
THE  - A -  MATRIX  IS  GIVEN  AS  FOLLOWS
```

$$
\begin{bmatrix}
-.4000E+01 & .5000E+01 & .1000E+01 & -.3000E+01 \\
-.8000E+01 & -.4000E+01 & .3000E+01 & -.5000E+01 \\
-.1200E+02 & .3000E+01 & -.4000E+01 & .2000E+01 \\
.4000E+01 & .6000E+01 & -.5000E+01 & .3000E+01
\end{bmatrix}
$$

```
X1 = -.9000E+01

DELTA  1 = -.9000E+01

X2 = .1070E+03

DELTA  3 = -.9630E+03

DETERMINANT  OF  A = .1480E+04

DELTA  5 = -.1325E+07

X3 = .6000E+03

DETERMINANT  OF  A  HATH = .9792E+05

DELTA  7 = -.1307E+12

ALL  DELTAS  ARE  NEGATIVE  THAT  IS  DIAGONAL  MATRIX  IS  NEGATIVE  INNERWISE

THEREFORE  - A -  MATRIX  HAS  EIGENVALUES  WITH  NEGATIVE  REAL  PART.

FOLLOWING  IS  DOUBLE  TRIANGULARIZED  - A -  MATRIX
```

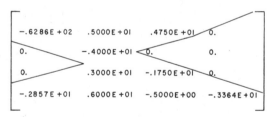

```
END  AT  LINE  67.  49  STATEMENTS  EXECUTED  IN  0.071  SECONDS.
```

Example 7.16

Step 1: Pivot at a_{22}, that is, -4. Delete second row and second column of A and store remaining 3×3 matrix as B_1:

$$
B_1 = \begin{bmatrix}
-4 & 1 & -3 \\
-12 & -4 & 3 \\
4 & -5 & 3
\end{bmatrix}
\tag{7.162b}
$$

Continue the double triangularization on A:

$$A_1 = \begin{bmatrix} -14 & 5 & \dfrac{19}{4} & -\dfrac{37}{4} \\ 0 & -4 & 0 & 0 \\ -18 & 3 & -\dfrac{7}{4} & -\dfrac{7}{4} \\ -8 & 6 & -\dfrac{1}{2} & -\dfrac{9}{2} \end{bmatrix} \qquad (7.162c)$$

Compute $x_2 = \tilde{a}_{22}(\tilde{a}_{11} + \tilde{a}_{33} + \tilde{a}_{44})$ where $\tilde{a}_{ii}$ are entries of A_1. Store x_2 as $x_{2\text{stored}} = -4(-14 - (7/4) - (9/2)) = 81$. Delete third row and third column of A_1 and store remaining 3×3 matrix as B_2:

$$B_2 = \begin{bmatrix} -14 & 5 & -\dfrac{37}{4} \\ 0 & -4 & 0 \\ -8 & 6 & -\dfrac{9}{2} \end{bmatrix} \qquad (7.162d)$$

Step 2: Complete the double triangularization of A_1:

$$A_2 = \begin{bmatrix} -62.86 & 5 & 4.75 & 0 \\ 0 & -4 & 0 & 0 \\ 0 & 3 & -1.75 & 0 \\ 2.857 & 6 & -0.5 & -3.364 \end{bmatrix} \qquad (7.162e)$$

Compute

$$x_4 = |A_2| = -62.86 \cdot -4 \cdot -1.75 \cdot -3.364 = 1480 = |A| \qquad (7.162f)$$
$$x_3 = \tilde{a}_{22}(\tilde{\tilde{a}}_{11}\tilde{\tilde{a}}_{33} + \tilde{\tilde{a}}_{33}\tilde{\tilde{a}}_{44}) = -4(-62.86 \cdot -1.75 + -1.75 \cdot -3.364)$$
$$= -468$$

Store x_3 as $x_{3\text{stored}}$.

Step 3: Apply the double triangularization algorithm on B_1 and B_2 as described in Section 7.1:

$$\tilde{B}_1 = \begin{bmatrix} -7 & 1 & -2.5 \\ 0 & -4 & 0 \\ 19 & -5 & -0.5 \end{bmatrix} \qquad \tilde{\tilde{B}}_1 = \begin{bmatrix} -7 & 1 & 0 \\ 0 & -4 & 0 \\ 19 & -5 & -\dfrac{44}{7} \end{bmatrix}$$

$$(7.162g)$$

Compute

$$x_2 = x_{2\text{stored}} + \tilde{b}_{22}(\tilde{b}_{11} + \tilde{b}_{33}) = 81 - 4(-7 + 0.5) = 107$$

$$(7.162\text{h})$$

Compute $x_3 = x_{3\text{stored}} + \tilde{b}_{11} \cdot \tilde{b}_{21} \cdot \tilde{b}_{33} = -468 + (-7 \cdot -4 \cdot -(44/7)) = 644$, where $\tilde{b}_{ii}$ and $\tilde{\tilde{b}}_{ii}$ are the entries of $\tilde{B}_1$ and $\tilde{\tilde{B}}_1$, respectively. Store x_3 as $x_{3\text{stored}}$. Continue the double triangularization of (7.162d):

$$\bar{B}_2 = \begin{bmatrix} -14 & 5 & 0 \\ 0 & -4 & 0 \\ -8 & 6 & \dfrac{11}{14} \end{bmatrix} \qquad (7.162\text{i})$$

Compute $x_3 = x_{3\text{stored}} + \bar{b}_{11}\bar{b}_{22}\bar{b}_{33} = -644 + 44 = -600$ where $\bar{b}_{ii}$ are entries of $\bar{B}_2$. Store x_3 as $x_{3\text{stored}}$.

Substitute

$$x_3 = -x_{3\text{stored}} = 600 \qquad (7.162\text{j})$$

The condition x_2 of (7.162h) is equivalent to $|\Delta_7|/|\Delta_5|$ from (3.251h). Now calculate $|\Delta_3|$ (presented in the computer printout) as follows:

$$|\Delta_3| = x_2\,|\Delta_1| = 107 \times (-9) = -0.963 \times 10^3 \qquad (7.162\text{k})$$

Now we calculate $|\Delta_5|$ (presented in the computer printout) as follows:

$$|\Delta_5| = x_4\,|\Delta_3| = -0.1325 \times 10^7 \qquad (7.162\text{l})$$

We may note that x_3 of (7.162j) is equivalent to $|\Delta_3|/|\Delta_1|$ in (3.251h).

Since $|\Delta_5|$ in (7.162l) is negative, which constitutes another stability condition from (3.251h), we proceed to obtain $|\Delta_7|$ as follows.

Similar to (7.13), we can express the determinant of $\tilde{A}$ given in (3.251f) (by actual expansion) in terms of the earlier calculated conditions. This is given as

$$x_5 = |\tilde{A}| = x_1 x_2 x_3 - x_1^2 x_4 - x_3^2 = 9.792 \times 10^4 \qquad (7.162\text{m})$$

The above can also be ascertained from the Hurwitz determinant $|\Delta_{3H}|$ for $F(s) = |sI - A| = s^4 + x_1 s^3 + x_2 s^2 + x_3 s + x_4 = 0$. Now $|\Delta_7|$ (presented in the computer printout) is given by

$$|\Delta_7| = x_5\,|\Delta_5| = -1.30654 \times 10^{11} \qquad (7.162\text{n})$$

We may note that x_4 of (7.162f) is equivalent to $|\Delta_5|/|\Delta_3|$ in (3.251h).

Since $|\Delta_7|$ in (7.162l) is negative, all the $|\Delta_i|$'s, that is, $|\Delta_1|$, $|\Delta_3|$, $|\Delta_5|$, $|\Delta_7|$, are negative, which corresponds to the matrix Δ_7 of (3.249) being negative innerwise. This constitutes the stability condition of the A matrix. The summary of the above detailed procedure is given in the computer printout of Example 7.16.

Example 7.17 Testing the Rank of a Matrix [25]. In many applications of matrix theory to problems of system theory, we require the test of a matrix rank. For instance, in controllability and observability problems we need to test the rank of a certain matrix formed from the system matrices. Such a test would determine whether or not the system is controllable and observable. For example, if the linear system is described by

$$\dot{x} = Ax + Bu \qquad (7.163)$$

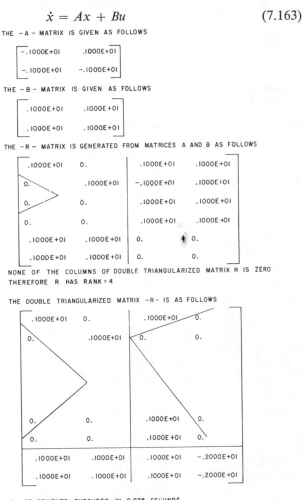

THE $-A-$ MATRIX IS GIVEN AS FOLLOWS

$$\begin{bmatrix} -.1000E+01 & .1000E+01 \\ -.1000E+01 & -.1000E+01 \end{bmatrix}$$

THE $-B-$ MATRIX IS GIVEN AS FOLLOWS

$$\begin{bmatrix} .1000E+01 & .1000E+01 \\ .1000E+01 & .1000E+01 \end{bmatrix}$$

THE $-R-$ MATRIX IS GENERATED FROM MATRICES A AND B AS FOLLOWS

NONE OF THE COLUMNS OF DOUBLE TRIANGULARIZED MATRIX R IS ZERO THEREFORE R HAS RANK = 4

THE DOUBLE TRIANGULARIZED MATRIX $-R-$ IS AS FOLLOWS

END AT LINE 69. 54 STATEMENTS EXECUTED IN 0.078 SECONDS.

Example 7.17

where $\qquad$ dimension $A = n \times n,$ $\qquad$ dimension $B = n \times \ell$ $\qquad$ (7.164)

$$Y = Cx + Du \qquad (7.165)$$

where $\qquad$ dimension cup $= \ell' \times n,$ $\qquad$ dimension $D = \ell' \times \ell$ $\qquad$ (7.166)

then the system is completely controllable and completely observable if certain matrices formed from the system matrices have the rank "n." This fact is illustrated in the accompanying example, which was programmed in the computer. It is shown that by applying the double-triangularization procedure to the matrix, we can determine whether its rank is $n^2 = 4$ or not. In this case, it is equal to 4, and hence the system described in the example is completely controllable. Similarly, we can ascertain the observability condition. The main idea of the double triangularization is to determine whether the matrix has linearly independent columns (or rows) or not, which consequently determines whether its rank is of n^2 or not.

Example 7.18 Root Distribution of an Even Polynomial (Continuous Case).

Following the discussion of the critical cases in Section 7.2, we will apply here the double-triangularization procedure to obtain the root distribution of an even polynomial. Let the $F(s)$ equation be given as

$$F(s) = 3.6s^{12} + 8.4s^{10} + 4.9s^8 - 1.2s^6 - 1.4s^4 + 0.1 = 0 \quad (7.167)$$

The polynomial of (7.167) is an even polynomial. Hence it constitutes the first critical case at the start. We generate the generalized Routh–Hurwitz matrix following (2.4) for this case.† This is shown in the accompanying computer printout. By doing the double triangularization, we arrive at another critical case of the first form. This indicates that roots exist which are of multiplicity 2 (if no further critical case of this type follows). From the inner determinants before the second critical case of the first form, we obtain one change of sign (taking into account the reversal of sign needed for exchanging the bottom with the top rows of the matrix). This single change of sign indicates the existence of at least one root in the right half of the s-plane. To proceed with the test after the second critical case of the first form, we generate the new even polynomial equation from the computer printout as indicated by the arrow. In this case, it is

$$F_1(s) = -0.3527s^6 - 0.4115s^4 + 0s^2 + 0.05879 = 0 \qquad (7.168)$$

Now we proceed to obtain the root distribution of (7.168) in a similar fashion. From the computer printout after the double triangularization, we

---

† Note that the matrix is normalized by dividing all its elements by the element a_{11} before applying the double triangularization. This alters the inners determinants values but not their sign.

obtain one change of sign of the inners determinants. This indicates another root in the right half-plane. Because of the repeated form of the first critical case, this root should be a multiple of the first root obtained earlier. Since $F(s)$ is an even polynomial, we have the same double roots in the left half-plane (symmetrical location). Hence the remaining eight roots are on the $j\omega$ axis. Because of the repeated form of the first critical case, these roots are of multiplicity 2. Therefore we have one real root in the right half-plane of multiplicity 2, one root of multiplicity 2 in the left half-plane, and four distinct roots of multiplicity 2 on the $j\omega$ axis. The total is $2 + 2 + 8 = 12$. Indeed, by factoring the polynomial of (7.167), we obtain

$$F(s) = 0.1 \, (1 + s^2)^2(1 + 2s^2)^2(1 - 3s^2)^2 \qquad (7.169)$$

By minor modification of the preceding method, we can readily determine the nonnegativity and positivity condition of any real even polynomial. This subject has been discussed in Chapter 5 in connection with the modified Routh table. For instance, if we replace s by $j\omega$ in (7.169), we obtain the root distribution of the following even polynomial.

$$\Pi(\omega^2) = 0.1 \, (1 - \omega^2)^2(1 - 2\omega^2)^2(1 + 3\omega^2)^2 \qquad (7.170)$$

This polynomial has positive real roots with even multiplicity† and thus it is a nonnegative polynomial [26].

Example 7.19 Root Distribution of a Reciprocal Polynomial (Discrete Case).

We obtain the root distribution of a reciprocal polynomial discussed in Chapter 5, Equation (5.117), by using the double-triangularization procedure. The discussion of this example closely follows the procedure indicated in Section 7.2 for the critical cases of reciprocal (or self-inversive) polynomials. Let $F(z)$ be given as in (5.117):

$$F(z) = z^8 - 2.172z^7 - 1.89z^6 + 1.16z^5 + 6.72z^4 + 1.16z^3 - 1.89z^2$$
$$- 2.172z + 1 = 0 \quad (5.117)$$

The polynomial in (5.117) represents the first critical case, as indicated in the accompanying printout. We obtain a new Schur–Cohn matrix and proceed. In setting up this matrix and those that follow, the conjugate operation in (7.29) was not used. This simply changes the sign of the inners determinants but does not affect the root distribution, since the polynomial is reciprocal. We reach another critical case of the first form. Before generating the new polynomial, we find one root inside the unit circle from the inners determinants sign. We generate the new polynomial from the rectangular

† This can also be ascertained from (5.62) and (5.63) as follows:

$$\pi_1 = \left(\frac{12 - 6}{2} - 1\right) - \left(\frac{6}{2} - 1\right) = 0 \qquad \pi_2 = \frac{6}{2} - 1 = 2$$

LET F(S) = 3.6 S 12 + 8.4 S10 + 4.9 S8 − 1.2 S6 − 1.4 S4 + 0.1 BE THE POLYNOMIAL.
SINGULARITY NO.1
THIS IS THE FIRST TYPE OF CRITICAL CASE (EVEN POLYNOMIAL) Z(S) = F(S) + F_0(S)
THIS IS THE GENERALIZED ROUTH–HURWITZ MATRIX GENERATED FROM COEFFICIENTS OF Z(S).

```
             -.8400E+01 0.     .4900E+01 0.      .1200E+01 0.     -.1400E+01 0.      0.          0.
600E+01 0.               -.8400E+01 0.     .4900E+01 0.       .1200E+01 0.     -.1400E+01 0.      0.
       .3600E+01 0.      -.8400E+01 0.     .4900E+01 0.       .1200E+01 0.     -.1400E+01 0.
       0.       .3600E+01 0.      -.8400E+01 0.     .4900E+01 0.      .1200E+01 0.     -.1400E+01
       0.       0.       .3600E+01 0.      -.8400E+01 0.     .4900E+01 0.      .1200E+01 0.
       0.       0.       0.       .3600E+01 0.      -.8400E+01 0.     .4900E+01 0.      .1200E+01
       0.       0.       0.       0.       .3600E+01 0.      -.8400E+01 0.     .4900E+01 0.
       0.       0.       0.       0.       0.       .3600E+01 0.      -.8400E+01 0.     .4900E+01
       0.       0.       0.       0.       0.       0.       .3600E+01 0.      -.8400E+01 0.
       0.       0.       0.       0.       0.       0.       0.       .3600E+01 0.      -.8400E+01
       0.       0.       0.       0.       0.       0.       0.       0.       .3600E+01 0.
       0.       0.       0.       0.       0.       0.       0.       0.       0.       .3600E+01
       0.       0.       0.       0.       0.       0.       0.       0.       0.       0.
       0.       0.       0.       0.       0.       0.       0.       0.       -.4320E+02 0.
       0.       0.       0.       0.       0.       0.       0.       -.4320E+02 0.      .8400E+02
       0.       0.       0.       0.       0.       0.       -.4320E+02 0.      .8400E+02 0.
       0.       0.       0.       0.       0.       -.4320E+02 0.      .8400E+02 0.      -.3920E+02
       0.       0.       0.       0.       -.4320E+02 0.      .8400E+02 0.      -.3920E+02 0.
       0.       0.       0.       -.4320E+02 0.      .8400E+02 0.      -.3920E+02 0.      -.7200E+01
       0.       0.       -.4320E+02 0.      .8400E+02 0.      -.3920E+02 0.      -.7200E+01 0.
       0.       -.4320E+02 0.      .8400E+02 0.      -.3920E+02 0.      -.7200E+01 0.      .5600E+01
     -.4320E+02 0.      .8400E+02 0.      -.3920E+02 0.      -.7200E+01 0.      .5600E+01 0.
0E+02 0.       .8400E+02 0.      -.3920E+02 0.      -.7200E+01 0.      .5600E+01 0.      0.
```

```
.1000E+00 0.        0.        0.        0.        0.        0.        0.        0.        0.        0.        0.
0.        .1000E+00 0.        0.        0.        0.        0.        0.        0.        0.        0.       0.
0.        0.        .1000E+00 0.        0.        0.        0.        0.        0.        0.       0.        0.
0.        0.        0.        .1000E+00 0.        0.        0.        0.        0.       0.        0.        0.
-.1400E+01 0.       0.        .1000E+00 0.        0.        0.        0.       0.        0.        0.        0.
0.        -.1400E+01 0.       0.        0.        .1000E+00 0.       0.        0.        0.        0.        0.
.1200E+01 0.        -.1400E+01 0.       0.        0.        .1000E+00 0.      0.        0.        0.        0.
0.        .1200E+01 0.        -.1400E+01 0.       0.        0.        .1000E+00 0.       0.        0.        0.
.4900E+01 0.        .1200E+01 0.        -.1400E+01 0.       0.        0.        .1000E+00 0.       0.        0.
0.        .4900E+01 0.        .1200E+01 0.        -.1400E+01 0.      0.        0.        .1000E+00 0.       0.
-.8400E+01 0.       .4900E+01 0.        .1200E+01 0.        -.1400E+01 0.      0.        0.        .1000E+00 0.
0.        -.8400E+01 0.       .4900E+01 0.        .1200E+01 0.        -.1400E+01 0.      0.        0.        .1000E+00
-.4320E+02 0.       .8400E+02 0.        -.3920E+02 0.       -.7200E+01 0.      .5600E+01 0.       0.        0.      Δ₂
0.        .8400E+02 0.        -.3920E+02 0.       -.7200E+01 0.      .5600E+01 0.       0.        0.      Δ₄
.8400E+02 0.        -.3920E+02 0.       -.7200E+01 0.      .5600E+01 0.       0.        0.        0.      Δ₆
0.        -.3920E+02 0.       -.7200E+01 0.       .5600E+01 0.       0.        0.        0.        0.      Δ₈
-.3920E+02 0.       -.7200E+01 0.       .5600E+01 0.       0.        0.        0.        0.        0.      Δ₁₀
0.        -.7200E+01 0.       .5600E+01 0.       0.        0.        0.        0.        0.        0.      Δ₁₂
-.7200E+01 0.       .5600E+01 0.       0.        0.        0.        0.        0.        0.        0.      Δ₁₄
0.        .5600E+01 0.       0.        0.        0.        0.        0.        0.        0.        0.      Δ₁₆
.5600E+01 0.        0.        0.        0.        0.        0.        0.        0.        0.        0.      Δ₁₈
0.        0.        0.        0.        0.        0.        0.        0.        0.        0.        0.      Δ₂₀
0.        0.        0.        0.        0.        0.        0.        0.        0.        0.        0.      Δ₂₂
0.        0.        0.        0.        0.        0.        0.        0.        0.        0.        0.      Δ₂₄
```

$,\Delta_2| = -.1200000E+02$

$|\Delta_4| = .5600000E+02$

$|\Delta_6| = -.2032593E+03$

$|\Delta_8| = -.4065185E+03$

$|\Delta_{10}| = .1843621E+03$

$|\Delta_{12}| = -.6503074E+02$

CRITICAL CASE

```
.1000E+01 0.        -.2333E+01 0.         .1361E+01 0.         .3333E+00 0.        -.3889E+00 0.          0.         0.
0.         .1000E+01 0.        -.2333E+01 0.         .1361E+01 0.         .3333E+00 0.        -.3889E+00 0.         0.
0.        0.         .1000E+01 0.        -.2333E+01 0.         .1361E+01 0.         .3333E+00 0.        -.3889E+00 0.
0.        0.        0.         .1000E+01 0.        -.2333E+01 0.         .1361E+01 0.         .3333E+00 0.        -.3889E+00
0.        0.        0.        0.         .1000E+01 0.        -.2333E+01 0.         .1361E+01 0.         .3333E+00 0.
0.        0.        0.        0.        0.         .1000E+01 0.        -.2333E+01 0.         .1361E+01 0.         .3333E+00
0.        0.        0.        0.        0.        0.         .1000E+01 0.        -.2333E+01 0.         .1361E+01 0.
0.        0.        0.        0.        0.        0.        0.         .1000E+01 0.        -.2333E+01 0.         .1361E+01
0.        0.        0.        0.        0.        0.        0.        0.         .1000E+01 0.        -.2333E+01 0.
0.        0.        0.        0.        0.        0.        0.        0.        0.         .1000E+01 0.        -.2333E+01
0.        0.        0.        0.        0.        0.        0.        0.        0.        0.         .1000E+01 0.
0.        0.        0.        0.        0.        0.        0.        0.        0.        0.        0.         .1000E+01
0.        0.        0.        0.        0.        0.        0.        0.        0.        0.        0.        -.1200E+02
0.        0.        0.        0.        0.        0.        0.        0.        0.        0.        -.1200E+02 0.         .2333E+02
0.        0.        0.        0.        0.        0.        0.        0.        0.        -.1200E+02 0.         .2333E+02 0.
0.        0.        0.        0.        0.        0.        0.        0.        -.1200E+02 0.         .2333E+02 0.        -.1089E+02
0.        0.        0.        0.        0.        0.        -.1200E+02 0.         .2333E+02 0.        -.1089E+02 0.
0.        0.        0.        0.        0.        -.1200E+02 0.         .2333E+02 0.        -.1089E+02 0.        -.2000E+01 0.
0.        0.        0.        0.        -.1200E+02 0.         .2333E+02 0.        -.1089E+02 0.        -.2000E+01 0.         .1556E+01
0.        0.        0.        -.1200E+02 0.         .2333E+02 0.        -.1089E+02 0.        -.2000E+01 0.         .1556E+01 0.
0.        -.1200E+02 0.         .2333E+02 0.        -.1089E+02 0.        -.2000E+01 0.         .1556E+01 0.         0.
```

```
.2778E-01 0.      .5401E-01 0.      -.1300E+01 0.     -.3948E-01 0.      .5401E-02 0.      .6301E-02 0.

0.        .2778E-01 0.      -.4555E+00 0.      .2094E-32 0.      -.4192E-01 0.      .5401E-02 0.      .6650E-02

0.        0.        -.1207E+01 0.      .5130E-01 0.      .1991E+01 0.      -.3274E+00 0.      -.3305E+00 0.

0.        -.9074E+00 0.      -.8333E-01 0.      .5770E+01 0.      -.4741E+01 0.      -.3274E+00 0.      .6312E+00

-.3889E+00 0.      .3025E+00 0.      .4869E+01 0.      -.1495E+01 0.      .2468E+00 0.      .2485E+00 0.

0.        .3889E+00 0.      .1708E+01 0.      -.4583E+01 0.      .3848E+01 0.      .2468E+00 0.      -.5149E+00

.3333E+00 0.      .4582E+01 0.      -.3972E+01 0.      0.        0.        0.        0.        0.        0.

0.        .3509E+01 0.      -.1075E+01 0.      0.        0.        0.        0.        0.        0.

.1361E+01 0.      -.4431E+01 0.      0.        0.        0.        0.        0.        0.        0.

0.        -.4083E+01 0.      0.        0.        0.        0.        0.        0.        0.

-.2333E+01 0.      0.        0.        0.        0.        0.        0.        0.        0.

0.        0.        0.        0.        0.        0.        0.        0.        0.        0.     Δ2

-.1200E+02 0.      0.        0.        0.        0.        0.        0.        0.        0.

0.        -.4667E+01 0.      0.        0.        0.        0.        0.        0.        0.     Δ4

.2333E+02 0.      -.3630E+01 0.      0.        0.        0.        0.        0.        0.     Δ6

0.        .4356E+02 0.      .2000E+01 0.      0.        0.        0.        0.        0.     Δ8

-.1089E+02 0.      .5093E+02 0.      -.4535E+00 0.      0.        0.        0.        0.     Δ10

0.        -.2741E+02 0.      .1368E+02 0.      -.3527E+00 0.      .4115E+00 0.      .1421E-13 0.      -.5879E-01   Δ12

-.2000E+01 0.      -.3692E+02 0.      .4713E+02 0.      .7958E-12 0.      -.1421E-13 0.      -.1563E-12 0.

0.        -.3111E+01 0.      -.1457E+02 0.      .5455E+02 0.      -.4570E+02 0.      -.2962E+01 0.      .6110E+01

.1556E+01 0.      -.3327E+01 0.      -.4051E+02 0.      .1794E+02 0.      -.2962E+01 0.      -.2982E+01 0.

0.        .3630E+01 0.      -.3920E+00 0.      -.4835E+01 0.      .3949E+02 0.      .2777E+01 0.      -.5252E+01

0.        0.        .4940E+01 0.      -.2358E+01 0.      -.1692E+02 0.      .2777E+01 0.      .2807E+01 0.

0.        0.        0.        .1952E+01 0.      -.2519E+01 0.      .2261E+01 0.      .1194E+00 0.      -.3057E+00
```

```
SINGULARITY NO. 2
THIS IS THE FIRST TYPE OF CRITICAL CASE ( EVEN POLYNOMIAL.)
NULL INNERWISE MATRIX
```

```
.1000E+01 0.      -.2333E+01 0.      .1361E+01 0.      .3333E+00 0.      -.3889E+00 0.          0.          0.
0.        .1000E+01 0.      -.2333E+01 0.      .1361E+01 0.      .3333E+00 0.      -.3889E+00 0.          0.
0.        0.        .1000E+01 0.      -.2333E+01 0.      .1361E+01 0.      .3333E+00 0.      -.3889E+00 0.
0.        0.        0.        .1000E+01 0.      -.2333E+01 0.      .1361E+01 0.      .3333E+00 0.      -.3889E+00
0.        0.        0.        0.        .1000E+01 0.      -.2333E+01 0.      .1361E+01 0.      .3333E+00 0.
0.        0.        0.        0.        0.        .1000E+01 0.      -.2333E+01 0.      .1361E+01 0.      .3333E+00
0.        0.        0.        0.        0.        0.        .1000E+01 0.      -.2333E+01 0.      .1361E+01 0.
0.        0.        0.        0.        0.        0.        0.        .1000E+31 0.      -.2333E+01 0.      .1361E+01
0.        0.        0.        0.        0.        0.        0.        0.        .1000E+01 0.      -.2333E+01 0.
0.        0.        0.        0.        0.        0.        0.        0.        0.        .1000E+01 0.      -.2333E+01
0.        0.        0.        0.        0.        0.        0.        0.        0.        0.        .1000E+01 0.
0.        0.        0.        0.        0.        0.        0.        0.        0.        0.        0.        -.1200E+02 0.
0.        0.        0.        0.        0.        0.        0.        0.        0.        0.        -.1200E+02 0.
0.        0.        0.        0.        0.        0.        0.        0.        0.        -.1200E+02 0.      .2333E+02
0.        0.        0.        0.        0.        0.        0.        0.        -.1200E+02 0.      .2333E+02 0.
0.        0.        0.        0.        0.        0.        0.        -.1200E+02 0.      .2333E+02 0.      -.1089E+02
0.        0.        0.        0.        0.        0.        -.1200E+02 0.      .2333E+02 0.      -.1089E+02 0.
0.        0.        0.        0.        0.        -.1200E+02 0.      .2333E+02 0.      -.1089E+02 0.      -.2000E+01
0.        0.        0.        0.        -.1200E+02 0.      .2333E+02 0.      -.1089E+02 0.      -.2000E+01 0.
0.        0.        0.        -.1200E+02 0.      .2333E+02 0.      -.1089E+02 0.      -.2000E+01 0.      .1556E+01
0.        0.        -.1200E+02 0.      .2333E+02 0.      -.1089E+02 0.      -.2000E+01 0.      .1556E+01 0.
0.        -.1200E+02 0.      .2333E+02 0.      -.1089E+02 0.      -.2000E+01 0.      .1556E+01 0.          0.
```

```
.2778E-01 0.      .5401E-01 0.      -.1300E+01 0.      -.3948E-01 0.      .5401E-02 0.      .6301E-02 0.
0.      .2778E-01 0.      -.4555E+00 0.      .2094E-32 0.      -.3948E-01 0.      .5401E-02 0.      .6301E-02
0.      0.      -.1207E+01 0.      .5130E-01 0.      .1991E+01 0.      -.3274E+00 0.      -.3305E+00 0.
0.      -.9074E+00 0.      -.8333E-01 0.      .5773E+01 0.      .1991E+01 0.      -.3274E+00 0.      -.3305E+00
-.3889E+00 0.      .3025E+00 0.      .4869E+01 0.      -.1495E+01 0.      .2468E+00 0.      .2485E+00 0.
0.      .3889E+00 0.      .1708E+01 0.      -.4580E+01 0.      -.1495E+01 0.      .2468E+00 0.      .2485E+00
.3333E+00 0.      .4582E+01 0.      -.3972E+01 0.      0.      0.      0.      0.      0.      0.
0.      .3509E+01 0.      -.1075E+01 0.      0.      0.      0.      0.      0.      0.
.1361E+01 0.      -.4431E+01 0.      0.      0.      0.      0.      0.      0.      0.
0.      -.4083E+01 0.      0.      0.      0.      0.      0.      0.      0.
-.2333E+01 0.      0.      0.      0.      0.      0.      0.      0.      0.
0.      0.      0.      0.      0.      0.      0.      0.      0.      0.
-.1200E+02 0.      0.      0.      0.      0.      0.      0.      0.      0.
            Δ2
0.      -.4667E+01 0.      0.      0.      0.      0.      0.      0.      0.
            Δ4
.2333E+02 0.      -.3630E+01 0.      0.      0.      0.      0.      0.      0.      0.
                  Δ6
0.      .4356E+02 0.      .2000E+01 0.      0.      0.      0.      0.      0.      0.
                        Δ8
-.1089E+02 0.      .5093E+02 0.      -.4535E+00 0.      0.      0.      0.      0.      0.
                              Δ10
0.      -.2741E+02 0.      .1368E+02 0.      -.3527E+00 0.      0.      0.      0.      0.
                                    Δ12
-.2000E+01 0.      -.3692E+02 0.      .4713E+02 0.      .7958E-12 0.      -.1421E-13 0.      -.1563E-12 0.
                                          Δ14
0.      -.3111E+01 0.      -.1457E+02 0.      .5455E+02 0.      .1794E+02 0.      -.2962E+01 0.      -.2982E+01
.1556E+02 0.      -.3327E+01 0.      -.4051E+02 0.      .1794E+02 0.      -.2962E+01 0.      -.2982E+01 0.
0.      .3630E+01 0.      -.3920E+00 0.      -.4835E+02 0.      -.1692E+02 0.      .2777E+01 0.      .2807E+01
0.      0.      .4940E+01 0.      -.2358E+01 0.      -.1692E+02 0.      .2777E+01 0.      .2807E+01 0.
0.      0.      0.      .1952E+01 0.      -.2519E+01 0.      -.6780E+00 0.      .1194E+00 0.      .1141E+00
```

THIS IS THE GENERALIZED ROUTH–HURWITZ MATRIX GENERATED FROM COEFFICIENTS OF NEW Z(S).

$$|\Delta_2^l| = -.7463132E+00$$

$$|\Delta_4^l| = .2165725E+00$$

$$|\Delta_6^l| = -.4888770E-01$$

$$|\Delta_8^l| = -.6081612E-02$$

$$|\Delta_{10}^l| = .1715799E-03$$

$$|\Delta_{12}^l| = -.3765549E-05$$

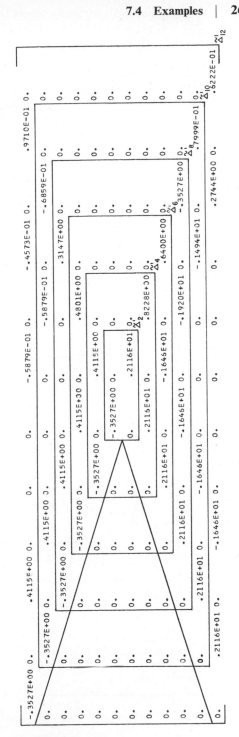

Example 7.18

LET F(Z) = 79-2.17Z7-1.89Z6+1.16Z5+6.72Z4+1.15Z3-1.89Z2-2.172Z+1 BE THE POLYNOMIAL

THIS IS THE SCHUR-COHN MATRIX GENERATED FROM COEFFICIENTS OF ABOVE POLYNOMIAL.

SINGULARITY NO. 1

THIS THE FIRST TYPE OF CRITICAL CASE.

FULL INVERSE MATRIX

1.00	0.	0.	0.	0.	0.	0.	0.	0.	0.	0.	0.	0.	0.	0.	1.00
-2.17	1.00	0.	0.	0.	0.	0.	0.	0.	0.	0.	0.	0.	0.	1.00	-2.17
-1.89	-2.17	1.00	0.	0.	0.	0.	0.	0.	0.	0.	0.	0.	1.00	-2.17	-1.89
1.16	-1.89	-2.17	1.00	0.	0.	0.	0.	0.	0.	0.	0.	1.00	-2.17	-1.89	1.16
6.72	1.16	-1.89	-2.17	1.00	0.	0.	0.	0.	0.	0.	1.00	-2.17	-1.89	1.16	6.72
1.16	6.72	1.16	-1.89	-2.17	1.00	0.	0.	0.	0.	1.00	-2.17	-1.89	1.16	6.72	1.16
-1.89	1.16	6.72	1.16	-1.89	-2.17	1.00	0.	0.	1.00	-2.17	-1.89	1.16	6.72	1.16	-1.89
-2.17	-1.89	1.16	6.72	1.16	-1.89	-2.17	1.00	1.00	-2.17	-1.89	1.16	6.72	1.16	-1.89	-2.17
-2.17	-1.89	1.16	6.72	1.16	-1.89	-2.17	1.00	1.00	-2.17	-1.89	1.16	6.72	1.16	-1.89	-2.17
-1.89	1.16	6.72	1.16	-1.89	-2.17	1.00	0.	0.	1.00	-2.17	-1.89	1.16	6.72	1.16	-1.89
1.16	6.72	1.16	-1.89	-2.17	1.00	0.	0.	0.	0.	1.00	-2.17	-1.89	1.16	6.72	1.16
6.72	1.16	-1.89	-2.17	1.00	0.	0.	0.	0.	0.	0.	1.00	-2.17	-1.89	1.16	6.72
1.16	-1.89	-2.17	1.00	0.	0.	0.	0.	0.	0.	0.	0.	1.00	-2.17	-1.89	1.16
-1.89	-2.17	1.00	0.	0.	0.	0.	0.	0.	0.	0.	0.	0.	1.00	-2.17	-1.89
-2.17	1.00	0.	0.	0.	0.	0.	0.	0.	0.	0.	0.	0.	0.	1.00	-2.17
1.00	0.	0.	0.	0.	0.	0.	0.	0.	0.	0.	0.	0.	0.	0.	1.00

The central 2×2 block (rows and columns 8–9) is boxed and labeled Δ_2.

NEWLY GENERATED SCHUR-COHN MATRIX

$|\Delta'_2| = .9263E+00$

$|\Delta'_4| = -.1191E+00$

$|\Delta'_6| = -.4128E+01$

$$
\begin{array}{cccccccccccccc}
1.00 & -1.90 & 1.90 & .72 & 3.35 & .73 & -.47 & -.13 & .50 & -277.88 & 325.18 & 477.94 & 349.36 & -288.35 \\
0. & 1.90 & -1.90 & -1.90 & .72 & 3.35 & .43 & .12 & 8.15 & -189.41 & 192.90 & 202.90 & 218.45 & -175.17 \\
0. & 1.00 & -1.90 & -1.90 & -1.42 & .72 & 3.36 & .91 & 5.25 & 46.27 & -50.54 & -70.10 & -49.73 & 44.24 \\
0. & 1.00 & -1.90 & -1.42 & -1.42 & .72 & .20 & .20 & -2.49 & 207.70 & -219.81 & -322.00 & -236.13 & 197.16 \\
0. & 0. & 1.00 & -1.90 & -1.42 & -1.42 & -.38 & -5.88 & 0. & 0. & 0. & 0. & 0. & 0. \\
0. & 0. & 0. & 0. & 1.00 & -1.90 & -.52 & 0. & 0. & 0. & 0. & 0. & 0. & 0. \\
0. & 0. & 0. & 0. & 0. & 1.00 & .53 & 0. & 0. & 0. & 0. & 0. & 0. & 0. \\
0. & 0. & 0. & 0. & 0. & -.27 & -.13 & 0. & 0. & 0. & 0. & 0. & 0. & 0. \\
0. & 0. & 0. & -.27 & -.47 & .43 & .12 & -.62 & 34.66 & -37.43 & -53.13 & -37.63 & 34.66 \\
0. & 0. & -.27 & -.47 & .43 & 3.36 & .91 & 4.58 & -28.15 & 29.02 & 44.12 & 33.45 & -25.29 \\
0. & 0. & -.47 & .43 & 3.36 & .72 & .20 & 8.49 & -168.23 & 177.11 & 261.19 & 193.01 & -157.27 \\
0. & -.27 & .43 & 3.36 & .72 & -1.42 & -.28 & .15 & -202.48 & 319.94 & 468.71 & 343.83 & -284.81 \\
0. & -.47 & 3.36 & .72 & -1.42 & -1.90 & -.52 & -5.25 & 15.64 & -15.43 & -25.41 & -20.98 & 15.16 \\
-.27 & -.47 & .72 & -1.42 & -1.90 & & & & & & & & \\
\end{array}
$$

$\tilde{\Delta}_2 \qquad \tilde{\Delta}_4 \qquad \tilde{\Delta}_6$

SINGULARITY NO. 2

THIS THE FIRST TYPE OF CRITICAL CASE.

WILL INFLUENCE MATRIX

1.00	-1.90	-1.42	.72	3.36	.43	-.47	-.13	.50	-317.88	-13.24	-206.36	12.57	227.47		
0.	1.00	-1.90	.72	.72	3.36	.43	.12	8.19	-109.41	2.03	30.43	-.69	-32.42		
0.	0.	1.00	-1.42	-1.42	.72	3.36	.91	5.25	46.27	10.53	165.04	-11.55	-181.38		
0.	0.	0.	1.00	-1.90	-1.42	.72	.20	-2.49	237.70	0.	0.	0.	0.		
0.	0.	0.	0.	1.00	-1.90	-1.42	-.38	-5.88	0.	0.	0.	0.	0.		
0.	0.	0.	0.	0.	1.00	-1.90	-.52	0.	0.	0.	0.	0.	0.		
0.	0.	0.	0.	0.	0.	1.00	.93	-.13	0.	0.	0.	0.	0.		
0.	0.	0.	0.	-.27	-.27	-.77	-.13	-.62	34.66	0.	0.	0.	0.		
0.	0.	0.	-.27	-.47	-.47	-.47	.12	4.58	-23.15	0.	0.	0.	0.		
0.	0.	-.27	-.47	.43	.43	.43	.91	8.49	-168.23	-2.36	-44.77	0.	0.		
0.	-.27	-.47	.43	3.36	3.36	3.36	.20	.15	-302.48	-10.96	-171.33	3.14	49.20		
0.	-.47	.43	3.36	.72	.72	.72	-.38	15.94	15.94	-17.66	-10.96	11.60	188.02		
-.27	.43	3.36	.72	-1.42	-1.42	-1.42	-.52	-5.29	-5.29	-17.66	-275.68	18.38	301.77		

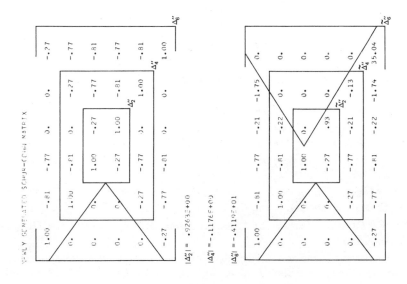

Example 7.19

box denoted by the arrow. We proceed with the double-triangularization procedure for the new Schur–Cohn matrix of this new polynomial. We obtain one change of sign of inner determinants, hence another root inside the unit circle. Since the polynomial equation $F(z) = 0$ is a reciprocal one, we have two roots outside the unit circle. Because of two repeated forms of first form critical case, these roots are of multiplicity 2. The remaining four roots are on the unit circle and are also of multiplicity 2. Indeed, the factored form of $F(z)$ in (5.127) verifies this root distribution.

A minor extension of the double-triangularization procedure, and following the discussions of Chapter 5, we can readily test the reciprocal polynomial for nonnegativity or positivity.† These last two examples indicate that the nonnegativity and positivity conditions of Chapter 5 can be obtained in a unified way by using one algorithm based, as shown above, on the double-triangularization procedure. A computer flowchart for both continuous and discrete cases is shown in Figures 7.2 and 7.3.

REFERENCES

[1] C. F. GAUSS, "C. F. Gauss Works Gottingen: konigliches Gesellschaft dern Wissenschaften, 1863."

[2] F. R. GANTMACHER, *The Theory of Matrices*. New York: Chelsea, 1959.

[3] E. I. JURY and S. M. AHN, "A computational algorithm for inners," *IEEE Trans. on Automatic Control*, **AC-17** (4), 541–543 (Aug. 1972).

[4] E. I. JURY, *Theory and Application of the z-Transform Method*. New York: Wiley, 1964, Chap. 3.

[5] M. MARDEN, *Geometry of Polynomials*. Providence, R.I.: American Mathematical Society, 1 and 66, Chap. X.

[6] D. K. CHANG, *Analysis of Linear Systems*. Reading, Mass.: Addison-Wesley, 1959.

[7] C. HERMITE, "Sur le nombre des racines d'une équation algébrique comprise entre des limites donnés," *J. Reine Angewandte Mathematik*, **52,** 39–51 and Ouvres 1, 397–419, 854.

[8] R. J. DUFFIN, "Algorithms for classical stability problems," *SIAM Rev.*, **11** (2), 136–213 (April 1969).

[9] B. D. O. ANDERSON and E. I. JURY, "A simplified Schur–Cohn test," *IEEE Trans. on Automatic Control*, **AC-18** (2), 157–163 (April 1973).

† For instance, in this example π_1 and π_2 (number of positive real roots with multiplicity 1 and 2, respectively) can be obtained by using (5.108) as follows:

$$\pi_1 = (4 - 2) - (4 - 2) = 0 \qquad \pi_2 = 4 - 2 = 2$$

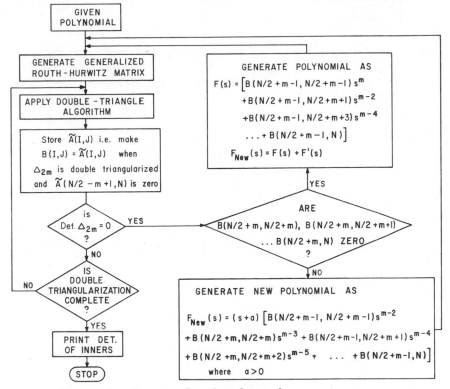

Figure 7.2. **Computer flow chart for continuous systems.**

[10] C. J. GREAVES et al., "Evaluations of integral appearing in minimization problems of discrete data systems," *IEEE Trans. on Automatic Control*, **AC-11** (1), 145–148 (Jan. 1960).

[11] R. M. GOLDEN, "Digital filters synthesis by sample-data transformation," *IEEE Trans. on Audio and Electroacoustics*, **AU-16** (3), 321–327 (Sept. 1968).

[12] A. M. BUSH and D. C. FIELDER, "Simplified algebra for the bilinear and related transformations," *IEEE Trans. on Audio and Electroacoustics*, 127–128 (April 1973).

[13] M. MANSOUR, "Über die Stabilitat Lineares Abtastsystemes," *Regelungstechnik*, **12**, 523–534 (1964).

[14] M. H. POWER, "The mechanics of the bilinear transformation," *IEEE Trans. on Education* (Correspondence), **E-10**, 114–116 (June 1967).

[15] D. C. FIELDER, "Some classroom comments on bilinear transformations," *IEEE Trans. on Education (Correspondence)*, **E-13** (2), 105–107 (Aug. 1970).

[16] C. A. HALIJAK and M. L. MOĖ, "Effective computation of coefficients of $(1 - z)^m (1 + z)^n$," *Proceedings 14th Midwest Symposium on Circuit Theory*, University of Denver, Denver, Colo., May 1971, pp. 1.5-1–1.5-5.

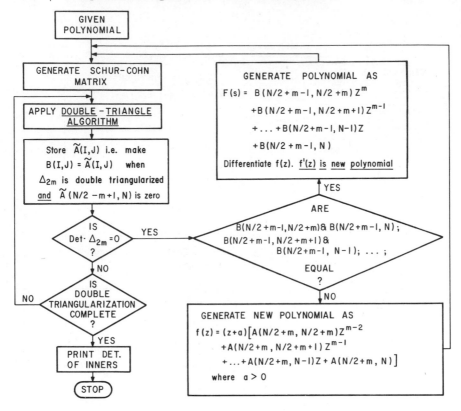

Figure 7.3. Computer flow chart for discrete systems.

[17] E. I. Jury, "Remarks on the mechanics of bilinear transformation," *Trans. on Audio and Electroacoustics*, **AU-21** (4), 380–382 (Aug. 1973).

[18] E. I. Jury and O. W. C. Chan, "Combinatorial rules for some useful transformations," *IEEE Trans. on Circuit Theory*, **CT-20,** 476–480 (Sept. 1973).

[19] J. H. Wilkinson, *The Algebraic Eigenvalue Problem*. New York: Oxford University Press, 1965.

[20] S. K. Mitra and R. J. Sherwood, "Canonic realizations of digital filters using the continued fraction expansion," *IEEE Trans. on Audio and Electroacoustics*, **AU-20** (3), 185–194 (Aug. 1972).

[21] E. I. Jury and S. M. Ahn, "The theory of inners as applied to networks (part VI)," Electronics Res. Laboratory, University of California, Berkeley, Dec. 1970. Also, *IEEE, 1970, Int. Symp. on Circuit Theory*, pp. 183–184.

[22] L. Weinberg, *Network Analysis and Synthesis*. New York: McGraw-Hill, 1962.

[23] A. T. FULLER, "Stability criteria for linear systems and realizability criteria for RC networks," *Proc. Camb. Phil. Soc.*, **53** (1957).

[24] E. I. JURY and S. GUTMAN, "The inner formulation of the total square integral (sum)," *Proc. IEEE*, 395–397 (March 1973).

[25] E. I. JURY, "Inner algorithm test for controllability and observability," *IEEE Trans. on Automatic Control*, **18,** 682–683 (Dec. 1973).

[26] J. KARMARKER, "On Šiljak's absolute stability test," *Proc. IEEE*, 817–819 (May 1970).

[27] G. W. STEWART III, "On Lehmer's method for finding the zeros of a polynomial," *Mathematics of Computation*, **23** (108), 829–835 (Oct. 1969).

[28] I. S. PACE and S. BARNETT, "Numerical comparison of root-location algorithms for constant linear systems," *in* D. J. Bell (Ed.), *Recent Mathematical Developments in Control.* London and New York: Academic Press, 1973, pp. 373–392.

[29] J. J. H. MILLER, "On the location of zeros of certain classes of polynomials with applications to numerical analysis," *J. Inst. Math. Appl.*, **8,** 397–406 (1971).

[30] P. S. Kamat, "On Computational Algorithm for Inners," *IEEE Trans. on Aut. Contr.*, AC–**19** (2), (April 1974), 154–155.

Tables of
Chebyshev Functions

Chebyshev functions have already been mentioned in several sections in the text. For instance, in Section 2.2 Chebyshev functions are used to obtain conditions on the relative stability of continuous systems. In Section 3.4 the Chebyshev theorem, in relation to Hankel matrices, is discussed. In addition, Chebyshev functions are used in transforming the unit circle into a segment of the real axis between zero and unity [1]. Thus one can apply the Sturm sequence theorem to obtain conditions on positivity and hence on the stability conditions discussed in Chapter 6. Many other applications of Chebyshev functions exist in the scientific literature. In this appendix we will present the Chebyshev functions of the first kind, $T_k(\zeta)$, and of the second kind, $U_k(\zeta)$, in an innerwise matrix form. Doing so is feasible because of the recursive form of these functions [2]. The innerwise matrix form enables us to apply the double-triangularization procedure discussed in Chapter 7 to evaluate the Chebyshev functions. These functions are presented in Tables A.1 and A.2 for $T_k(\zeta)$ and $U_k(\zeta)$ for certain k's and can be readily extended to other values of k.

A.1   INNERWISE MATRIX FOR $T_k(\zeta)$

We can readily construct an innerwise matrix for $T_k(\zeta)$ for any k, based on the following relationships obtained from the basic definitions of T_0 and T_1:

$$|\Delta_1| = T_0 = 1 \tag{A.1}$$

$$|\Delta_3| = T_1 = \zeta \tag{A.2}$$

$$|\Delta_5| = 2\zeta |\Delta_3| - |\Delta_1| = 2\zeta^2 - 1 = T_2 \tag{A.3}$$

$$|\Delta_7| = 2\zeta |\Delta_5| - |\Delta_3| = 2\zeta T_2 - T_1 = 2\zeta(2\zeta^2 - 1) - \zeta = 4\zeta^3 - 3\zeta = T_3 \tag{A.4}$$

$$|\Delta_9| = 2\zeta |\Delta_7| - |\Delta_5| = 2\zeta T_3 - T_2 = 2\zeta(4\zeta^3 - 3\zeta) - 2\zeta^2 + 1$$
$$\vdots \qquad\qquad \vdots \qquad\qquad\qquad\qquad \vdots \quad = 8\zeta^4 - 8\zeta^2 + 1 = T_4 \tag{A.5}$$

ξ	T_0	T_1	T_2	T_3	T_4	T_5	T_6	T_7	T_8	T_9	T_{10}
.	1.00000000	0.	-1.00000000	0.	1.00000000	0.	-1.00000000	0.	1.00000000	0.	-1.00000000
.05000000	1.00000000	.05000000	-.99500000	-.14950000	.98005000	.24750500	-.95529950	-.34303495	.92099600	.43513455	-.87748255
.10000000	1.00000000	.10000000	-.98000000	-.29600000	.92080000	.48016000	-.82476800	-.66511360	.69574528	.78426266	-.53889275
.15000000	1.00000000	.15000000	-.95500000	-.43650000	.82405000	.68371500	-.61893550	-.86939565	.35811681	.97683069	-.06506760
.20000000	1.00000000	.20000000	-.92000000	-.55800000	.69280000	.84512000	-.35475200	-.98702080	-.04005632	.97099827	.42845563
.25000000	1.00000000	.25000000	-.87500000	-.68750000	.53125000	.95312500	-.05468750	-.98046875	-.43554687	.76269531	.81689453
.30000000	1.00000000	.30000000	-.82000000	-.79200000	.34480000	.99888000	.25452800	-.84616320	-.76222592	.38882765	.99552251
.35000000	1.00000000	.35000000	-.75500000	-.87850000	.14005000	.97653500	.54352450	-.59606785	-.96077199	-.07647255	.90724121
.40000000	1.00000000	.40000000	-.68000000	-.94400000	-.07520000	.88384000	.78227200	-.25802240	-.98866992	-.53292954	.56234629
.45000000	1.00000000	.45000000	-.59500000	-.98550000	-.29195000	.72274500	.94242050	.12543345	-.82953040	-.87201081	.04472067
.50000000	1.00000000	.50000000	-.50000000	-1.00000000	-.50000000	.50000000	1.00000000	.50000000	-.50000000	-1.00000000	-.50000000
.55000000	1.00000000	.55000000	-.39500000	-.98450000	-.68795000	.22775500	.93848050	.80457355	-.05344960	-.86336810	-.89625532
.60000000	1.00000000	.60000000	-.28000000	-.93600000	-.84320000	-.07584000	.75219200	.97847040	.42197248	-.47210342	-.98849659
.65000000	1.00000000	.65000000	-.15500000	-.85150000	-.95195000	-.38603550	.45010450	.97117085	.81241760	.08497204	-.70195396
.70000000	1.00000000	.70000000	-.02000000	-.72800000	-.99920000	-.67088000	.05996800	.75483520	.99680128	.64066659	-.09998005
.75000000	1.00000000	.75000000	.12500000	-.56250000	-.96875000	-.89062500	-.36718750	.33984375	.87695313	.97558594	.58647578
.80000000	1.00000000	.80000000	.29000000	-.35200000	-.84320000	-.99712000	-.75219200	-.20638720	.42197248	.89154317	.98849659
.85000000	1.00000000	.85000000	.44500000	-.09350000	-.60395000	-.93321500	-.98251550	-.73706135	-.27048879	.27723040	.74178047
.90000000	1.00000000	.90000000	.62000000	.21600000	-.23120000	-.63216000	-.90668800	-.99987840	-.89309312	-.60768922	-.20074747
.95000000	1.00000000	.95000000	.80500000	.57950000	.29605000	-.01700500	-.32835950	-.60687805	-.82470880	-.96006866	-.99942166
1.00000000	1.00000000	1.00000000	1.00000000	1.00000000	1.00000000	1.00000000	1.00000000	1.00000000	1.00000000	1.00000000	1.00000000

Table A.1

Table A.2

ξ	U_0	U_1	U_2	U_3	U_4	U_5	U_6	U_7	U_8	U_9	U_{10}
.	0.	1.00000000	0.	-1.00000000	0.	1.00000000	0.	-1.00000000	0.	1.00000000	0.
.05000000	0.	1.00000000	.10000000	-.99000000	-.19900000	.97010000	.29961000	-.94049900	-.39005990	.90149301	.48020920
.10000000	0.	1.00000000	.20000000	-.96000000	-.39200000	.88160000	.56832000	-.76793600	-.72190720	.62355456	.84661911
.15000000	0.	1.00000000	.30000000	-.91000000	-.57300000	.73810000	.79443000	-.49977100	-.94436130	.21646261	1.00930008
.20000000	0.	1.00000000	.40000000	-.84000000	-.73600000	.54560000	.95424000	-.16390400	-1.01980160	-.24401664	.92219494
.25000000	0.	1.00000000	.50000000	-.75000000	-.87500000	.31250000	1.03125000	.20312500	-.92968750	-.66796875	.59570313
.30000000	0.	1.00000000	.60000000	-.64000000	-.98400000	.06960000	1.01376000	.55865600	-.67856640	-.95795584	.09908890
.35000000	0.	1.00000000	.70000000	-.51000000	-1.05700000	-.22990000	.89607000	.85714900	-.29605570	-1.05639499	-.44901079
.40000000	0.	1.00000000	.80000000	-.35000000	-1.08800000	-.51040000	.67968000	1.05414400	.16363520	-.92323584	-.90222387
.45000000	0.	1.00000000	.90000000	-.19000000	-1.07100000	-.77390000	.37449000	1.11094100	.62535690	-.54811977	-1.11886471
.50000000	0.	1.00000000	1.00000000	-.00000000	-1.00000000	-1.00000000	.00000000	1.00000000	1.00000000	-.00000000	-1.00000000
.55000000	0.	1.00000000	1.10000000	.21000000	-.86900000	-1.15590000	-.41349000	.71106100	1.19565710	.60416181	-.53107911
.60000000	0.	1.00000000	1.20000000	.44000000	-.67200000	-1.24640000	-.82368000	.25798400	1.13326080	1.10192896	.18905395
.65000000	0.	1.00000000	1.30000000	.69000000	-.40300000	-1.21390000	-1.17507000	-.31369100	.76727170	1.31114421	.93721577
.70000000	0.	1.00000000	1.40000000	.95000000	-.05600000	-1.03840000	-1.39776000	-.91846400	.11191040	1.07513856	1.39328358
.75000000	0.	1.00000000	1.50000000	1.25000000	.37500000	-.68750000	-1.40625000	-1.42187500	-.72656250	.33203125	1.22460938
.80000000	0.	1.00000000	1.60000000	1.56000000	.89600000	-.12640000	-1.09824000	-1.63078400	-1.51101440	-.78683904	.25207194
.85000000	0.	1.00000000	1.70000000	1.89000000	1.51300000	.68210000	-.35343000	-1.28293100	-1.82755270	-1.82390853	-1.27309190
.90000000	0.	1.00000000	1.80000000	2.24000000	2.23200000	1.77760000	.96768000	-.03577600	-1.03207680	-1.82196224	-2.24745523
.95000000	0.	1.00000000	1.90000000	2.61000000	3.05900000	3.20210000	3.02499000	2.54538100	1.81123390	.89596341	-.10890942
1.00000000	0.	1.00000000	2.00000000	3.00000000	4.00000000	5.00000000	6.00000000	7.00000000	8.00000000	9.00000000	10.00000000

$$|\Delta_{2k+1}| = 2\zeta\,|\Delta_{2k-1}| - |\Delta_{2k-3}| \tag{A.6}$$

Similarly, we can generate the other $|\Delta_{2k+1}|$ determinants that are related to the corresponding $T_k(\zeta)$.† The innerwise matrix is represented in

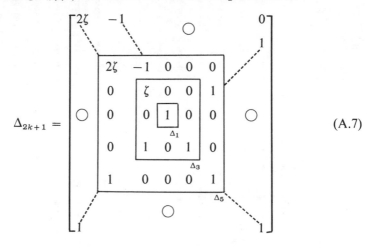

$$\Delta_{2k+1} = \tag{A.7}$$

A.2 INNERWISE MATRIX FOR $U_k(\zeta)$

The innerwise matrix for $U_k(\zeta)$ is similar to that of $T_k(\zeta)$ except for the fact that $|\Delta_3| = U_2 = 2\zeta$ and $U_0 \triangleq 0$. The relationships between the inners determinants $|\Delta_{2k-1}|$ and $U_k(\zeta)$ are given as

$$|\Delta_1| = 1 = U_1 \tag{A.8}$$

$$|\Delta_3| = 2\zeta = U_2 \tag{A.9}$$

$$|\Delta_5| = 2\zeta(2\zeta) - 1 = 4\zeta^2 - 1 = U_3 \tag{A.10}$$

$$|\Delta_7| = 2\zeta(4\zeta^2 - 1) - 2\zeta = 8\zeta^3 - 4\zeta = U_4 \tag{A.11}$$

$$|\Delta_9| = 2\zeta(8\zeta^3 - 4\zeta) - 4\zeta^2 + 1 = 16\zeta^4 - 12\zeta^2 + 1 = U_5 \tag{A.12}$$

$$\vdots \qquad\qquad \vdots$$

$$|\Delta_{2k-1}| = 2\zeta\,|\Delta_{2k-3}| - |\Delta_{2k-5}| \tag{A.13}$$

Similarly, other $|\Delta_{2k-1}|$ determinants are related to $U_k(\zeta)$‡ through the following innerwise matrix for generating the $U_k(\zeta)$:

† The general form of $T_k(\zeta)$ is given in various books [2].
‡ The general form of $U_k(\zeta)$ is given in various books [2].

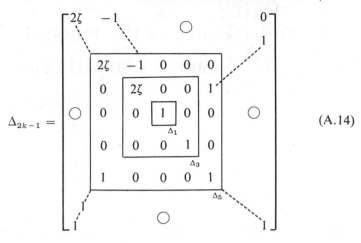

$$\text{(A.14)}$$

REFERENCES

[1] B. D. O. ANDERSON and E. I. JURY, "Stability test for two-dimensional recursive filters," *IEEE Trans. on Audio and Electroacoustic*, **AU-21** (4), 366–372 (Aug. 1973).

[2] D. D. ŠILJAK, *Nonlinear Systems*. New York: Wiley, 1969, pp. 388–390. See also Appendix A.

Proof of the reduced Schur–Cohn Criterion

In Section 3.2 we presented the reduced Schur–Cohn criterion for stability of linear discrete systems without a proof. Although a proof based on geometric distribution of the roots exists in the literature [1], in this appendix we will present a more concise proof based on algebraic consideration of the root distribution. The basic idea of the proof is based on the bilinear transformation from the unit circle in the z-plane into the left half of the complex s-plane. By doing so, we will first show the connection between Hermite and Schur–Cohn criteria, and then, using this connection, we will show the connection between the reduced Hermite and the reduced Schur–Cohn criteria. The former criterion is proven in Section 3.1, and with the developed transformation, a proof of the reduced Schur–Cohn criterion is presented. The Hermite–Schur–Cohn and the reduced Hermite and the reduced Schur–Cohn transformations are also of importance in establishing the connection between Hurwitz matrices and the determinantal criterion, as well as between the Liénard–Chipart criterion and the simplified determinantal criterion for the discrete case. Hence a unified proof for stability of both the continuous and the discrete systems can be established [2].

B.1 SCHUR–COHN–HERMITE TRANSFORMATION

In this transformation we will relate the root distribution relative to the unit circle with the root distribution relative to the imaginary axis. Let $\phi(z) = \sum_{i=0}^{n} a_i z^i$ be a polynomial of n-degree with real or complex coefficients; we are interested in its zero distribution relative to the unit circle. By means of the bilinear transformation (7.72)

$$z = \frac{s+1}{s-1} \quad \text{and} \quad s = \frac{z+1}{z-1} \tag{7.72}$$

we obtain the n-degree polynomial $f(s) = \sum_{i=0}^{n} b_i s^i$ from $\phi(z)$. We are interested in the zero distribution of $f(s)$ relative to the imaginary axis. The Möbius mapping of (7.72) leads to the following relationships [3] between the polynomials $f(s)$ and $\phi(z)$.

$$f(s) = 2^{-n/2}(s - 1)^n \phi\left(\frac{s + 1}{s - 1}\right) \tag{B.1}$$

$$f(-s) = (-1)^n 2^{-n/2}(s + 1)^n \phi\left(\frac{s - 1}{s + 1}\right) \tag{B.2}$$

The relation between the coefficients of $\phi(z)$ and $f(s)$ is given by [2]

$$b' = a'\Gamma^{(n+1)} \tag{B.3}$$

where

$$b' = [b_0 b_1 \cdots b_n] \tag{B.4}$$

$$a' = [a_0 a_1 \cdots a_n] \tag{B.5}$$

and $\Gamma^{(n+1)}$ is an $(n + 1) \times (n + 1)$ matrix of arrays Γ_{ij} and is related to the Q matrix of (7.79) and (7.80) by

$$2^{(n/2)}\Gamma^{(n+1)} = R_{n+1}Q'R_{n+1} \tag{B.6}$$

where R_{n+1} is a matrix of $n + 1$ rows and columns that, when used as a premultiplier, reverses the ordering of rows and, when used as a postmultiplier, reverses the ordering of columns. Thus

$$R_{n+1} = \begin{bmatrix} 0 & 0 & \cdots & 0 & 1 \\ 0 & 0 & \cdots & 1 & 0 \\ \vdots & & & & \\ 0 & & & & \\ & 1 & & & \\ 1 & 0 & \cdots & & 0 \end{bmatrix} \tag{B.7}$$

The matrix Q' denotes the transpose of Q.

If we let S denote the Schur–Cohn matrix as presented in (3.16) and if H denotes the Hermite matrix as presented in (3.2), then, following Fujiwara's [4] transformation, we can state the definitions of S and H.

DEFINITION OF S. *The entries s_{ij} are defined by*

$$\frac{\phi(z)w^n\bar{\phi}(w^{-1}) - \phi(w)z^n\bar{\phi}(z^{-1})}{z - w} = \sum_{i,j=1}^{n} z^{i-1}s_{ij}w^{n-j} \tag{B.8}$$

DEFINITION OF H. *The entries h_{ij} of H are defined by*

$$\frac{f(s)\bar{f}(-t) - f(t)\bar{f}(-s)}{s - t} = \sum_{i,j=1}^{n} s^{i-1}h_{ij}(-t)^{j-1} \tag{B.9}$$

Note that in both definitions the ¯ denotes the conjugates form with respect to coefficients. Using (7.72), (B.1), and (B.2), and setting $w = (t + 1)/(t - 1)$, we obtain for the left-hand side of (B.8) the following

$$\frac{\phi(z)w^n\bar{\phi}(w^{-1}) - \phi(w)z^n\bar{\phi}(z^{-1})}{z - w}$$

$$= \frac{\phi\left(\dfrac{s + 1}{s - 1}\right)\left(\dfrac{t + 1}{t - 1}\right)^n\bar{\phi}\left(\dfrac{t - 1}{t + 1}\right) - \phi\left(\dfrac{t + 1}{t - 1}\right)\left(\dfrac{s + 1}{s - 1}\right)^n\bar{\phi}\left(\dfrac{s - 1}{s + 1}\right)}{\left(\dfrac{s + 1}{s - 1}\right) - \left(\dfrac{t + 1}{t - 1}\right)}$$

$$= \frac{(-1)^{n-1}2^{n-1}}{(s - 1)^{n-1}(t - 1)^{n-1}}\frac{f(s)\bar{f}(-t) - f(t)\bar{f}(-s)}{s - t} \qquad \text{(B.10)}$$

Noting the right-hand side of (B.9), we obtain from (B.10) the following:

$$\frac{\phi(z)w^n\bar{\phi}(w^{-1}) - \phi(w)z^n\bar{\phi}(z^{-1})}{z - w}$$

$$= \frac{(-1)^{n-1}2^{n-1}}{(s - 1)^{n-1}(t - 1)^{n-1}}\sum_{k,\ell=1}^{n} s^{k-1}h_{k\ell}(-t)^{\ell-1} \qquad \text{(B.11)}$$

The right-hand side of (B.8) yields

$$\sum_{i,j=1}^{n} z^{i-1}s_{ij}w^{n-j}$$

$$= \sum_{i,j=1}^{n}\left(\frac{s + 1}{s - 1}\right)^{i-1}s_{ij}\left(\frac{t + 1}{t - 1}\right)^{n-j}$$

$$= \frac{1}{(s - 1)^{n-1}(t - 1)^{n-1}}\sum_{i,j=1}^{n}(s + 1)^{i-1}(s - 1)^{n-i}s_{ij}(t + 1)^{n-j}(t - 1)^{j-1}$$

$$\text{(B.12)}$$

Let $\Gamma^{(n)}$ be the $n \times n$ matrix of arrays Γ_{ij}, mapping coefficients of $(n - 1)$-dimensional polynomials into one another, similar to (B.3). We also note that the matrix Γ_{ij} is defined by [3]

$$2^{-n/2}(w + 1)^i(w - 1)^{n-i} = \sum_{j=0}^{n}\Gamma_{ij}w^j$$

$$i = 0, 1, \ldots, n \qquad \text{(B.13)}$$

Using (B.13), we have

$$2^{-(n-1)/2}(s - 1)^{n-i}(s + 1)^{i-1} = 2^{-(n-1)/2}(s - 1)^{(n-1)-(i-1)}(s + 1)^{i-1}$$

$$= \sum_{k=1}^{n}\Gamma_{ik}^{(n)}s^{k-1} \qquad i = 1, 2, \ldots, n \qquad \text{(B.14)}$$

and also

$$2^{-(n-1)/2}(t+1)^{n-j}(t-1)^{j-1} = 2^{-(n-1)/2}(-1)^{n-1}[(-t)-1]^{n-j}[(-t+1)]^{j-1}$$

$$= (-1)^{n-1} \sum_{\ell=1}^{n} \Gamma_{j\ell}^{(n)}(-t)^{\ell-1}$$

$$j = 1, 2, \ldots, n \quad \text{(B.15)}$$

Using (B.14) and (B.15) in the right-hand side of (B.12), we obtain for the left-hand side

$$\sum_{i,j=1}^{n} z^{i-1} s_{ij} w^{n-j}$$

$$= \frac{2^{n-1}(-1)^{n-1}}{(s-1)^{n-1}(t-1)^{n-1}} \sum_{i,j=1}^{n} \left[\sum_{k=1}^{n} \Gamma_{ik}^{(n)} s^{k-1} \right] s_{ij} \left[\sum_{\ell=1}^{n} \Gamma_{j\ell}^{(n)}(-t)^{\ell-1} \right]$$

$$= \frac{2^{n-1}(-1)^{n-1}}{(s-1)^{n-1}(t-1)^{n-1}} \sum_{k,\ell=1}^{n} s^{k-1} \left[\sum_{i,j=1}^{n} \Gamma_{ik}^{(n)} s_{ij} \Gamma_{j\ell}^{(n)} \right] (-t)^{\ell-1}$$

$$\text{(B.16)}$$

Comparing (B.11) and (B.16), and noting (B.8), we have

$$h_{k\ell} = \sum_{i,j=1}^{n} \Gamma_{ik}^{(n)} s_{ij} \Gamma_{j\ell}^{(n)} \tag{B.17}$$

or
$$H = \Gamma^{(n)'} S \Gamma^{(n)} \tag{B.18}$$

Noting the symmetry in the expression

$$\sum_{i=1}^{n} \sum_{j=1}^{n} a_i \Gamma_{ij} \Gamma_{jk} = a_k \tag{B.19}$$

and since the a_i are arbitrary, it follows that $\Gamma^{(n)}$ is idempotent [3]. Therefore (B.18) gives

$$S = \Gamma^{(n)'} H \Gamma^{(n)} \tag{B.20}$$

Equations (B.18) and (B.19) establish the connection between H and S, and vice versa. If one of these matrices is positive definite, so is the other.

The formulas of (B.18) and (B.20) show that any zero distribution property for $f(s)$ that is describable in terms of the signs of the eigenvalues of H gives rise to a parallel property for $\phi(z)$, describable in terms of the signs of the eigenvalues [4] of S.

Illustration Example B.1. To explain the various relationships derived earlier, we will present the following example for $n = 4$: Let

$$\phi(z) = a_4 z^4 + a_3 z^3 + a_2 z^2 + a_1 z + a_0 \tag{B.21}$$

From (B.1), we have

$$f(s) = 2^{-2}\phi\left(\frac{s+1}{s-1}\right)(s-1)^4 = \sum_{i=0}^{4} b_i s^i \tag{B.22}$$

From (B.3), we have

$$[b_0 b_1 b_2 b_3 b_4] = [a_0 a_1 a_2 a_3 a_4]\Gamma^{(5)} \tag{B.23}$$

The matrix $\Gamma^{(5)}$ is obtained from (7.87) and (B.6) to give
$\Gamma^{(5)} = 2^{-2}[R_5 Q' R_5]$

$$= 2^{-2}\begin{bmatrix} & & & & 1 \\ & & & 1 & \\ & & 1 & & \\ & 1 & & 0 & \\ 1 & & & & \end{bmatrix}\begin{bmatrix} 1 & 4 & 6 & 4 & 1 \\ 1 & 2 & 0 & -2 & -1 \\ 1 & 0 & -2 & 0 & 1 \\ 1 & -2 & 0 & 2 & -1 \\ 1 & -4 & 6 & -4 & 1 \end{bmatrix}\begin{bmatrix} & & & & 1 \\ & & & 1 & \\ & & 1 & & \\ & 1 & & 0 & \\ 1 & & & & \end{bmatrix}$$

$$= 2^{-2}\begin{bmatrix} 1 & 4 & 6 & 4 & 1 \\ -1 & 2 & 0 & -2 & 1 \\ 1 & 0 & -2 & 0 & 1 \\ -1 & -2 & 0 & 2 & 1 \\ 1 & 4 & 6 & 4 & 1 \end{bmatrix} \tag{B.24}$$

The b_i's (within a constant) are obtained from (B.23) and (B.24) to give

$$\begin{aligned} b_0 &= a_0 - a_1 + a_2 - a_3 + a_4 \\ b_1 &= -4a_0 + 2a_1 - 2a_3 + 4a_4 \\ b_2 &= 6a_0 - 2a_2 + 6a_4 \\ b_3 &= -4a_0 - 2a_1 + 2a_3 + 4a_4 \\ b_4 &= a_0 + a_1 + a_2 + a_3 + a_4 \end{aligned} \tag{B.25}$$

Noting (3.2) and that $P = R_n H R_n$, we have for $f(s)$

$$H = \begin{bmatrix} b_0 b_1 & 0 & b_0 b_3 & 0 \\ 0 & -b_0 b_3 + b_1 b_2 & 0 & b_1 b_4 \\ b_0 b_3 & 0 & -b_1 b_4 + b_2 b_3 & 0 \\ 0 & b_1 b_4 & 0 & b_3 b_4 \end{bmatrix} \tag{B.26}$$

Now to obtain S using (B.20), we must first obtain $\Gamma^{(4)}$. This matrix is obtained in the same way as $\Gamma^{(5)}$ except that in this case the Q matrix obtained from (7.80) is as follows:

$$
Q = \begin{bmatrix} 1 & 1 & 1 & 1 \\ 3 & 1 & -1 & -3 \\ 3 & -1 & -1 & 3 \\ 1 & -1 & 1 & -1 \end{bmatrix}
\tag{B.27}
$$

The corresponding $\Gamma^{(4)}$ yields

$$
\Gamma^{(4)} = 2^{-3/2} \begin{bmatrix} -1 & 3 & -3 & 1 \\ 1 & -1 & -1 & 1 \\ -1 & -1 & 1 & 1 \\ 1 & 3 & 3 & 1 \end{bmatrix}
\tag{B.28}
$$

Substituting (B.26) and (B.28) in (B.20), we obtain (within a constant)

$$
S = \Gamma^{(4)\prime} H \Gamma^{(4)} = \begin{bmatrix} a_4^2 - a_0^2 & a_3 a_4 - a_1 a_0 & a_2 a_4 - a_2 a_0 & a_1 a_4 - a_3 a_0 \\ a_3 a_4 - a_1 a_0 & a_4^2 + a_3^2 - a_1^2 - a_0^2 & a_3 a_4 + a_2 a_3 - a_2 a_1 - a_1 a_0 & a_2 a_4 - a_2 a_0 \\ a_2 a_4 - a_2 a_0 & a_3 a_4 + a_2 a_3 - a_2 a_1 - a_1 a_0 & a_4^2 + a_3^2 - a_1^2 - a_0^2 & a_3 a_4 - a_1 a_0 \\ a_1 a_4 - a_3 a_0 & a_2 a_4 - a_2 a_0 & a_3 a_4 - a_1 a_0 & a_4^2 - a_0^2 \end{bmatrix}
\tag{B.29}
$$

Noting (3.16), the above matrix is the Schur–Cohn matrix S corresponding to the fourth-degree polynomial $\phi(z)$.

B.2 REDUCED SCHUR–COHN TO REDUCED HERMITE RELATIONSHIPS

In the following discussion we will assume that the polynomials $f(s)$ and $\phi(z)$ are real. By utilizing the Schur–Cohn to Hermite transformation shown earlier, we will prove that the reduced Schur–Cohn matrices $B > 0$, $A > 0$ of (3.21) and (3.20) correspond to the reduced Hermite matrices $C > 0$ and $D > 0$ of (3.3) and (3.4), and vice versa. Such relationships enable us to complete the proof of the reduced Schur–Cohn criterion.

Let $f(s) = \sum_{i=0}^{n} b_i s^i$ be the real polynomial whose zeros in the open left half-plane are under consideration. The matrix $H = R_n P R_n$ corresponding to $f(s)$ can be written, following (B.9) and (3.2) as follows:

$$
h_{ij} = \sum_{k=1}^{i} b_{k-1} b_{i+j-k} (-1)^{k+i} \qquad \begin{array}{l} i \le j, \quad i+j \text{ even} \\ \end{array}
$$
$$
= 0 \qquad\qquad\qquad\qquad i + j = \text{odd} \tag{B.30}
$$

Define a matrix M by

$$M' = \begin{bmatrix} 1 & 0 & 0 & & \cdots & & 0 & \cdots \\ 0 & 0 & 1 & 0 & \cdots & & 0 & \cdots \\ 0 & 0 & 0 & 0 & 1 & \cdots & 0 & \cdots \\ \multicolumn{8}{c}{\cdots\cdots\cdots\cdots\cdots\cdots\cdots\cdots} \\ 0 & 1 & 0 & & \cdots & & 0 & \cdots \\ 0 & 0 & 0 & 1 & & \cdots & 0 & \cdots \\ \vdots & \vdots & \vdots & \vdots & & & \vdots \end{bmatrix} \qquad \text{(B.31)}$$

When $n = 2m$, the top and bottom halves of M' have n rows each. When $n = 2m + 1$, the top part has $m + 1$ rows, the bottom part m rows. Evidently premultiplication of an $n \times n$ matrix by M' moves the odd-numbered rows to the first m (or $m + 1$) rows and the even-numbered rows to the last m rows Postmultiplication by M produces the same operation on the columns. Thus

$$M'HM = \left[\begin{array}{c|c} \tilde{C} & \\ \hline & \tilde{D} \end{array}\right] \qquad \text{(B.31a)}$$

where $\tilde{D}$ is an $m \times m$ matrix and $\tilde{C}$ either an $(m + 1) \times (m + 1)$ matrix when n is odd or an $m \times m$ matrix when n is even. The matrices $\tilde{C}$ and $\tilde{D}$ are simply related to the reduced Hermite ones discussed in (3.3) and (3.4).

Let $\phi(z) = \sum_{i=0}^{n} a_i z^i$ be a real polynomial whose roots inside the unit circle are to be established. Following (B.8) and (3.16), the elements of the Schur–Cohn matrix are given by

$$S_{ij} = \sum_{p=1}^{\min(i,j)} a_{n-i+p}a_{n-j+p} - a_{i-p}a_{j-p}$$

$$i, j = 1, 2, \ldots, n \quad \text{(B.32)}$$

We define R_m as in (B.7) and I_m as the identity matrix, and by using the matrices A and B as in (3.20) and (3.21), we can write

$$S = \frac{1}{2} \begin{bmatrix} I_m & -I_m \\ R_m & R_m \end{bmatrix} \begin{bmatrix} A & 0 \\ 0 & B \end{bmatrix} \begin{bmatrix} I_m & R_m \\ -I_m & R_m \end{bmatrix} \qquad \text{(B.33)}$$

and

$$S = \frac{1}{2} \begin{bmatrix} I_m & 0 & -I_m \\ 0 & 1 & 0 \\ R_m & 0 & R_m \end{bmatrix} \left[\begin{array}{c|c} A & 0 \\ \hline 0 & B \end{array}\right] \begin{bmatrix} I_m & 0 & R_m \\ 0 & 1 & 0 \\ -I_m & 0 & R_m \end{bmatrix}$$

$$\text{for } n = 2m + 1 \quad \text{(B.34)}$$

The matrices A and B are the reduced Schur–Cohn matrices. Based on the symmetric and skew-symmetric properties of the matrix Q as shown in Section 7.3, equations (7.80), (7.90), (7.91) and the relationship of (B.6), it follows that the matrix $\Gamma^{(n)}$ has the same properties. Thus for $\Gamma^{(n)}$ we can write

$$2^{(n-1)/2}M'\,\Gamma^{(n)\prime} \triangleq \left[\begin{array}{c|c} -X & XR_m \\ \hline Y & YR_m \end{array}\right] \quad \text{for } n = 2m \qquad \text{(B.35)}$$

Since M and $\Gamma^{(n)}$ are nonsingular, clearly X and Y are also.

Now we are ready to show the reduced Schur–Cohn to reduced Hermite connections.

(a) Even case, $n = 2m$. From (B.18) and (B.31), we can write

$$H = \Gamma^{(n)\prime}S\Gamma^{(n)} \qquad \text{(B.36)}$$

and $$M'HM = M'\Gamma^{(n)\prime}S\Gamma^{(n)}M \qquad \text{(B.37)}$$

Noting (B.31), (B.33), and (B.35), we have

$$\left[\begin{array}{c|c} \tilde{C} & 0 \\ \hline 0 & \tilde{D} \end{array}\right] = \beta \left[\begin{array}{c|c} -X & XR_m \\ \hline Y & YR_m \end{array}\right]\left[\begin{array}{cc} I_m & -I_m \\ R_m & R_m \end{array}\right]$$

$$\left[\begin{array}{cc} A & 0 \\ 0 & B \end{array}\right]\left[\begin{array}{cc} I_m & R_m \\ -I_m & R_m \end{array}\right]\left[\begin{array}{cc} -X' & Y' \\ R_mX' & R_mY' \end{array}\right]$$

$$\qquad \text{(B.38)}$$

or

$$\left[\begin{array}{c|c} \tilde{C} & 0 \\ \hline 0 & \tilde{D} \end{array}\right] = \beta \left[\begin{array}{cc} 0 & X \\ Y & 0 \end{array}\right]\left[\begin{array}{cc} A & 0 \\ 0 & B \end{array}\right]\left[\begin{array}{cc} 0 & Y' \\ X' & 0 \end{array}\right] \qquad \text{(B.39)}$$

or

$$\left[\begin{array}{c|c} \tilde{C} & 0 \\ \hline 0 & \tilde{D} \end{array}\right] = \left[\begin{array}{cc} \beta XBX' & 0 \\ 0 & \beta YAY' \end{array}\right] \qquad \text{(B.40)}$$

Hence

$$\tilde{C} > 0 \Leftrightarrow B > 0 \qquad \text{(B.41)}$$

$$\tilde{D} > 0 \Leftrightarrow A > 0 \qquad \text{(B.42)}$$

(b) Odd case, $n = 2m + 1$. Using the symmetry of Γ in rows and columns, we have

$$
M'\Gamma' \triangleq \left[\begin{array}{c|c} X_{m+1} & X_{m+1}\begin{bmatrix} R_m \\ 0 \end{bmatrix} \\ \hline [-Y_m \quad \vdots \quad 0] & Y_m R_m \end{array} \right]
\tag{B.43}
$$

and thus for $n = 2m + 1$

$$
M'HM = \left[\begin{array}{c|c} \tilde{C} & 0 \\ \hline 0 & \tilde{D} \end{array} \right] = M'\Gamma'S\Gamma M
\tag{B.44}
$$

Hence $\tilde{C}$ is an $(m + 1) \times (m + 1)$ matrix and $\tilde{D}$ an $m \times m$ matrix. Using (B.34) and (B.43) in (B.44), we obtain

$$
\left[\begin{array}{c|c} \tilde{C} & 0 \\ \hline 0 & \tilde{D} \end{array} \right] = \left[\begin{array}{c|c} X_{m+1}\begin{bmatrix} \alpha I_m & 0 \\ 0 & 1 \end{bmatrix} & 0 \\ \hline 0 & Y_m \end{array} \right] \left[\begin{array}{c|c} A & 0 \\ \hline 0 & B \end{array} \right] \left[\begin{array}{c|c} \begin{bmatrix} \alpha I_m & 0 \\ 0 & 1 \end{bmatrix} X'_{m+1} & 0 \\ \hline 0 & Y'_m \end{array} \right]
\tag{B.45}
$$

where α is a positive constant.

The above equation can also be written as

$$
\left[\begin{array}{c|c} \tilde{C} & 0 \\ \hline 0 & \tilde{D} \end{array} \right] = \left[\begin{array}{c|c} X_{m+1}\begin{bmatrix} \alpha I_m & 0 \\ 0 & 1 \end{bmatrix} A \begin{bmatrix} \alpha I_m & 0 \\ 0 & 1 \end{bmatrix} X'_{m+1} & 0 \\ \hline 0 & Y_m B Y'_m \end{array} \right]
\tag{B.46}
$$

The matrices X_{m+1} and Y_{m+1} are nonsingular for the same reasons as X_m and Y_m, and hence

$$
\tilde{C} > 0 \Leftrightarrow A > 0
\tag{B.47}
$$

$$
\tilde{D} > 0 \Leftrightarrow B > 0
\tag{B.48}
$$

The formulas (B.41), (B.42), (B.47), and (B.48) establish the connection between the reduced Schur–Cohn matrices and the reduced Hermite matrices. In particular, if one pair is positive definite, so is the other. Therefore by proving one of these two criteria for stability, the other follows directly. Since in Chapter 3 we proved the reduced Hermite criterion for the open left half of the s-plane, we have now completed the proof of the reduced Schur–Cohn criterion for stability within the unit circle in the z-plane.

Example B.2. The following example illustrates the reduced Hermite to reduced Schur–Cohn connection for n-even. Let $f(s)$ be given as

$$f(s) = \sum_{i=0}^{4} b_i s^i \tag{B.49}$$

From (B.31) and (B.40), we can write

$$M'HM = \left[\begin{array}{c|c} \tilde{C} & 0 \\ \hline 0 & \tilde{D} \end{array}\right] = \beta \left[\begin{array}{c|c} XBX' & 0 \\ \hline 0 & YAY' \end{array}\right] \tag{B.50}$$

The matrix M' for this case is given from (B.31) as follows:

$$M' = \begin{bmatrix} 1 & 0 & 0 & 0 \\ 0 & 0 & 1 & 0 \\ 0 & 1 & 0 & 0 \\ 0 & 0 & 0 & 1 \end{bmatrix} \tag{B.51}$$

The matrix $\Gamma^{(4)}$ is given in (B.28). Hence

$$M'\Gamma^{(4)'} = 2^{-3/2} \left[\begin{array}{cc|cc} -1 & 1 & -1 & 1 \\ -3 & -1 & 1 & 3 \\ \hline 3 & -1 & -1 & 3 \\ 1 & 1 & 1 & 1 \end{array}\right] \tag{B.52}$$

From (B.35), we have

$$-X = \begin{bmatrix} -1 & 1 \\ -3 & -1 \end{bmatrix} \tag{B.53}$$

or

$$X = \begin{bmatrix} 1 & -1 \\ 3 & 1 \end{bmatrix} \tag{B.53}$$

Now we can check $\tilde{C} = \beta XBX'$. For $\phi(z) = \sum_{i=0}^{4} a_i z^i$ we have from (3.23) the following:

$$B = \begin{bmatrix} a_4^2 - a_4 a_1 + a_0 a_3 - a_0^2 & a_4 a_3 - a_4 a_2 + a_0 a_2 - a_0 a_1 \\ a_4 a_3 - a_4 a_2 + a_0 a_2 - a_0 a_1 & a_4^2 - a_4 a_3 + a_3^2 - a_3 a_2 - a_1^2 + a_1 a_2 - a_0^2 + a_0 a_1 \end{bmatrix} \tag{B.54}$$

Using (B.54) and (B.53), we obtain

$$2XBX' = \begin{bmatrix} b_0 b_1 & b_0 b_3 \\ b_0 b_3 & -b_1 b_4 + b_2 b_3 \end{bmatrix} \tag{B.55}$$

where the b_i's are those given in (B.25). Noting (3.4), the above is equivalent to R_2DR_2. Hence

$$B > 0 \Leftrightarrow D > 0 \tag{B.56}$$

Similarly noting (3.3), we can verify that

$$R_2CR_2 = \tilde{D} = 2YAY', \quad \text{or} \quad A > 0 \Leftrightarrow C > 0$$

where Y is obtained from (B.52) and (B.35) as follows:

$$Y = \begin{bmatrix} 3 & -1 \\ 1 & -1 \end{bmatrix}$$

and A is obtained from (3.22) for $\phi(z) = \sum_{i=0}^{4} a_i z^i$.

REFERENCES

[1] B. D. O. ANDERSON and E. I. JURY, "A simplified Schur–Cohn test," *IEEE Trans. on Automatic Control*, **AC-18** (2), 157–163 (April 1973).

[2] B. D. O. ANDERSON and E. I. JURY, "On the reduced Hermite and reduced Schur–Cohn matrix relationships," *Int. J. Control, England*, May (1974). **19** (5), 877–890.

[3] R. J. DUFFIN, "Algorithms for classical stability problems," *SIAM Rev.*, **11** (2), 196–213 (1969).

[4] M. FUJIWARA, "Über die Algebraischen Gleichungen deren Wurzeln in einem Kreise oder in einer Halbebene liegen," *Math. Z.*, **24**, 160–169 (1926).

Proof of Stability Conditions for the A Matrix

In this appendix we will present the proof of Theorem 3.15 discussed in Chapter 3. The proof follows that of Fuller [1] and is based on a theorem by Routh [2] on stability and a theorem by Stephanos [3] on the characteristic roots of the A matrix. Also, in this appendix we will present a simple procedure [4] for generating the $\tilde{A}$ matrix of (3.234). Finally, conditions for the eigenvalues of the A matrix to be distinct and negative real (aperiodicity condition of Section 2.2) and to have magnitude less than unity (stability within the unit circle, Section 2.3) will be presented. Also discussed is the relative stability of the A matrix.

THEOREM C.1 (ROUTH [2]). *Let the equation*

$$p_n \lambda^n + p_{n-1} \lambda^{n-1} + \cdots + p_0 = 0 \qquad (C.1)$$

have real coefficients and let $p_n > 0$. Let

$$s_m \mu^m + s_{m-1} \mu^{m-1} + \cdots + s_0 = 0 \qquad (C.2)$$

be the equation of root pair sums of (C.1)—that is, let the roots of (C.2) be the $m = \frac{1}{2}n(n-1)$ values

$$\mu = \lambda_i + \lambda_j$$
$$i = 2, 3, \ldots, n, \quad j = 1, 2, \ldots, i-1 \quad (C.3)$$

where $\lambda_1, \lambda_2, \ldots, \lambda_n$ are the roots of (C.1). Let $s_m > 0$. Then for the roots of (C.1) to have their real parts negative, it is necessary and sufficient that the coefficients $p_0, p_1, \ldots, p_{n-1}$ and $s_0, s_1, \ldots, s_{m-1}$ should all be positive.

Proof: Suppose that the real roots of (C.1) are $\alpha_1, \alpha_2, \ldots$ and the complex roots are $\beta_1 \pm j\gamma_1, \beta_2 \pm j\gamma_2, \ldots$. Then (C.1) is

$$p_n(\lambda - \alpha_1)(\lambda - \alpha_2)\cdots(\lambda^2 - 2\beta_1\lambda + \beta_1^2 + \gamma_1^2)$$
$$\times (\lambda^2 - 2\beta_2\lambda + \beta_2^2 + \gamma_2^2)\cdots = 0 \quad (C.4)$$

If $\alpha_1, \alpha_2, \ldots$ and $\beta_1, \beta_2, \ldots$ are all negative, the factors in (C.4) are all polynomials (of degree 0, 1, or 2) with positive coefficients. Thus when multiplied out, (C.4) is a polynomial with positive coefficients. Hence it is necessary that $p_0, p_1, \ldots, p_{n-1}$ should be positive. If the real parts of the roots of (C.1) are all negative, then, from (C.3), so are the real parts of the roots of (C.2). Hence, by the same reasoning, it is necessary that $s_0, s_1, \ldots,$ s_{m-1} should be positive.

To prove sufficiency, note that if $p_0, p_1, \ldots, p_{n-1}$ are all positive, the left side of (C.1) is positive for $\lambda \geq 0$. That is, (C.1) has no positive or zero real roots. Similarly, if $s_0, s_1, \ldots, s_{m-1}$ are all positive, (C.2) has no positive or zero real roots. But from (C.3) the real roots of (C.2) include the values $2\beta_1, 2\beta_2, \ldots,$ that is, twice the real parts of the complex roots of (C.1). Hence if $p_0, p_1, \ldots, p_{n-1}, s_0, s_1, \ldots, s_{m-1}$ are all positive, the real parts of the roots of (C.1) are all negative. This completes the proof of Theorem C.1.

Note that the number of stability conditions of the theorem is $\frac{1}{2}n(n + 1)$. Hence a certain redundancy exists, for there should be n conditions. If we try to reduce the number of constraints, we obtain the equivalent of the Liénard–Chipart criterion, since, as Routh [2] showed, $s_0, \ldots, s_{n-1}$ can be expressed as functions of $p_0, p_1, \ldots, p_n$. Thus for the stability test of (C.1), the Liénard–Chipart criterion is the simplest form and yields about n conditions. However, for the stability of the A matrix, the Routh theorem is of much importance when we proceed to prove Theorem 3.15.

THEOREM C.2 (STEPHANOS [3]). *The characteristic roots of the matrix*

$$\tilde{A} = 2A \cdot I_n \tag{C.5}$$

are the $\frac{1}{2}n(n - 1)$ values

$$\lambda_i + \lambda_j$$
$$i = 2, 3, \ldots, n, \quad j = 1, 2, \ldots, i - 1 \tag{C.6}$$

(where the $\cdot$ denotes the bialternate product of $2A$ and I_n). The bialternate product is defined as follows. Let A be an n-dimensional matrix (a_{ij}) and B an n-dimensional matrix (b_{ij}). Let F be an $m = \frac{1}{2}n(n - 1)$-dimensional matrix $(f_{pq,rs})$ whose rows are labeled pq $(p = 2, 3, \ldots, n, q = 1, 2, \ldots, p - 1)$,

whose columns are labeled rs $(r = 2, 3, \ldots, n, s = 1, 2, \ldots, r - 1)$, and whose elements are

$$f_{pq,rs} = \frac{1}{2}\left[\begin{vmatrix} a_{pr} & a_{ps} \\ b_{qr} & b_{qs} \end{vmatrix} + \begin{vmatrix} b_{pr} & b_{ps} \\ a_{qr} & a_{qs} \end{vmatrix} \right] \tag{C.7}$$

Then F is the bialternate product of A and B.

Proof: Assume first that the matrix A has distinct characteristic roots $\lambda_1, \lambda_2, \ldots, \lambda_n$. For each λ_i there is a corresponding characteristic vector (eigenvector) [4] $(x_1^i, x_2^i, \ldots, x_n^i)$. That is,

$$\lambda_i x_p^i = a_{p1} x_1^i + a_{p2} x_2^i + \cdots + a_{pn} x_n^i$$
$$i = 1, 2, \ldots, n, \quad p = i, 2, \ldots, n \quad \text{(C.8)}$$

Let us define

$$y_{pq}^{ij} = \begin{vmatrix} x_p^i & x_p^j \\ x_q^i & b_q^j \end{vmatrix}$$
$$i = 2, 3, \ldots, n, \quad j = 1, 2, \ldots, i - 1$$
$$p = 1, 2, \ldots, n, \quad q = 1, 2, \ldots, n \quad \text{(C.9)}$$

Then

$$(\lambda_i + \lambda_j) y_{pq}^{ij} = \begin{vmatrix} \lambda_i x_p^i & x_p^j \\ \lambda_i x_q^i & x_q^j \end{vmatrix} + \begin{vmatrix} x_p^i & \lambda_j x_p^j \\ x_q^i & \lambda_j x_q^j \end{vmatrix} \quad \text{(C.10)}$$

Expanding the two determinants and then regrouping into two different determinants, we write (C.10) as

$$(\lambda_i + \lambda_j) y_{pq}^{ij} = \begin{vmatrix} \lambda_i x_p^i & \lambda_j x_p^j \\ x_q^i & x_q^j \end{vmatrix} + \begin{vmatrix} x_p^i & x_p^j \\ \lambda_i x_q^i & \lambda_j x_q^j \end{vmatrix} \quad \text{(C.11)}$$

Substitution from (C.8) in (C.11) yields

$$(\lambda_i + \lambda_j) y_{pq}^{ij} = \sum_{\ell=1}^{n} a_{p\ell} \begin{vmatrix} x_\ell^i & x_\ell^j \\ x_q^i & x_q^j \end{vmatrix} + \sum_{\ell=1}^{n} a_{q\ell} \begin{vmatrix} x_p^i & x_p^j \\ x_\ell^i & x_\ell^j \end{vmatrix} \quad \text{(C.12)}$$

Noting the definition of (C.3), we can write (C.12) as

$$(\lambda_i + \lambda_j) y_{pq}^{ij} = \sum_{\ell=1}^{n} a_{p\ell} y_{\ell q}^{ij} + \sum_{\ell=1}^{n} a_{q\ell} y_{p\ell}^{ij} \quad \text{(C.13)}$$

Let us define y as a column vector with $\frac{1}{2} n(n - 1)$ components y_{pq} ($p = 2, 3, \ldots, n, q = 1, 2, \ldots, p - 1$). Then (C.13) is one component of the vector equation

$$(\lambda_i + \lambda_j) y^{ij} = \tilde{A} y^{ij} \quad \text{(C.14)}$$

where $\tilde{A}$ is square matrix of dimension $\frac{1}{2} n(n - 1)$ and whose elements will be defined later on. Equation (C.14), which holds for $i = 2, 3, \ldots, n, j = 1, 2, \ldots, i - 1$, states that the characteristic vectors of $\tilde{A}$ are the y^{ij} and the characteristic roots of $\tilde{A}$ are $\lambda_i + \lambda_j$.

In order to specify the elements $a_{pq,rs}$ of $\tilde{A}$, we use the relations below; which follow from (C.9).

$$y_{rs}^{ij} = -y_{sr}^{ij} \quad \text{(C.15)}$$

and

$$y_{rr}^{ij} = 0 \quad \text{(C.16)}$$

Equations (C.15) and (C.16) enable us to write the right side of (C.13) as a linear combination of the components y_{rs}^{ij} with $r > 0$, in conformity with the above-stated definition of the vector y. Thus (C.13) is

$$(\lambda_i + \lambda_j) y_{pq}^{ij} = -\sum_{\ell=1}^{q-1} a_{p\ell} y_{q\ell} + \sum_{q+1}^{n} a_{p\ell} y_{\ell q}^{ij} + \sum_{\ell=1}^{p-1} a_{q\ell} y_{p\ell}^{ij} - \sum_{p+1}^{n} a_{q\ell} y_{\ell p}^{ij}$$

(C.17)

The coefficient of y_{rs}^{ij} on the right side of (C.17) is (with $p > q$ and $r > s$)

$$\tilde{a}_{pq,rs} = \begin{cases} -a_{ps} & \text{if } r = q \\ a_{pr} & \text{if } r \neq p \text{ and } s = q \\ a_{pp} + a_{qq} & \text{if } r = p \text{ and } s = q \\ a_{qs} & \text{if } r = p \text{ and } s \neq q \\ -a_{qr} & \text{if } s = p \\ 0 & \text{otherwise} \end{cases}$$

(C.18)

With the rows of $\tilde{A}$ labeled pq ($p = 2, 3, \ldots, n, q = 1, 2, \ldots, p - 1$) and the columns $\tilde{A}$ labeled rs ($r = 2, 3, \ldots, n, s = 1, 2, \ldots, r - 1$), (C.18) gives the element at the intersection of the pqth row and the rsth column. A simple procedure for generating the matrix $\tilde{A}$ from the matrix A will be presented later.

To complete the proof of Theorem C.2, we have to verify that the matrix $\tilde{A}$ determined by (C.18) is the same as the $\tilde{A}$ matrix defined in (C.5). To show this, we can write, following (C.7), the general element of $\tilde{A}$ of (C.5) as

$$\tilde{a}_{pq,rs} = \begin{vmatrix} a_{pr} & a_{ps} \\ \delta_{qr} & \delta_{qs} \end{vmatrix} + \begin{vmatrix} \delta_{pr} & \delta_{ps} \\ a_{qr} & a_{qs} \end{vmatrix}$$

(C.19)

where δ_{ij} is the Kronecker delta defined by

$$\delta_{ij} = \begin{cases} 0 & \text{if } i \neq j \\ 1 & \text{if } i = j \end{cases}$$

(C.20)

If $r = q$ in (C.19), δ_{qr} is unity and the other δ's are zero (since $s < r$ and $p > q$), so that (C.19) is then

$$\tilde{a}_{pq,rs} = -a_{ps} \qquad \text{if } r = q \quad \text{(C.21)}$$

Equation (C.21) checks with the first line of (C.18). In the same way, it may be checked by inspection that (C.19) is the same as (C.18) for all other cases listed in the latter equation. Therefore the matrix $\tilde{A}$ of the theorem is the same as that determined by (C.18). We have thus proved the theorem for the case when $\lambda_1, \lambda_2, \ldots, \lambda_n$ are distinct. Since the characteristic roots of a matrix are continuous functions of its elements, Theorem C.2 is also valid when $\lambda_1, \lambda_2, \ldots, \lambda_n$ are not distinct. This completes the proof of the theorem. It may be noted that (C.5) can also be written as

$$\tilde{A} = A \cdot I_n + I_n \cdot A \tag{C.22}$$

which indicates that the $\tilde{A}$ is the bialternate sum of A with itself.

By noting that the characteristic polynomials of A and $\tilde{A}$ are the polynomials in (C.1) and (C.2), respectively, it follows from Theorems C.1 and C.2 and from (C.18) that Theorem 3.15 is immediate.

GENERATION OF THE MATRIX $\tilde{A}$ FROM THE ELEMENTS OF A

In Chapter 4 we obtained a method of obtaining the matrix B of (4.68) and of dimension $\frac{1}{2}n(n + 1)$ from the matrix A. This has been shown for the example of (4.69), where B' is obtained from A. Following a procedure discussed by Barnett [4], we can readily obtain $\tilde{A}$ from B'. This procedure involves the following steps.

1. The rows and columns of B' numbered 1, 3, 6, 10, ..., $\frac{1}{2}n(n + 1)$ must be deleted. After performing this operation, the new matrix is of dimension $\frac{1}{2}n(n - 1)$, which is of the same dimension as $\tilde{A}$.

2. Insert negative signs symmetrically in the positions

row	columns	
1	3, 5, 8, 12, 17, 23, . . .	
2, 3	6, 9, 13, 18, 24, . . .	
4, 5, 6	10, 14, 19, 25, . . .	(C.22a)
7, 8, 9, 10	15, 20, 111	
⋮	⋮	

together with the same arrangement but with "rows" and "columns" interchanged.

Example C.1. To illustrate the above procedure, we apply steps 1 and 2 to B' as given in (4.69) and rewritten below.

$$B' = \begin{bmatrix} 2a_{11} & a_{12} & 0 & a_{13} & 0 & 0 \\ 2a_{21} & a_{11} + a_{22} & 2a_{12} & a_{23} & a_{13} & 0 \\ 0 & a_{21} & 2a_{22} & 0 & a_{23} & 0 \\ 2a_{31} & a_{32} & 0 & a_{11} + a_{33} & a_{12} & 2a_{13} \\ 0 & a_{31} & 2a_{32} & a_{21} & a_{22} + a_{33} & 2a_{33} \\ 0 & 0 & 0 & a_{31} & a_{32} & 2a_{33} \end{bmatrix}$$

$$\tag{4.69}$$

Step 1: Delete rows and columns 1, 3, and 6 to obtain

$$B_1' = \begin{bmatrix} a_{11} + a_{22} & a_{23} & a_{13} \\ a_{32} & a_{11} + a_{33} & a_{12} \\ a_{31} & a_{21} & a_{12} + a_{33} \end{bmatrix} \tag{C.23}$$

Step 2: Insert minus sign in row 1 and column 3 and column 1 and row 3 to obtain $\tilde{A}$ [as verified in (3.251)] as follows:

$$\tilde{A} = \begin{bmatrix} a_{11} + a_{22} & a_{23} & -a_{13} \\ a_{32} & a_{11} + a_{33} & a_{12} \\ -a_{31} & a_{21} & a_{22} + a_{33} \end{bmatrix} \tag{C.24}$$

APERIODICITY CONDITION FOR THE A MATRIX

This condition, as explained in Section 2.2 for the characteristic polynomial, requires that all the eigenvalues of A be distinct and negative real. This condition can be expressed by the following theorem due to Bose [5].

THEOREM C.3. *A necessary and sufficient condition for a real non-symmetric matrix A to have all distinct nonpositive real eigenvalues is that*

$$Z(s) = \frac{d/ds|A - sI|}{|A - sI|} \tag{C.25}$$

where I is the identity matrix of the same order as A, be an RC impedance function.

In Section 2.4d we presented the condition for physical realization of an RC passive network (or for being an impedance function). Furthermore, the aperiodicity conditions given in Section 2.2 can also be applied to the characteristic polynomial associated with the matrix A.

RELATIVE STABILITY CONDITION FOR THE A MATRIX

This condition, as explained in Section 2.2 for the characteristic polynomial, requires that all the eigenvalues of A lie in the region shown in Figure 2.1. This condition can be expressed by the following theorem due to Davison–Ramesh [6].

THEOREM C.3a. *A necessary and sufficient condition that the eigenvalues of a real matrix A lie within the relative stability region of Figure 2.1 is that the eigenvalues of the matrix*

$$A_1 = \begin{bmatrix} A\cos\delta & -A\sin\delta \\ A\sin\delta & A\cos\delta \end{bmatrix} \tag{C.26}$$

have negative real part. The angle δ is given by the expression

$$\delta = \theta - \frac{\pi}{2} \qquad \text{(C.27)}$$

Proof: Let the eigenvalues of A be denoted by λ_i, $i = 1, 2, \ldots, n$, and the eigenvalues of A_1 be $\overset{*}{\lambda}_i$, $i = 1, 2, \ldots, 2n$. Then

$$\{\overset{*}{\lambda}_i\} = \{\hat{\lambda}_i, \bar{\hat{\lambda}}_i\} \qquad \text{(C.28)}$$

where the bar denotes complex conjugate and

$$\hat{\lambda}_i = \lambda_i(\cos \delta + j \sin \delta) \qquad i = 1, 2, \ldots, n \quad \text{(C.29)}$$

Now Re $(\overset{*}{\lambda}_i) \le 0$, $i = 1, 2, \ldots, 2n$, if and only if Re $(\hat{\lambda}_i) \le 0$, Re $(\bar{\hat{\lambda}}_i) \le 0$, $i = 1, 2, \ldots, n$, or if and only if

$$\text{Re } [\lambda_i(\cos \delta \pm j \sin \delta)] \le 0 \quad i = 1, 2, \ldots, n \quad \text{(C.30)}$$

or if and only if

$$\{\lambda_i\} \in \Omega \qquad \text{(C.31)}$$

where Ω is the region of relative stability indicated in Figure 2.1.

STABILITY CONDITION FOR THE A MATRIX
WITHIN THE UNIT CIRCLE

The stability within the unit circle requires that all the eigenvalues of A be inside the unit circle or that the characteristic roots be less than unity in modulus. The necessary and sufficient condition can be readily ascertained by using the following bilinear matrix transformation [7] on the matrix A, which we denote as A_d:

$$A_c = (A_d + I)(A_d - I)^{-1} \qquad \text{(C.32)}$$

Equation (C.32) indicates that when A_c is a stability matrix (i.e., all eigenvalues have negative real part), then the matrix A_d is a stability matrix within the unit circle (i.e., all eigenvalues are less than unity in modulus). Hence we can apply the stability condition on A_c, as indicated in Theorem 3.15, to obtain the stability condition within the unit circle for the A_d or A matrix.

The computation involved in the bilinear matrix transformation of (C.32), is quite cumbersome, and it would be desirable if a direct stability test could be obtained from the A matrix. A test has been obtained [8] that avoids the matrix bilinear transformation, and it is presented by the following two theorems, the proof of which is similar to that of Theorem 3.15 and is available in the literature [8].

THEOREM C.4. *This theorem is the discrete counterpart of the Routh theorem proven in Theorem C.1. It is given as follows. Let*

$$\zeta_n \lambda^n + \zeta_{n-1} \lambda^{n-1} + \cdots + \zeta_0 = 0 \qquad \zeta_n > 0 \quad \text{(C.33)}$$

have real coefficients. Let

$$s_m \mu^m + s_{m-1} \mu^{m-1} + \cdots + s_0 = 0 \qquad s_m > 0 \quad \text{(C.34)}$$

have the following $m = \frac{1}{2}n(n-1)$ *roots:*

$$\mu = \lambda_i \lambda_j - 1$$
$$i = 2, 3, \ldots, n, \quad j = 1, 2, \ldots, i-1 \quad \text{(C.35)}$$

where $\lambda_1, \lambda_2, \ldots, \lambda_n$ *are the roots of* (C.33).

For the roots of (C.33) *to lie inside the unit circle, it is necessary and sufficient that:*

1. *The linear combinations* b_i *of* ζ_i *are all positive, where for n-odd*

$$b_i = \sum_{r=0}^{2s-1} \left[\sum_j (-1)^{r+i-j+1} \zeta_r \binom{r}{j} \binom{2s-1-r}{i-j} \right]$$
$$n = 2s - 1 \quad \text{(C.36)}$$

and for n-even

$$b_i = \sum_{r=0}^{2s} \left[\sum_j (-1)^{r+1-j} \zeta_r \binom{r}{j} \binom{2s-r}{i-j} \right] \qquad n = 2s \quad \text{(C.37)}$$

The summand over j is governed by $\max(0, 2s - r - i) \le j \le \min(i, r)$.
2. *The coefficients* $s_0, s_1, \ldots, s_{m-1}$ *should all be positive.*

It is of interest to note that (C.36) and (C.37) are the same as (2.44) and (2.48). They are repeated here for continuity. Furthermore, the $\frac{1}{2}n(n+1)$ conditions of Theorem C.4 represent an alternate form for the stability of the real polynomial inside the unit circle. These $\frac{1}{2}n(n+1)$ conditions can be reduced to n conditions, which then coincide exactly with the simplified determinantal criterion of Section 2.3a. A method for obtaining the s_k's in (C.34) is presented in Theorem C.8.

THEOREM C.8. *This theorem is the discrete counterpart of Theorem 3.15. It is given as follows. Let A be a real square matrix of dimension $n \times n$. Let $\tilde{A}$ be the bialternate product of A by itself. Then for the characteristic roots (eigenvalues) of A to lie inside the unit circle, it is necessary and sufficient that in*

(a) $|\lambda I - A|$ (C.38)

(b) $|\mu I - \tilde{K}|$

$$\tilde{K} = \tilde{A} - I, \quad \text{the dimension of } \tilde{K} \text{ is } m = \frac{1}{2}n(n-1) \quad \text{(C.39)}$$

the coefficients of μ^i $(i = 0, 1, 2, \ldots, m - 1)$ and the coefficients b_i $(i = 0, 1, \ldots, n)$ given by (C.36) *and* (C.37) *should all be positive.*

It may be noted that this theorem also yields $\frac{1}{2}n(n + 1)$ conditions, which when reduced to n gives the simplified determinantal criterion presented in Section 2.3a.

Remarks

1. The bialternate product† $\tilde{A} = A \cdot A$ can be readily obtained from the A matrix as given by Fuller [1] and Stephanos [3]. Also, there exists a simple procedure for generating it from the Lyapunov matrix K' or from the A matrix, as is presented in the literature [8].

2. Similar to the critical-case formulation presented in (3.239), we have for the discrete case

$$|I - A| \geq 0 \qquad |I + A| \geq 0 \qquad \text{(C.39)}$$

$$|\Delta_{n-1}^-| = (-1)^m |\tilde{K}| \triangleq (-1)^m |\tilde{A} - I| \geq 0$$
$$m = \tfrac{1}{2}n(n - 1) \quad \text{(C.40)}$$

where $|\Delta_{n-1}^-|$ is given by (2.42).

The preceding conditions can be readily verified and are proven in the literature [8].

REFERENCES

[1] A. T. FULLER, "Conditions for a matrix to have only characteristic roots with negative real parts," *J. Math. Anal. and Appl.*, **23** (1), 71–98 (July 1968).

[2] E. J. ROUTH, *Stability of a Given State of Motion*. London: Macmillan, 1877.

[3] C. STEPHANOS, "Sur une extension du calcul des substitutions linéaires," *J. Math. Pure Appl.*, **6**, 73–128 (1900).

† The bialternate product is defined as follows: Let A be an n-dimensional matrix (a_{ij}). Let $\tilde{A}$ be an $m = \frac{1}{2}n(n - 1)$ dimensional matrix $(\tilde{a}_{pq,rs})$ whose rows are labeled $pq(p = 2, 3, \ldots, n, q = 1, 2, \ldots, p - 1)$, whose columns are labeled $rs(r = 2, 3, \ldots, n, s = 1, 2, \ldots, r - 1)$, and whose elements are

$$\tilde{a}_{pq,rs} = \begin{vmatrix} a_{pr} & a_{ps} \\ a_{qr} & a_{qs} \end{vmatrix}$$

Then $\tilde{A}$ is the bialternate product of A by itself and is written as $\tilde{A} = A \cdot A$. The above definition can be obtained as a special case of (C.7).

To illustrate the above definition, assume A is given ay (3.247), then its bialternate product

$$\tilde{A} = \begin{bmatrix} a_{11}a_{22} - a_{12}a_{21} & a_{11}a_{23} - a_{21}a_{13} & a_{12}a_{23} - a_{22}a_{13} \\ a_{11}a_{32} - a_{12}a_{31} & a_{11}a_{33} - a_{13}a_{31} & a_{12}a_{33} - a_{13}a_{32} \\ a_{21}a_{32} - a_{22}a_{31} & a_{21}a_{33} - a_{23}a_{31} & a_{22}a_{33} - a_{23}a_{32} \end{bmatrix}$$

[4] S. BARNETT and C. STOREY, *Matrix Methods in Stability Theory*. New York: Barnes and Noble, 1970, p. 90.

[5] N. K. BOSE, "On real eigenvalues of real nonsymmetric matrices," *Proc. IEEE*, **56**, 1380 (Aug. 1968). *Ibid.*, **57** (4), 705 (April 1969).

[6] E. J. DAVISON and N. RAMESH, "A note on the eigenvalues of a real matrix," *IEEE Trans. on Automatic Control*, 252–253 (April 1970).

[7] S. BARNETT, *Matrices in Control Theory*. London: Van Nostrand Reinhold, 1971.

[8] E. I. JURY and S. GUTMAN, "On the stability of the A matrix inside the unit circle," Electronics Res. Lab. Memo–446, University of Calif., Berkeley, Calif., June, 1974.

Index

303